METAL IONS IN BIOLOGICAL SYSTEMS

METAL IONS IN BIOLOGICAL SYSTEMS

Edited by

Helmut Sigel
and Astrid Sigel

Institute of Inorganic Chemistry
University of Basel
CH-4056 Basel, Switzerland

VOLUME 30

Metalloenzymes Involving Amino Acid-Residue and Related Radicals

MARCEL DEKKER, INC. New York • Basel • Hong Kong

ISBN: 0-8247-9093-6

This book is printed on acid-free paper.

MARCEL DEKKER, INC.
270 Madison Avenue, New York, New York 10016

Current printing (last digit):
10 9 8 7 6 5 4 3 2 1

PRINTED IN THE UNITED STATES OF AMERICA

Preface to the Series

Recently, the importance of metal ions to the vital functions of living organisms, hence their health and well-being, has become increasingly apparent. As a result, the long-neglected field of "bioinorganic chemistry" is now developing at a rapid pace. The research centers on the synthesis, stability, formation, structure, and reactivity of biological metal ion-containing compounds of low and high molecular weight. The metabolism and transport of metal ions and their complexes is being studied, and new models for complicated natural structures and processes are being devised and tested. The focal point of our attention is the connection between the chemistry of metal ions and their role for life.

No doubt, we are only at the brink of this process. Thus, it is with the intention of linking coordination chemistry and biochemistry in their widest sense that the Metal Ions in Biological Systems series reflects the growing field of "bioinorganic chemistry." We hope, also, that this series will help to break down the barriers between the historically separate spheres of chemistry, biochemistry, biology, medicine, and physics, with the expectation that a good deal of future outstanding discoveries will be made in the interdisciplinary areas of science.

Should this series prove a stimulus for new activities in this fascinating "field," it would well serve its purpose and would be a satisfactory result for the efforts spent by the authors.

Fall 1973

Helmut Sigel
Institute of Inorganic Chemistry
University of Basel
CH-4056 Basel, Switzerland

Preface to Volume 30

Biochemical systems and free radicals somehow do not appear as going well together: until recently, free radical species have been regarded as unstable, toxic, and uncontrollable with respect to reaction course. Indeed, living systems developed sophisticated enzyme systems to cope with superoxide ($O_2^{\bullet -}$) or peroxide ($HOO^\bullet$, $ROO^\bullet$) radicals. However, in recent years a number of metalloenzymes have been discovered that also contain free radicals on their amino acid residues, and there is now growing awareness that enzymes evolved ways to use the reactivity inherent in a radical site to catalyze a series of stereospecific reactions.

The first enzymes shown to involve protein free radicals were ribonucleotide reductases, where a binuclear iron-tyrosyl free radical center plays a role, yet there is also the intriguing pyruvate-formate lyase whose catalytic mechanism includes a free radical at the α-carbon of a glycine residue. These and other enzymes involving amino acid-residue and related radicals are outlined and discussed in this volume, and experimental approaches for studies in this difficult area are indicated. The metal ions that appear in the metalloenzymes (or as cofactors) are manganese, iron, cobalt, and copper; e.g., in amine oxidases a Cu(I)-semiquinone radical is crucial. The volume closes with a somewhat different focus, i.e., the bacterial transport of and resistance to copper; the role of radicals in this respect still remains elusive, yet there is evidence for the participation of hydroxyl radicals in certain related aspects.

Helmut Sigel
Astrid Sigel

Contents

Contributors

Georg Auling Institut für Mikrobiologie, Universität Hannover, D-30167 Hannover, FRG (131)

Gerald T. Babcock Department of Chemistry, Michigan State University, Chemistry Building, East Lansing, MI 48824-1322, USA (FAX: 1-517-353-1793) (77)

Nigel L. Brown School of Biological Sciences, University of Birmingham, Edgbaston, Birmingham, B15 2TT, England (FAX: 44-21-4146557) (405)

David M. Dooley Department of Chemistry, Amherst College, Amherst, MA 01002, USA (361)

Roger E. Fenna Department of Biochemistry and Molecular Biology, University of Miami School of Medicine, P.O. Box 016129, Miami, FL 33101, USA (25)

Hartmut Follmann Biochemie, Fachbereich Biologie-Chemie, Universität Kassel, Heinrich-Plett-Strasse 40, D-34109 Kassel-Oberzwehren, FRG (FAX: 49-561-8044735) (131)

Astrid Gräslund Department of Medical Biochemistry and Biophysics, University of Umeå, S-90187 Sweden (FAX: 46-90-136310) (109)

Maris G. N. Hartmanis Kabi Pharmacia BioScience Center, Strandbergsgatan 49, S-11287 Stockholm, Sweden (FAX: 46-8-6188262) (201)

Curtis W. Hoganson Department of Chemistry, Michigan State University, Chemistry Building, East Lansing, MI 48824-1322, USA (FAX: 1-517-353-1793) (77)

Peter F. Knowles Department of Biochemistry and Molecular Biology, University of Leeds, Leeds LS2 9JT, England (FAX: 44-532-333167) (361)

John W. Kozarich Department of Chemistry and Biochemistry, University of Maryland, College Park, MD 20742, USA (FAX: 1-301-405-9376) (279)

Barry T. O. Lee Department of Genetics, University of Melbourne, Parkville, Victoria 3052, Australia (405)

Lawrence J. Marnett A. B. Hancock Jr. Memorial Laboratory for Cancer Research, Center in Molecular Toxicology, Departments of Biochemistry and Chemistry, Vanderbilt University School of Medicine, Nashville, TN 37232, USA (163)

Ei-Ichiro Ochiai Department of Chemistry, Juniata College, Huntingdon, PA 16652, USA (FAX: 1-814-643-3620) (1/255)

Thomas L. Poulos Departments of Molecular Biology and Biochemistry, and Physiology and Biophysics, University of California at Irvine, Irvine, CA 92717, USA (25)

Simon Silver Department of Microbiology and Immunology, University of Illinois College of Medicine, MC/90, Box 6998, Chicago, IL 60680, USA (FAX: 1-312-996-6415) (405)

William L. Smith Department of Biochemistry, Michigan State University, Biochemistry Building, Room 510, East Lansing, MI 48824-1319, USA (FAX: 1-517-353-9334) (163)

Lars Thelander Department of Medical Biochemistry and Biophysics, University of Umeå, S-90187 Umeå, Sweden (FAX: 46-90-136310) (109)

Tetsuo Toraya Department of Biotechnology, Faculty of Engineering, Okayama University, 3-1-1 Tsushima-Naka, Okayama 700, Japan (FAX: 81-862-535755) (217)

James W. Whittaker Department of Chemistry, Carnegie Mellon University, 4400 Fifth Avenue, Pittsburgh, PA 15213-3890, USA (FAX: 1-412-268-3206) (315)

Kenny K. Wong Department of Chemistry and Biochemistry, University of Maryland, College Park, MD 20742 (FAX: 1-301-405-9376) (279)

Contents of Previous Volumes

---

*Out of print.

Volume 11. Metal Complexes as Anticancer Agents

Volume 12. Properties of Copper

Volume 15. Zinc and Its Role in Biology and Nutrition

Volume 16. Methods Involving Metal Ions and Complexes in Clinical Chemistry

Volume 17. Calcium and Its Role in Biology

Volume 18. Circulation of Metals in the Environment

Volume 19. Antibiotics and Their Complexes

Volume 20. Concepts on Metal Ion Toxicity

Volume 21. Applications of Nuclear Magnetic Resonance to Paramagnetic Species

Volume 22. ENDOR, EPR, and Electron Spin Echo for Probing Coordination Spheres

Volume 23. Nickel and Its Role in Biology

Volume 24. Aluminum and Its Role in Biology

Volume 25. Interrelations Among Metal Ions, Enzymes, and Gene Expression

Volume 26. Compendium on Magnesium and Its Role in Biology, Nutrition, and Physiology

Volume 29. Biological Properties of Metal Alkyl Derivatives

Other volumes are in preparation.

Comments and suggestions with regard to contents, topics, and the like for future volumes of the series are welcome.

The following Marcel Dekker, Inc. books are also of interest for any reader dealing with metals or other inorganic compounds:

Handbook on Toxicity of Inorganic Compounds
edited by Hans G. Seiler and Helmut Sigel, with Astrid Sigel
In 74 chapters, written by 84 international authorities, this book covers the physiology, toxicity, and levels of tolerance, including prescriptions for detoxification, for all elements of the Periodic Table (up to atomic number

103). The book also contains short summary sections for each element, dealing with the distribution of the elements, their chemistry, technological uses, and ecotoxicity as well as their analytical chemistry.

Handbook on Metals in Clinical and Analytical Chemistry
edited by Hans G. Seiler, Astrid Sigel, and Helmut Sigel
This book is written by 80 international authorities and covers over 3500 references. The first part (15 chapters) focuses on sample treatment, quality control, etc., and on the detailed description of the analytical procedures relevant for clinical chemistry. The second part (43 chapters) is devoted to a total of 61 metals and metalloids; all these contributions are identically organized covering the clinical relevance and analytical determination of each element as well as, in short summary sections, its chemistry, distribution, and technical uses.

METAL IONS IN BIOLOGICAL SYSTEMS

1
Free Radicals and Metalloenzymes: General Considerations

Ei-Ichiro Ochiai

Department of Chemistry, Juniata College,
Huntingdon, PA 16652, USA

1. INTRODUCTION: A SURVEY OF BIOLOGICALLY RELEVANT FREE RADICALS

A number of metalloenzymes have been discovered in recent years that also contain free radicals on their amino acid residues. The free radical on an amino acid residue plays a catalytic role in these enzymes. Examples of such enzymes include the small subunit of iron-dependent ribonucleotide reductase, galactose oxidase, and prostaglandin endoperoxide synthetase. B_{12}-coenzyme–dependent enzymes involve an adenosyl free radical which forms upon addition of a substrate. B_{12} contains a cobalt atom. This volume is devoted to a discussion of some of the better characterized systems that contain both metals and free radicals on amino acid residues or on other intrinsic parts. Stubbe recently contributed two review articles [1] on the subject of radicals in biological catalysis.

Let us first survey briefly the free radical species involved in biochemical systems. They can be classified in four categories: (1) the intermediates of quinones and flavins; (2) oxygen or dioxygen radicals ($O^{\bullet-}$ or $HO^{\bullet}$; $O_2^{\bullet-}$ or $HO_2^{\bullet}$) or its analogs ($RO^{\bullet}$ or $RO_2^{\bullet}$ which are derived from ROOH); (3) radicals on some amino acid residues such as phenoxy (of tyrosine) or thiol (of cysteine); and (4) alkyl radicals $R^{\bullet}$. The last group (4) includes the free radical intermediates in enzymatic reactions that are formed through H abstraction from a substrate, as well as the adenosyl free radical in B_{12}-coenzyme–dependent enzymatic reactions.

Quinones and flavins are the only organic compounds that can relatively easily participate in one-electron transfer reactions. This is because semiquinone, the intermediate in the reduction of quinone or oxidation of hydroquinone, is relatively stable. Flavins have the same type of chemistry as quinones [see reaction (1) below]. The semiquinone is a free radical. Quinones are employed as electron transfer agents in

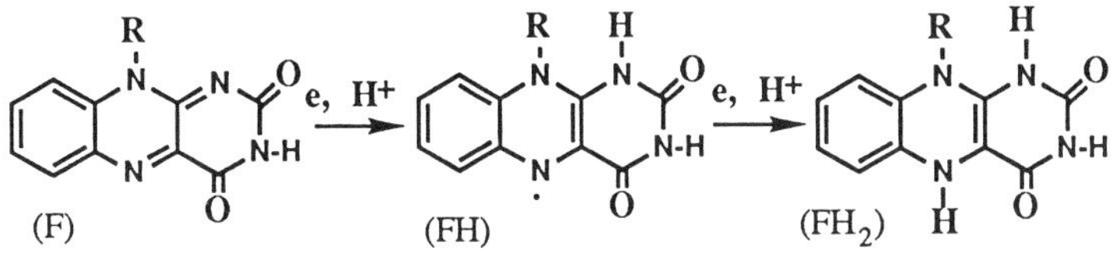 (1)

the mitochondrial or photosynthetic electron transfer systems, and flavins [such as flavin adenine dinucleotide (FAD) and flavin mononucleotide (FMN)] function as an electron transfer intermediary in a number of enzymes which contain iron-sulfur units and molybdenum. The cofactor of mitochondrial amine oxidase has recently been determined to be a kind of quinone (see Sec. 4.5).

Oxygen and dioxygen radicals are often involved in monooxygenation and dioxygenation reactions. For example, cytochrome P-450-dependent monooxygenation is believed to involve the so-called ferryl ion $(FeO)^{3+}$, which is hypothesized to abstract an H atom from the substrate [2], thus giving rise to the reaction sequence (2):

$$[(FeO)^{3+} \Leftrightarrow (Fe^{IV}, O^{\bullet -})] + RCH_2{-}H \rightarrow$$
$$[(Fe^{IV}, OH^-, RCH_2^{\bullet}) \Leftrightarrow (Fe^{III}, HO^{\bullet}, RCH_2^{\bullet})] \rightarrow$$
$$Fe^{III} + RCH_2OH \qquad (2)$$

The oxygen radical postulated here needs to be associated with the heme iron throughout the reaction because an electron is assumed to be transferred between the O moiety and the iron center as described. This hypothesis has not unequivocally been substantiated. Other enzymes of this category include methane monooxygenase and numerous other iron- or copper-dependent oxygenases and peroxidases (see also Chap. 2 of this volume), and, perhaps, isopenicillin N synthetase. The enzymes of this group, however, are too numerous to be treated here. Besides, they have been well documented already.

Oxygen and dioxygen radicals are also believed to participate in many physiological processes including the aging process of the cell membranes and the bacteriocidal effect of macrophage cells. Readers are referred to, e.g., the series *Free Radicals in Biology* [3a], and Halliwell and Gutteridge [3b] for further information on this issue.

Radicals of the aforementioned categories 3 and 4 are in the focus of this volume. The phenoxy radical (of tyrosine) and the thiol radical (of cysteine) have been identified as the essential radical moieties in several metalloenzymes. Adenosylcobalamin (abbreviated as AdoCbl = vitamin B_{12}-coenzyme) is believed to produce the adenosyl radical by splitting the Co-C bond upon binding a substrate to the enzyme protein. There are also some indications in certain enzyme systems, such as ethanolamine ammonia lyase, that the radical on the adenosyl residue is actually transferred to an amino acid residue. The amino acid is thought to be a cysteine. The thiol radical thus created participates in the catalytic activity, abstracting an H atom from the substrate (see Chaps. 8 and 9 for details).

Pyruvate formate lyase itself is a metal-free enzyme, but is activated by an iron-requiring enzyme containing adenosylmethionine and dihydroflavodoxin as cofactors. The activated enzyme has been demonstrated to have a free radical at the α carbon of a glycine residue (Gly734), which is at the catalytic site [4].

A fair number of enzymes are now known to involve a free radical as a catalytic entity. This implies that certain kinds of biochemical reactions would not take place without the participation of a free radical. Therefore, next we shall ask "why free radical?" and "why are metal ions associated with free radicals?" and then describe briefly what is known about the reactivities of free radicals.

2. NECESSITY OF FREE RADICALS AND METAL IONS

2.1. Why Radicals?

Chemical reactions fall into two broad categories: the acid–base type and oxidation–reduction type. Reactions in biological systems are no exception.

Reactions of the acid–base type involve the heterolytic cleavage or formation of chemical bonds. The substrates that are subject to the action of enzymes of this kind have bonds that are relatively polar and thus susceptible to the action of an acid/base. Biochemical examples of this type include hydrolysis of bonds (peptidase, phosphorylase, and similar reactions), synthetase reactions (condensation reactions), and kinase reactions (phosphate transfer).

Reactions of the oxidation–reduction type involve transfer of electrons. They can also occur as atom transfer reactions such as O-atom transfer (as in oxygenase reactions) or H atom(s) or H^- transfer (as in dehydrogenase reactions). Biochemical examples of this type include electron transfer reactions (as in-between cytochromes or iron-sulfur proteins); various oxidation and oxygenation reactions involving iron, molybdenum, and copper enzymes; and dehydrogenase reactions.

Free radical reactions can be regarded as a special case of the oxidation–reduction type in the sense that they involve a transfer of an entity containing an odd number of electrons, and a homolytic bond cleavage and formation. The most common entity of this kind is an H atom. Substrates that have no polar bonds, such as saturated hydrocarbons, are not very reactive and cannot easily be made to react in an acid–base manner. For example, a heterolytic cleavage of a C-H bond in hydrocarbons to form a carbonium ion is virtually unknown. The only feasible means to render such a C-H bond reactive appears to be a free radical abstraction of the H atom, leaving an unpaired electron on the carbon atom. This free radical site on the carbon atom is very reactive. It can abstract an H atom from another moiety, recombine with another free radical species, or undergo some other reactions. Indeed, many of the biologically important free radicals function as a hydrogen atom abstractor from a rather inert substrate such as a hydrocarbon or hydrocarbon-like compound. Frey emphasized this same point in a review article [5].

It was recently demonstrated [4] that pyruvate formate lyase has a free radical at the α-carbon of a glycine residue (Gly734) and catalyzes the following reaction:

$$CH_3COCOOH + CoA\text{-}SH \rightarrow CoA\text{-}S{-}(C{=}O)CH_3 + HCOOH \quad (3)$$

The regular acetyl-CoA formation from pyruvate by another enzyme, pyruvate dehydrogenase, requires NAD^+ and produces CO_2 as a product. This is an oxidative reaction. In contrast, reaction (3) appears to be of an acid–base type; i.e., the thiol of $CoA\text{-}S^-$ seems to attack the carbonyl carbon of pyruvate and displaces $-COOH$ ($-COOH + H^+ \rightarrow HCOOH$). However, the reaction seems to involve a free radical instead. A possible reaction scheme might be:

$$\begin{aligned} &R^{\bullet}\ (\text{glycine radical}) + H{-}S\text{-}CoA \rightarrow R{-}H + {}^{\bullet}S\text{-}CoA \\ &{}^{\bullet}S\text{-}CoA + CH_3COCOOH \rightarrow CH_3CO{-}S\text{-}CoA + {}^{\bullet}COOH \qquad (4) \\ &{}^{\bullet}COOH + RH \rightarrow HCOOH + R^{\bullet} \end{aligned}$$

2.2. Free Radical–Metal Ion Relationships

The descriptions presented so far have clearly indicated that free radicals are necessary in certain types of biochemical reactions, and that the free radicals often occur in association with transition metal ions, particularly iron, cobalt, and copper. The only exception to this general observation is the $O_2^{\bullet-}$ formation by flavin [either its dihydro or semiquinone form; see Equation (1)].

How are free radicals produced in general? Chemistry textbooks tell us that the most often used routes to a free radical are (a) photochemical reactions e.g., $Br_2 + h\nu \rightarrow 2Br^{\bullet}$) or (b) thermal decompositions of peroxides (e.g., $R(C{=}O)O{-}O(O{=}C)R \rightarrow 2R(C{=}O)O^{\bullet}$). Neither of these reactions seems to be readily applicable to biological systems.

Here lies the necessity for transition elements. Two important characteristics of these elements in this regard are first the relatively weak bond between a transition element and a carbon atom in their organometallic compounds, and second the variability of their oxidation states:

1. The homolytic cleavage of the cobalt-to-carbon bond in AdoCbl is a typical example. The bond dissociation energy of the cobalt-to-carbon bond in AdoCbl has been determined to be about 120 kJ/mol, which is not very large (compared with, e.g., 410kJ/mol for the C-C bond). The mechanism of this homolytic bond cleavage has eluded an elucidation as discussed below and in Chaps. 10 and 11, but it can be seen from the dissociation energy value that the bond can be broken relatively easily. It has been shown that the homolytic bond cleavage is easier in terms of bond energy than a heterolytic bond cleavage such as $Co{-}C \rightarrow Co(III) + C^{-}$ or $Co(I) + C^{+}$. The adenosyl free radical thus formed then abstracts an H atom from a substrate, thus initiating the subsequent reaction. AdoCbl is the only naturally occurring organometallic compound that has been found so far, but it seems possible that other organometallic compounds may be playing roles in biological systems.

2. The most conspicuous feature of transition elements and their compounds, as compared to organic compounds, is the variability of the oxidation state. Because of that, transition elements readily participate

in electron transfer reactions. Numerous iron- and copper-containing biocompounds are known to be involved in electron transfer processes. Likewise, transition elements can exchange electrons with oxygen, dioxygen, and their free radicals as mentioned earlier, though electron exchange partners are not limited to the oxygen entities. The oxygen radicals ($O_2^{\bullet -}$, $O^{\bullet -}$, etc.) then create an organic free radical by abstracting an H atom. In this regard, iron and copper appear to be the most useful [6].

3. REACTIVITIES OF FREE RADICALS

Free radical reactions in general are not understood as well as reactions of polarized bonds (involving heterolytic bond cleavage and formation). Useful summaries of free radical chemistry are found in Refs. 7–10.

Let us first examine the H-atom abstraction reaction by a free radical. A highly reactive radical species abstracts H atoms rather indiscriminately, while a mildly reactive one does so selectively. For example, Table 1 gives the relative reaction rates of H abstraction from hydrocarbons by several radicals [7]. The variation in rate is rather small (except for methane) in the case of reactive free radicals such as

TABLE 1

Relative Rates of Hydrogen Abstraction by Free Radicals

R-H	$Me^{\bullet}$	t-$BuO^{\bullet}$	$Cl^{\bullet}$	$Br^{\bullet}$	$Cl_3C^{\bullet}$
RCH_3	0.4	0.1	0.8	0.0001	
R_2CH_2	4.0	1.2	3	0.005	5
R_3CH	9.4	4.4	3.6	0.02	160
$PhCH_3$	4.1	3.2	2.5	17	50
$(Ph)RCH_2$	13	6.9	5.5	37	260
$(Ph)R_2CH$		4.7	2.0	10	50
Ph_2CH_2		9.6	7.2	18	160

Source: After Ref. 7.

$Cl^{\bullet}$, $Me^{\bullet}$, and tertiary butoxyl (t-$BuO^{\bullet}$), while the selectivity is evident in the case of the moderately reactive $Cl_3C^{\bullet}$ and $Br^{\bullet}$ species. The variations in the activation energy of H-atom abstraction by moderately reactive free radicals parallel somewhat the variation in the bond dissociation energy of C-H, as shown in Fig. 1 [7]. In this case the H-abstraction rate is determined to a large extent by the dissociation energy of the bond to be cleaved. That is, a linear relationship seems to hold between the activation energy and the enthalpy of reaction for these moderately reactive species.

The activation energies for the H abstraction from ethane by reactive radical species [reaction (5)] are given in Table 2 [8]. The

$$X^{\bullet} + H{-}CH_2CH_3 \rightarrow X{-}H + {}^{\bullet}CH_2CH_3 \quad (5)$$

activation energy is very low for $F^{\bullet}$ and $Cl^{\bullet}$. The activation energy does not correlate well with the heat of reaction (which is the difference in bond energy between C−H and X−H). The $HO^{\bullet}$ (as well as the $O^{\bullet -}$)

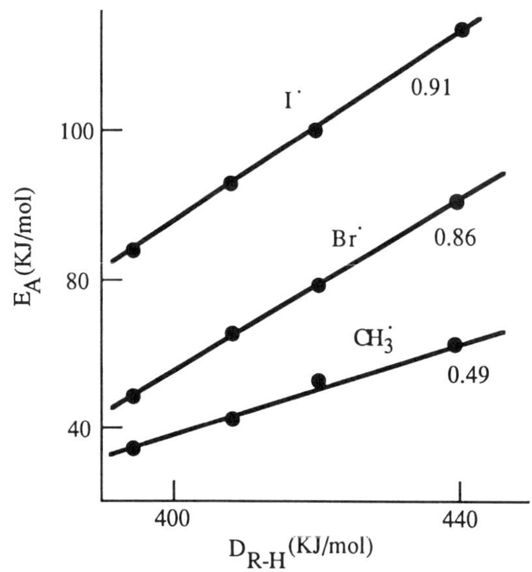

FIG. 1. Reactivity–selectivity relationship between the activation energy E_A for hydrogen abstraction from R−H and the bond dissociation energy D_{R-H}. The numbers given (0.91, 0.86, 0.49) are the slopes, which are a measure of the selectivity. (After Ref. 6.)

TABLE 2

Activation Energies for Reaction (5)

$X^\bullet$ (X=)	F	Cl	Br	MeO	H	Me
Activation energy (kJ/mol)	1.3	4.2	55	31	38	49

Source: After Ref. 7.

radical also behaves as a highly reactive indiscriminate radical [9]. The $^\bullet CH_3$ radical requires a substantial amount of activation energy in abstracting an H atom, but appears as reactive as $Cl^\bullet$ or *t*-$BuO^\bullet$ (see Table 1). That is, a hydrocarbon free radical as represented by a methyl (Me) radical is highly reactive in hydrogen abstraction despite its high activation energy.

The bond dissociation energy could provide some indication as to which bond would be most susceptible to H abstraction, or whether or not a certain H abstraction would be feasible. Table 3 lists the bond dissociation energy values for several biologically relevant bonds [10].

TABLE 3

Bond (Dissociation) Energies of Some Biologically Relevant Compounds

Bond	Energy (kJ/mol)	Bond	Energy (kJ/mol)
$H-CH_3$	435	$H-OCH_3$	427
$H-CH_2C_6H_5$	356	$H-SCH_3$	390
H−(C=O)H	368	H−Cl	431
H−OH	498	$H-C_6H_5$	431
H−SH	377	$H-CCl_3$	402
H−F	569	$CH_3CO-COCH_3$	347
$H-CH_2CH_3$	410	$H-OC_6H_5$	356
$H-CH_2OH$	389	H−NHR	444
CH_3-CH_3	368	(AdoCbl)Co−C	126

Source: Data compiled from Ref. 10 and various other sources.

For example, a tyrosyl radical (a phenoxy radical) plays an important role in many biological systems (as will be seen in later chapters). Its O-H bond dissociation energy is 356 kJ/mol, whereas the H-OH bond energy is 498 kJ/mol. Therefore, the H abstraction from the OH group of a tyrosine residue by an $^{\bullet}$OH radical to form the tyrosyl radical is thermodynamically favorable ($\Delta H = -142$ kJ/mol). The formation of a thiol radical is also easy, as the S-H bond energy (390 kJ/mol for CH_3SH) is relatively low. However, a reaction in which a tyrosyl radical abstracts an H atom from a substrate having a C-H bond would be thermodynamically unfavorable because ΔH has a positive value; e.g., approximately +54 kJ (for the H abstraction from C−H in ethane by a tyrosyl radical). In the case of a sulfhydryl radical, the corresponding ΔH value is around +20 kJ. Nonetheless, this type of reaction is postulated in many enzymatic systems. Is such an assumption reasonable?

Thermodynamic arguments assume that the reaction products remain in proximity immediately after a reaction and so a reverse reaction can take place; if this reverse reaction is thermodynamically favored, no forward reaction appears to take place. However, in case that reactants are supplied with enough energy to overcome the activation barrier and the products are immediately separated or the reverse reaction is somehow prevented, then the reaction could proceed in the forward direction even if it is not thermodynamically favored. The activation barrier is not exceedingly high in the cases of the H abstraction by free radicals in general, and it is almost nil in the case of reactive radicals such as $F^{\bullet}$, $Cl^{\bullet}$, or $HO^{\bullet}$, as seen in Table 2. However, should these conditions (i.e., a relatively low activation energy and a prevention of the reverse reaction) not be met, the free radical H-abstraction reaction would very likely be controlled thermodynamically.

Another issue is how the reactivity of a molecule would change when an H atom is abstracted from it, i.e., when it is converted to a free radical. For example, would the presence of an unpaired electron on a carbon affect the reactivity of the moiety bound to the same carbon atom or the one adjacent to it? Not much has been studied in this regard, unfortunately. One general effect that has been observed is that the acidity of an alcoholic OH group increases when the alcohol becomes a free radical [9,11]. The pK_a values of primary alcohols RCH_2OH are about 20 (i.e., $K_a = 10^{-20}$), while the pK_a values of RH(C$^{\cdot}$)OH fall in the range of 11–12. This means that the K_a value of an alcohol increases by a factor of 10^8–10^9 when an H atom is abstracted from the α carbon. The increase in K_a values of alcohol-OH is in the order of 10^4–10^5 when the unpaired electron is on the β carbon [8].

An alcohol RCH_2OH does not bind strongly to a metal cation such as Cu(II), but it may bind like RH(C$^{\cdot}$)O$^-$–Cu(II) when it becomes a free radical. Besides, in this state, an intramolecular electron transfer may become facile, i.e., RH(C$^{\cdot}$)O$^-$–Cu(II) → RCH=O + Cu(I). This reaction will be promoted because of the presence of an unpaired electron on the carbon atom and the double-bond formation.

The skeletal rearrangement reactions of radical species will be discussed in relation to the 1,2-shift reaction found in AdoCbl-dependent enzymes (Chaps. 10 and 11). It is evident that much more study is needed to better understand the free radical reactions in biological systems.

4. SUMMARY OF FREE RADICAL–DEPENDENT METALLOENZYMES

As an introduction to the subsequent chapters, some comments on the better characterized enzymes of this category are given below.

4.1. B_{12}-Coenzyme–Dependent Enzymes (Except Ribonucleotide Reductases)

4.1.1. Mutases, Diol Dehydratase, and Ethanolamine Ammonia Lyase

Enzymes that are dependent on B_{12}-coenzyme (= AdoCbl) include diol dehydratase, glutamate mutase, L-β-lysine aminomutase, methylmalonyl-CoA mutase, ethanolamine ammonia lyase, and ribonucleotide reductase. The first three enzymes consist of two subunits. Subunit E binds AdoCbl, and component S contains sulfhydryl groups. The last three enzymes are, however, of a single polypeptide, which contains several cysteine sulfhydryl groups. Both sulfhydryl groups and AdoCbl are believed to be essential for the enzymatic activity. The details of these enzymes are found in Chaps. 10 and 11 of this volume. AdoCbl contains a cobalt-to-carbon (C-5′ of the adenosyl group) bond.

Essentially two different issues await elucidation regarding the enzymatic mechanisms of these enzymes. They are (1) the mechanism by which the cobalt-to-carbon bond splits homolytically upon binding a substrate, and (2) and (3) the mechanisms of the subsequent rearrangement reactions as indicated below.

1. *Homolytic cleavage of the cobalt-to-carbon bond upon binding a substrate.* Researchers agree on the first step of the enzymatic reactions. That is the homolytic splitting of the cobalt–carbon bond when the enzyme binds a proper substrate. However, the mechanism of this homolytic cleavage is far from being well established.

A most popular mechanism has it that upon binding a substrate the corrin ring (of AdoCbl) changes its conformation in such a way that the steric interaction between the ring and the carbon atom (of Ado) bound to the cobalt atom becomes severe. Thus the steric interaction will force the cobalt–carbon bond to split ("mechanochemical mechanism" [12]).

The cobalt–carbon bond dissociation energy of AdoCbl has been estimated to be approximately 120 kJ/mol [13]. The Co-C bond dissociation energies of the so-called model compounds (of simpler structures) have been determined to be 84–126 kJ/mol. None of these values is exceedingly large for a bond dissociation energy. Therefore, the change in the hypothesized steric interaction has been thought by some to be sufficient to break such a weak bond. An estimation of the magnitude of the increase of such a steric interaction suggests that it would amount to about one third of the 120 kJ/mol under a most favorable condition [15]. That is, it is not nearly enough even under a most favorable circumstance likely to occur in the real system. This means that the steric effect may not be the major cause for the homolytic bond cleavage, though it may affect the bond splitting somewhat.

A suggestion has been made that the major portion of the binding energy of AdoCbl to the enzyme protein is utilized to weaken (lengthen) the Co-C bond [14,15]. This weakened bond would then readily be split with an addition of an extra energy due to the binding of a substrate [15]. Chapter 11 should be consulted for further details.

2. *Hydrogen abstraction from substrates.* The reactions that take place after the Co-C bond cleavage are believed to be an H abstraction from the substrate and the 1,2-shift rearrangement. That means,

$$\mathrm{Co{-}CH_2(Ado) \rightarrow Co(II) + {}^{\bullet}CH_2(Ado)}$$
$$\mathrm{{}^{\bullet}CH_2(Ado) + RCH_2{-}CH_2X \rightarrow CH_3(Ado) + RH({}^{\bullet}C){-}(C)H_2X} \quad (6a)$$
$$\mathrm{RH({}^{\bullet}C){-}(C)H_2X \rightarrow RHX(C){-}({}^{\bullet}C)H_2\ [+CH_3(Ado)] \rightarrow}$$
$$\mathrm{RHXCCH_3\ [+{}^{\bullet}CH_2(Ado)]} \quad (6b)$$

where $^{\bullet}CH_2(Ado)$ is the adenosyl radical. In this mechanism, $^{\bullet}CH_2(Ado)$ itself is assumed to participate directly in the catalytic process.

An alternative mechanism which has been suggested for ethanolamine ammonia lyase [16] and ribonucleotide reductase postulates that the radical entity is first transferred to an amino acid residue, probably cysteine. That is:

$$\mathrm{{}^{\bullet}CH_2(Ado) + HS(Cys) \rightarrow CH_3(Ado) + {}^{\bullet}S(Cys)} \quad (7)$$

This cysteine radical abstracts an H atom from the substrate and the product radical then picks an H atom from either H$-$S(Cys) or H$-$CH$_2$(Ado) [16]. A problem here is whether the thiol radical $^{\bullet}$S(Cys) can abstract an H atom from the H-C bond of a substrate such as diol or glutamate. This is the issue already discussed in Sec. 3. The H abstraction of H$-$S(Cys) by $^{\bullet}CH_2(Ado)$ would be exothermic (around -20 kJ). However, the following reaction, i.e., the H abstraction from the substrate by $^{\bullet}$S(Cys), would then be endothermic (around $+20$ kJ), though the ΔH is not exceedingly large. Thus the idea that the thiol radical acts as a catalytic radical may be feasible from the energy point of view. However, much more research is needed for a fuller understanding of the issue.

3. *The 1,2 shift or other reactions of substrate free radicals.* The major problem of these enzymatic reactions is the final step, i.e., how the substrate that has lost an H atom might transform into the product. Three questions arise: (1) Are all the reactions (perhaps, except that of ribonucleotide reductase) of a 1,2 shift type? (2) What is the mechanism of the 1,2 shift reaction itself? (3) Is the metal (Co) involved in the 1,2-shift reaction?

Glutamate mutase, methylmalonyl-CoA mutase, and other mutase reactions are definitely of a 1,2-shift reaction:

$$\mathrm{CH_2(OH){-}CH_2(OH) \rightarrow {}^{\bullet}CH(OH){-}CH_2(OH) \rightarrow CH(OH)_2{-}CH_2^{\bullet}}$$
$$\mathrm{CH(OH)_2{-}CH_2^{\bullet} + X{-}H \rightarrow CH(OH)_2{-}CH_3\ (+{}^{\bullet}X)} \quad (8)$$
$$\mathrm{\rightarrow O{=}CH{-}CH_3\ (+H_2O)}$$

or

$$\begin{aligned}&CH_2(OH)-CH_2(NH_2) \rightarrow {}^{\bullet}CH(OH)-CH_2(NH_2)\\&\rightarrow CH(OH)(NH_2)-CH_2^{\bullet} \rightarrow CH(OH)(NH_2)-CH_3\\&\rightarrow O{=}CH-CH_3\ (+\ NH_3)\end{aligned} \tag{9}$$

However, the dehydration of the free radical of ethylene glycol has been shown to be spontaneous and very fast in aqueous media [17]. This dehydration does not necessarily require a 1,2 shift as described by Eq. (8). Thus, the mechanism could be [17]:

$$\begin{aligned}&{}^{\bullet}CH(OH)-CH_2(OH) \rightarrow {}^{\bullet}CH(OH)-CH_2(OH_2)^{+}\\&\rightarrow {}^{\bullet}CH(OH)-CH_2^{+}\ (+\ H_2O) \rightarrow O{=}CH-CH_2^{\bullet}\ (+\ H^{+})\end{aligned} \tag{10}$$

Alternatively [16]:

$$\begin{aligned}&{}^{\bullet}CH(OH)-CH_2(OH) \rightarrow {}^{\bullet}CH(O^{-})-CH_2(OH)\ (+\ H^{+})\\&\rightarrow O{=}CH-CH_2^{\bullet}\ (+\ H_2O))\end{aligned} \tag{11}$$

The driving force here could be the stability of $O{=}CH-CH_2^{\bullet}$, which has an electronic structure equivalent to that of an allyl radical. A similar mechanism may be envisaged for the ethanolamine ammonia lyase reaction:

$$\begin{aligned}&{}^{\bullet}CH(OH)-CH_2(NH_2) \rightarrow {}^{\bullet}CH(OH)-CH_2(NH_3^{+})\\&\rightarrow {}^{\bullet}CH(OH)-CH_2^{+}\ (+NH_3) \rightarrow O{=}CH-CH_2^{\bullet}\end{aligned} \tag{12}$$

Thus it is possible that the enzymatic mechanisms of diol dehydratase and the ethanolamine ammonia lyase reaction are indeed different from those of mutases.

1,2-Shift (rearrangement) reactions are known to be facile in compounds with a carbonium ion, but are somewhat rare and slow in the case of free radicals. Any organic chemistry textbook gives a plausible reason for that on the basis of a bridged mechanism (see Sec. 3.2 in Chap. 11). Therefore, the 1,2-shift reaction as postulated for glutamate mutase and other AdoCbl-dependent mutases is rather unique. How the 1,2-shift reaction might proceed in these enzymatic reactions is virtually unexplored, though a number of so-called model reaction studies have been attempted to gain insight into the mechanism [18]. These studies have focused on reactions of diol dehydratase and ethanolamine ammonia lyase [18]. The mechanism of these reactions, however, could well be different from those of mutase reactions, as mentioned in the previous paragraph. Therefore, the results of the model studies discussed in Ref. 18 might be irrelevant for mutase reactions. Recent model studies by Dowd and coworkers [19] do indicate that the mechanisms of glutamate mutase and methylmalonyl-CoA mutase are indeed different from those of diol dehydratase in that they seem to involve the cobalt moiety whereas the diol dehydratase reaction occurs independent of the cobalt once the radical species is induced (see Sec. 3.2. in Chap. 11).

4.1.2. Lysine 2,3-Aminomutase

This mutase does not depend on AdoCbl unlike L-β-lysine aminomutase and instead requires Co(II), an Fe(II)-sulfur protein, S-adenosylmethionine, and pyridoxal phosphate [5]. It may be postulated that this system indeed forms an adenosyl-Co compound, which then participates in the enzymatic reaction like AdoCbl.

4.2. Ribonucleotide Reductases

The reaction of ribonucleotide reductase is represented by:

$$\begin{aligned} &-\mathrm{HC^{3'}(OH)-C^{2'}(OH)-}\ \text{(ribose of nucleotide)} + 2\mathrm{HSR} \\ &\rightarrow -\mathrm{HC(OH)-CH-}\ \text{(deoxyribose)} + \mathrm{RSSR} + \mathrm{H_2O} \qquad (13) \end{aligned}$$

There are three different types of ribonucleotide reductase: AdoCbl-dependent ones found in some fungi and anaerobic bacteria [20], Mn-dependent ones found in some bacteria [20], and iron-dependent ones (see Chaps. 4 and 5). The enzymes of the last category are found in a variety of organisms from virus and bacteria to mammals [21,22].

4.2.1. B_{12}-Coenzyme–Dependent Ribonucleotide Reductase

The enzyme from *Lactobacillus leichmannii* consists of a single polypeptide of a molecular weight of 76000. It contains a number of cysteine residues. The first step of the enzymatic reaction is believed to be the homolytic cleavage of the Co-C bond of AdoCbl, as in the case of the other AdoCbl-dependent enzymes discussed in the previous section. Isotopic studies with [3′-^{3}H]UTP ([3′-^{3}H]ATP) as the substrate have established that the adenosyl radical formed in the first step does not directly abstract an H atom from the substrate [23]. A suggestion has been made [23] that the radical character is transferred from the adenosyl group to perhaps a thiol group, which then acts as the true catalytic site.

4.2.2. Iron-Dependent Ribonucleotide Reductases

Iron-dependent ribonucleotide reductase (see Chap. 4) consists of $\alpha_2\beta_2$ subunits. The larger subunit α_2 (R1) contains the substrate binding sites as well as allosteric sites. Three Cys's 225, 462, and 439 in *E. coli* R1 are suggested to be located near the interface between R1 and R2 (= β_2) and to be directly involved in the reduction of ribose [23]. It has been discovered [23] that R1 from *E. coli* and *L. leichmannii*'s AdoCbl-dependent enzyme contains similar sequences containing two cysteine residues: –GC(745)ESGAC(759)KI– in *E. coli* R1 and –CEGGACPIK– in *L. leichmannii*. These sequences are considered to be recognized by thioredoxin [23].

Each of the small subunits β_2 (R2) contains one essential tyrosyl radical as well as a pair of antiferromagnetically coupled Fe(III). This tyrosyl radical was the first recognized free radical associated with metalloenzymes [20–22]. B2 subunit (R2 subunit of *E. coli* enzyme) has been extensively studied and its structure has been determined by X-ray crystallography [24]. The tyrosyl radical is located closely (about 500 pm) to the Fe(III) pair and is unusually stable.

The mechanism of tyrosyl radical formation has been a focus of study in recent years. A prevailing idea [24] before 1990 was based on the chemistry of monooxygenation by cytochrome P-450. It assumed an addition of O_2 to the dinuclear Fe(II) pair (analogous to that of hemerythrin) and then a splitting of the O-O bond in the following manner:

$$\mathrm{Fe(III)}-(\mathrm{O}-\mathrm{O})^{2-}-\mathrm{Fe(III)} \rightarrow (\mathrm{FeO})^{3+}-\mathrm{O}^{2-}-\mathrm{Fe(III)} \tag{14}$$

$(FeO)^{3+}$ is the analog of the ferryl ion postulated in cytochrome P-450 monooxygenases, and it was supposed to abstract an H atom from the tyrosine O–H group. This would leave Fe(IV) behind, which needs to be reduced. This last step was ignored in the mechanism.

Ochiai et al. proposed the following mechanism based on a careful study of the stoichiometric relationships of the tyrosyl radical formation [25]. The major point of this mechanism is the assumption that an

$$Fe(II)_2 + O_2 \Leftrightarrow O_2^{2-}{-}Fe(III)_2 \qquad (15a)$$

$$O_2^{2-}{-}Fe(III)_2 + Fe(II) \rightarrow O^{2-}{-}Fe(III)_2 + Fe(III) + O^{\bullet-} \qquad (15b)$$

$$O^{\bullet-} + PhOH \rightarrow HO^- + PhO^\bullet \qquad (15c)$$

extraneous electron is required to split the $(O{-}O)^{2-}$ bond and that this electron is supplied by a third Fe(II) ion. This mechanism has been confirmed by further studies [26,27].

4.2.3. Mechanism of Reduction of Ribose

The tyrosyl radical in *E. coli* R2 and the adenosyl radical in *L. leichmannii* enzyme are, however, not believed to be directly involved in the enzymatic reaction. As mentioned above, the radical entity is transferred to an amino acid residue, probably a cysteine residue. Evidence is strong that this thiol radical first abstracts an H atom from C-3′:

$$\begin{aligned}&-(HO)C^{3'}D-C^{2'}H(OH)- + (Cys)S^\bullet \rightarrow \\ &\quad -(HO)C^{3'\bullet}-C^{2'}H(OH)- + (Cys)SD \qquad (16)\end{aligned}$$

The mechanism of the subsequent reaction has not been well elucidated despite an intensive study. One that has been proposed [23] assumes:

$$\begin{aligned}&-(HO)C^{3'\bullet}-C^{2'}H(OH)- (+H^+) <==> -(HO)(C^\bullet)-C^+H- (+H_2O)\rightarrow \\ &\quad -(O{=}C)-CH_2- \rightarrow -(HO)(C^\bullet)-CH_2- (+(Cys)SD) \rightarrow \\ &\quad -(HO)CD-CH_2- (+(Cys)S^\bullet) \qquad (17)\end{aligned}$$

A pair of thiol groups is assumed to be involved in the electron transfer reactions necessary for steps 2 and 3 above, as well as an acid (to provide a proton in the first step). The last step in this mechanism is a hydrogen atom transfer from the catalytic thiol group back to the substrate (now in the form of product). This is to conform to an experimental finding that there was no hydrogen isotopic exchange at position C-3′ after a completion of one reaction cycle. When the substrate is exhausted, $(Cys)S^\bullet$ would abstract a hydrogen atom from the adenosine, resulting in the reformation of the adenosyl radical which then recombines with the Co(II) of cobalamin, completing the reaction. This reaction sequence implies the presence of a scheme of elaborate conformational changes in the enzyme that guarantees the specific hydrogen transfers between the important players at specific stages of the reaction.

4.3. Galactose Oxidase

Galactose oxidase (see also Chap. 6) catalyzes the oxidation of primary alcohols as well as galactose to produce the corresponding aldehydes and hydrogen peroxide:

$$RCH_2OH + O_2 \rightarrow RCH{=}O + H_2O_2 \tag{18}$$

A recent X-ray crystallographic determination [28] revealed that the active site contains a copper ion coordinated by two histidine nitrogens, one tyrosyl oxygen (Tyr272), and a water molecule. This tyrosine is covalently bonded through sulfur to a cysteine residue on the protein. A native enzyme from *Dactylium dendroides* can be activated by $Fe(CN)_6{}^{3-}$, and the active form can be reduced back to an inactive form by $Fe(CN)_6{}^{4-}$. The active form, though containing Cu(II), is electron paramagnetic resonance (EPR)-silent and shows an absorption spectrum indicative of a tyrosyl radical [29]. The inactive reduced state shows a Cu(II) EPR signal. Addition of a substrate to the active enzyme under anaerobic condition abolishes both the Cu(II) EPR signal and the tyrosyl radical optical absorption signal [29]. As a result of these studies, the following mechanism has been proposed:

$$\begin{aligned}&[\mathrm{Cu(II)}, \mathrm{PhO}^{\bullet}]\ (\text{EPR-silent}) + RCH_2OH \rightarrow \\ &\qquad [\mathrm{Cu(I)}, \mathrm{PhOH}] + RCH{=}O + H^+\end{aligned} \tag{19}$$

$PhO^{\bullet}$ would first abstract an H atom from the substrate. The resulting $RH(C^{\bullet})OH$ is a stronger acid than the original alcohol, and can thus reduce Cu(II) (see Sec. 3):

$$\begin{aligned}&PhO^{\bullet} + RCH_2OH \rightarrow PhOH + RH(C^{\bullet})OH \\ &\qquad RH(C^{\bullet})OH + \mathrm{Cu(II)} \rightarrow RCH{=}O + \mathrm{Cu(I)} + H^+\end{aligned} \tag{20}$$

The resulting Cu(I) moiety can readily be oxidized by O_2 to Cu(II), and $O_2^{\bullet-}$ may abstract an H atom from the tyrosine, recreating the active catalytic entity:

$$[Cu(I), PhOH] + O_2 \rightarrow [Cu(II), PhOH, O_2^{\bullet -}]$$
$$\rightarrow [Cu(II), PhO^{\bullet}] + HO_2^- \quad (21)$$

4.4. Prostaglandin Endoperoxide Synthetase and Cytochrome c Peroxidase

Prostaglandin endoperoxide synthetase (PGHS) catalyzes the first step in the conversion of arachidonic acid (AA) to prostaglandin derivatives (see also Chap. 8). This step consists of two reactions: (i) a cyclooxygenase reaction and (ii) a peroxidase reaction [Eq. (22)]. Both of these

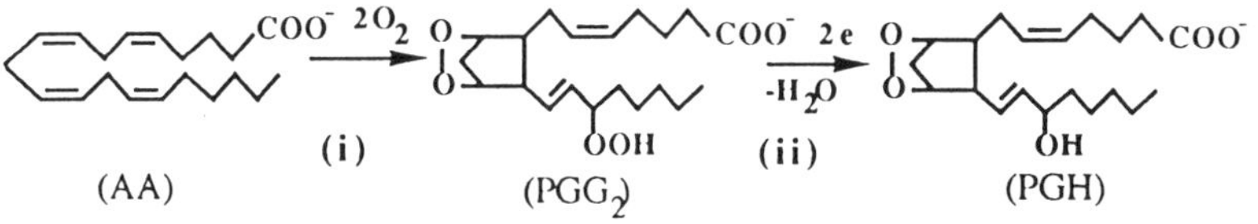

(22)

reactions are believed to be catalyzed by the single enzyme (PGHS), which contains a heme group [30]. The peroxidase activity (ii) is similar to that of the heme-containing cytochrome c peroxidase (see also Chap. 2) in terms of mechanism. It is believed to involve:

$$\text{P-Fe(III)-Por} + \text{ROOH} \rightarrow {}^{\bullet}\text{P-Fe(IV)-Por} + \text{ROH} \quad (23)$$

where P = protein and Por = porphyrin. ROOH accepts an electron from Fe(III) and then splits the RO-OH bond to produce RO^- and $HO^{\bullet}$. The latter abstracts an H atom from an amino acid residue. This residue was recently identified as tyrosine-385 [31,32].

The mechanism of cyclooxygenase activity (i) is not well understood. It has been demonstrated that the EPR spectrum changed little when a substrate was added to the enzyme under an anaerobic condition, but that the EPR obtained upon addition of the substrate under an aerobic condition was very similar to that obtained when the product (PHG) was added instead. This EPR signal was reminiscent of a tyrosyl radical [31]. This tyrosyl radical was thought to be essential to the cyclooxygenase activity [32]. However, a recent study [33] indicates that the tyrosyl radical is not catalytically competent in cyclooxygenase activity. It may be speculated that a free radical derived from a reaction such as Eq. (23) abstracts an H atom from C-13 of AA and turns AA to a radical, which is then transferred to C-11. O_2 can add to C-11 and forms the endoperoxide. This will result in the formation of a radical at C-15. O_2 adds to C-15, and the resulting ROO$^{\bullet}$ would abstract the H atom back from the tyrosine. This completes a reaction cycle. The crucial issue here is how the first hydroperoxide may be produced. Once a small amount of hydroperoxide is formed, the reactions, both (i) and (ii), will proceed.

As mentioned earlier, cytochrome c peroxidase is believed to produce also $^{\bullet}$P-Fe(IV)-Por as a result of the reaction with ROOH. The amino acid residue that forms the radical has been suggested to be tryptophan-191 [34]. This is the first and so far only protein that forms a tryptophan radical (see also Chap. 2).

4.5. Amine Oxidases

The cofactor other than a copper atom had long been unknown for this enzyme, but it was recently identified as 6-hydroxydopaquinone [35] (see also Chap. 7). The active enzyme exhibits an EPR spectrum of a Cu(II) entity. This signal did not change significantly when the enzyme was incubated with a substrate amine under an anaerobic condition. This had led to a belief that the copper did not function by changing its valence like Cu(II)-Cu(I).

There is a general basic problem in interpreting the EPR spectral data in regard to enzymatic mechanisms. The problem stems from the fact that EPR measurements are typically done at very low temperature such as 20 K, while the enzymatic reactions are carried out at around room temperature, say, 300 K. There is a huge temperature gap between these experiments. If the chemical entity of interest shows a strong temperature dependence, the situation at a low temperature (e.g., 20 K) may not necessarily reflect that at 300 K. One ought to keep this fact in mind when interpreting and extrapolating the data from low temperature to room temperature [6].

It turned out that the EPR signal does change when the EPR measurement is done at room temperature [36]. That is, the Cu(II) signal diminishes and a free radical signal appears when a substrate amine is added to the enzyme under an anaerobic condition [36]. This radical signal is likely to be due to the semiquinone form of 6-hydroxydopaquinone. The reaction can thus be described by Eq. (24):

$$[\text{Cu(II)}, \text{O}{=}\Phi{=}\text{O}] + \text{RCH}_2\text{NH}_2 \rightarrow [\text{Cu(I)}, \text{HO-}\Phi\text{-O}^{\bullet}] + \text{RCH}{=}\text{NH} + \text{H}^+ \quad (24)$$

It can be conjectured that (Cu(I), semiquinone) is in equilibrium with (Cu(II), dihydroquinone) and that the equilibrium would shift toward (Cu(II), dihydroquinone) at lower temperatures. If this is correct, (Cu(I), semiquinone) may not be observable at very low temperatures.

Methylamine dehydrogenase whose cofactor had been presumed to be PQ (2,7,9–tricarboxypyrroloquinolinequinone) has recently been demonstrated to contain another novel cofactor, tryptophan tryptophylquinone [37]. Two tryptophan moieties are covalently bound, and one of them is converted to an orthoquinone. This quinone is believed to participate in the redox reaction as in the amine oxidase reaction mentioned above.

4.6. Other Examples

The diol dehydratase obtained from *Clostridium glycocalycum* does not contain AdoCbl, unlike the enzyme from the other sources (see also Chap. 9). It exhibits an EPR signal at $g = 2.02$, which is sensitive to the action of hydroxyurea. Reagents that act on thiol groups diminished both the enzymatic activity and the radical signal intensity at the same time. This observation suggested that the radical is on a cysteine residue [38].

Pyruvate formate lyase from *E. coli* is activated by the action of an enzyme which depends on Fe(II). The enzyme thus activated shows an EPR signal at $g = 2.0037$ [4], which has been assigned to a free radical at the α carbon of Gly734. It is conceivable that the activation process involves an $O^{\bullet}$ radical formation as in Eq. (15) or a similar reaction involving Fe(II) and O_2.

It might be pointed out that factor $Z^{\bullet+}$ in Photosystem II has been identified as a tyrosyl radical [39] (see Chap. 3 for details).

ABBREVIATIONS

AA arachidonic acid
AdoCbl adenosylcobalamin

ATP	adenosine 5′-triphosphate
t-BuO	tertiary butoxyl
CoA	coenzyme A
EPR	electron paramagnetic resonance
FAD	flavin adenine dinucleotide
FMN	flavin mononucleotide
Me	methyl
MeO	methoxyl
NAD^+	nicotinamide adenine dinucleotide
PGG_2	prostaglandin G_2
PGH	prostaglandin H_2
PGHS	prostaglandin endoperoxide synthetase
Ph, Φ	phenyl group
UTP	uridine 5′-triphosphate

REFERENCES

1. (a) J. A. Stubbe, *Biochemistry*, *27*, 3893 (1988); (b) J. A. Stubbe, *Ann. Rev. Biochem.*, *58*, 257 (1989).
2. (a) T. J. McMurry and J. T. Groves, in *Cytochrome P-450* (P. R. Ortiz de Montellano, ed.), Plenum Press, New York, 1986, Chap. 11; (b) see Ref. 13; (c) P. R. Ortiz de Montellano and R. A. Stearns, *J. Am. Chem. Soc.*, *109*, 3415 (1987).
3. W. A. Pryor, ed., *Free Radicals in Biology*, Academic Press, New York, 1976–(series); (b) B. Halliwell and J. M. C. Gutteridge, *Methods in Enzymology*, Vol. 186 (L. Packer and A. N. Glazer, eds.), Academic Press, New York, 1990, p. 1 ff.
4. A. F. Volker Wagner, M. Frey, F. A. Neugebauer, W. Schaefer, and J. Knappe, *Proc. Natl. Acad. Sci.*, *89*, 996 (1992).
5. P.A. Frey, *Chem. Rev.*, *90*, 1343 (1990).
6. E.-I. Ochiai, *General Principles of Biochemistry of the Elements*, Plenum Press, New York, 1987, Chap. 4.
7. N. S. Isaacs, *Physical Organic Chemistry*, Longman, London, 1987, Chap. 15.
8. J. March, *Advanced Organic Chemistry*, 3rd ed., John Wiley and Sons, New York, 1985, Chaps. 5 and 14.

9. P. Neta, *Adv. Phys. Org. Chem., 12*, 223 (1976).
10. T. H. Lowry and K. S. Richardson, *Mechanism and Theory in Organic Chemistry*, 2nd ed., Harper and Row, New York, 1981, Chap. 9.
11. K.-D. Asmus, A. Henglein, A. Wigger, and G. Beck, *Ber. Bunsenges. Phys. Chem., 70*, 756 (1966).
12. J. Halpern, *Science, 227*, 869 (1985).
13. B. P. Hay and R. Finke, *J. Am. Chem. Soc., 109*, 8012 (1987).
14. (a) W. Holloway, H. A., White, K. N. Joblin, A. W. Johnson, W. F. Lappert, and O. C. Wallis, *Eur. J. Biochem., 82*, 143 (1978); (b) A. Gaudemer, J. Zybler, N. Zybler, M. Baran-Marszac, W. E. Hull, M. Fountoulakis, A. Koenig, K. Woelf, and J. Retey, *Eur. J. Biochem., 119*, 279 (1981).
15. E.-I. Ochiai and C. E. Loope, *J. Inorg. Chem., 36*, 194 (1989) (4th Int. Conf. Bioinorg. Chem., Boston).
16. J. R. O'Brien, J. A. Fox, M. G. Kopczynski, and B. M. Babior, *J. Biol. Chem., 260*, 16131 (1985).
17. K. M. Bansal, L. K. Patterson, A. Henglein, and E. Janata, *J. Phys. Chem., 77*, 16 (1973).
18. R. G. Finke, D. A. Schiraldi, and B. J. Mayer, *Coord. Chem. Rev., 54*, 1 (1984).
19. (a) P. Dowd, in *Selective Hydrocarbon Activation: Principles and Progress* (J. J. Davies, P. L. Watson, A. Greenberg, and J. F. Liebman, eds.), VCH, Weinheim, 1990, Chapt. 9 (p. 265 ff); (b) G. Choi, S.-C. Choi, A. Galan, B. Wilk, and P. Dowd, *Proc. Natl. Acad. Sci. USA, 87*, 3174 (1990).
20. L. Thelander and P. Reichard, *Ann. Rev. Biochem., 43*, 133 (1979).
21. M. Lammers and H. Follmann, *Struc. Bondg., 54*, 27 (1983).
22. G. J. Mann, A. Gräslund, E.-I. Ochiai, R. Ingemarson, and L. Thelander, *Biochemistry, 30*, 1939 (1991).
23. (a) J. A. Stubbe, *Ann. Rev. Biochem., 58*, 257 (1989); (b) J. A. Stubbe, *J. Biol. Chem., 265*, 5329 (1990); (c) J. A. Stubbe, *Adv. Enzymol., 63*, 349 (1990).
24. P. Nordlund, B-M. Sjöberg, and H. Eklund, *Nature*, 345, 593 (1990).
25. E.-I. Ochiai, G. J. Mann, A. Gräslund, and L. Thelander, *J. Biol. Chem., 265*, 15758 (1990).

26. E.-I. Ochiai, S. Swierczynski, A. Gräslund, and L. Thelander, *J. Inorg. Biochem.*, *43*, 531 (1991) (5th Int. Conf. Bioinorg. Chem., Oxford).
27. J. M. Bollinger, Jr., D. E. Edmonson, B. H. Huyn, J. Filley, J. R. Norton, and J. A. Stubbe, *Science*, *253*, 292 (1991).
28. N. Ito, S. E. V. Phillips, C. Stevens, Z. B. Orgel, M. J. McPherson, J. N. Keen, K. S. O. Yadav, and P. F. Knowles, *Nature*, *350*, 87 (1991)
29. (a) M. M. Whittaker and J. M. Whittaker, *J. Biol. Chem.*, *263*, 6074 (1988); (b) M. M. Whittaker and J. M. Whittaker, *J. Biol. Chem.*, *265*, 9610 (1990).
30. (a) T. Miyamoto, S. Yamamoto, and O. Hayaishi, *Proc. Natl. Acad. Sci. USA*, *71*, 3645 (1974); (b) T. Miyamoto, N. Ogino, S. Yamamoto, and O. Hayaishi, *J. Biol. Chem.*, *251*, 2629 (1976).
31. T. Katherin, R. Dietz, W. Nastainczyk, and H. H. Ruff, *Eur. J. Biochem.*, *171*, 313 (1990).
32. T. Shimokawa, R. J. Kulmacz, D. L. Dewitt, and W. L. Smith, *J. Biol. Chem.*, *265*, 20073 (1990).
33. G. Lassmann, R. Odenwaller, J. F. Curtis, J. A. DeGray, R. P. Mason, L. J. Marnett, and T. E. Eling, *J. Biol. Chem.*, *266*, 20045 (1991).
34. M. Sivaraja, D. B. Goodwin, M. Smith, and B. M. Hoffman, *Science*, *245*, 738 (1989).
35. S. M. Janes, D. Mu, D. Wemmer, A. K. Smith, S. Kaut, D. Maltby, A. L. Burlingame, and J. P. Klinman, *Science*, *248*, 981 (1990).
36. D. M. Dooley, M. A. McGuirl, D. E. Brown, P. N. Turowski, W. S. McIntire, and P. F. Knowles, *Nature*, *349*, 262 (1991).
37. W. S. McIntire, D. E. Wemmer, A. Chistoserdov, and M. E. Lidstrom, *Science*, *252*, 817 (1991).
38. M. G. N. Hartmanis and T. C. Stadtman, *Proc. Natl. Acad. Sci. USA*, *74*, 76 (1987).
39. (a) B. A. Barry and G. T. Babcock, *Proc. Natl. Acad. Sci. USA*, *84*, 7099 (1987); (b) B. A. Barry, M. K. El-Deeb, P. O. Sandusky, and G. T. Babcock, *J. Biol. Chem.*, *265*, 20139 (1990).

2

Peroxidases: Structure, Function, and Engineering

Thomas L. Poulos[1] and Roger E. Fenna[2]

[1]Departments of Molecular Biology and Biochemistry, and Physiology and Biophysics, University of California at Irvine, Irvine, California, USA 92717

[2]Department of Biochemistry and Molecular Biology, University of Miami Medical School, P.O. Box 016129, Miami, Florida USA 33101

1. INTRODUCTION

1.1. Biological Roles

The purpose of this chapter is to review structure–function relationships for the three heme peroxidases for which detailed crystallographic information is available: cytochrome c peroxidase (CCP), myeloperoxidase (MPO), and lignin peroxidase (LiP). Peroxidases catalyze the peroxide-dependent oxidation of various organic molecules and other protein redox centers. These enzymes are ubiquitous in nature although the most thoroughly studied peroxidases are from plants. In plants peroxidases are considered key components in plant cell wall biosynthesis [1] including lignin formation. Peroxidases catalyze formation of coniferyl, sinapyl, and *p*-coumaryl alcohol radicals which condense in a nonenzymatic reaction to give lignin [2,3]. Peroxidases also participate in plant hormone biosynthesis including indole acetic acid [1] and ethylene [4].

Of particular concern in this chapter is the role that peroxidases play in lignin degradation. This is an important reaction since lignin is the second most abundant polymer on earth [5] and provides a nearly impervious barrier around cellulose in woody plants. Thus, the degradation of lignin is critical to the carbon cycle. White rot basidiomycete fungi are one of the few organisms known to degrade lignin to water and carbon dioxide. *Phanerochaete chrysosporium* is the best studied of these organisms and secretes two peroxidases, lignin peroxidase, or LiP, and manganese peroxidase, which are major components in lignin degradation [6,7]. LiP is also able to attack substrates that cannot be oxidized by plant peroxidases such as nonphenolic aromatics [6–9]. This makes *P. chrysosporium* and its peroxidases potentially important tools in the degradation of aromatic pollutants [10–12].

Heme-containing peroxidases also are produced in a wide variety of mammalian tissues where they are largely involved in defense mechanisms that protect against pathogenic bacteria, viruses, fungi, and larger metazoan parasites. Myeloperoxidase and eosinophil peroxidase (EPO), produced by certain types of white blood cells, catalyze the oxidation of halide and thiocyanate ions to the powerful oxidizing agents: HOCl, HOBr and HOSCN which are effective in killing pathogenic organisms through a variety of secondary reactions. For a recent review of the biological activities of the MPO-peroxide-halide system see Ref. [13]. MPO has also been shown capable of inactivating human α-1-antiproteinase, thereby promoting proteolytic tissue degradation in emphysema [14,15]. Several studies have suggested a role for MPO as a modulator of the inflammatory response through its ability to catalyze the oxidative inactivation of certain leukoattractants as well as several enzymes released by polymorphonuclear leukocytes [16,17].

Lactoperoxidase (LPO), found abundantly in the milk of several mammals, utilizes hydrogen peroxide and thiocyanate to generate HOSCN that effectively kills bacteria capable of producing potentially lethal mammary infections. The addition of exogenous thiocyanate and sodium percarbonate, as a source of H_2O_2, has been used to facilitate the preservation of milk for human consumption through the utilization of endogenous LPO to kill pathogenic bacteria [18]. Peroxidases found in saliva, the intestine and uterus are believed to play similar protective roles. Thyroid peroxidase (TPO) is an exception in that it is specifically involved in the biosynthesis of a hormone: thyroxine, through its catalysis of both tyrosine iodination and the iodotyrosine coupling reaction. Initially, TPO catalyses the iodination of tyrosines on thyroglobulin to mono- and di-iodotyrosines which are subsequently coupled through phenoxy-ether formation to yield the thyroid hormones thyroxine (T_4) and triiodothyronine (T_3). For a recent review of thyroid peroxidase see Ref. [19].

Seven complete sequences and one partial sequence have now been determined for four different mammalian peroxidases: human and mouse MPO [20–22]; human EPO [23,24]; human, pig, and rat TPO [25–28]; bovine LPO [29,30]; and a partial sequence for human LPO [29]. These studies have revealed high levels of sequence identity (40–70%) providing evidence for the existence of an extensive gene family of mammalian peroxidases. A more distant relative, prostaglandin H synthase, exhibits both heme-dependent hydroperoxidase and cyclooxygenase activities in catalyzing the biosynthesis of prostaglandins and thromboxane from arachidonic acid. Alignment of the amino acid sequences of human MPO and prostaglandin H (PGH) synthase indicates 22% identical residues [31].

Despite what we do know about the biological role of peroxidases, it remains an irony that the in vivo function of the one peroxidase for which we have the most detailed structural information, cytochrome c peroxidase (CCP), remains an enigma. This enzyme is produced in yeast as a proenzyme with a 68-residue leader sequence [32]. The leader peptide is processed off as the enzyme is transported across the outer mitochondrial membrane [33] with the holoenzyme remaining between the inner and outer mitochondrial membranes [34]. Although there is evidence suggesting that CCP can support oxidative phosphorylation in the absence of a functioning cytochrome oxidase [35], it has not been demonstrated that this is the role of CCP in vivo. Another proposed role for CCP is as a protective peroxide scavenger but, again, this is only a suggestion and has not been demonstrated. Nevertheless, CCP has proven to be an excellent system for studying the mechanism of peroxide reduction, interprotein electron transfer reactions, and for developing the area of heme enzyme mutagenesis and engineering.

1.2. Biosynthesis of Mammalian Peroxidases

Extensive studies of the biosynthesis of MPO have been carried out using the human promyelocytic cell line HL-60. The primary translation product is an 80-kD polypeptide that includes a signal sequence and an amino terminal pro sequence [36]. Cotranslational cleavage of the signal peptide is accompanied by the addition of five high mannose carbohydrate chains all of which are linked to asparagines. The resultant 92-kD polypeptide undergoes rapid partial deglycosylation to an 89-kD form of proMPO [37]. Further processing involves removal of the 116-amino-acid pro sequence, the excision of six residues to generate the separate small and large polypeptides, and removal of the carboxy terminal serine. Little is known about the mechanism of heme incorporation, although this is known to occur at the level of the pro enzyme [38]. The formation of dimers is believed to occur at a later stage following incorporation into the secretory granules [39].

Other members of the mammalian peroxidase family are believed to undergo similar posttranslational processing from their proenzyme forms to the mature enzymes. However, while eosinophil peroxidase (EPO) resembles MPO in that processing involves cleavage into small and large polypeptides, the mature enzyme forms of LPO and TPO are both single polypeptides. In addition, MPO is the only member of the family known to form covalent dimers through disulfide bridge formation.

1.3. Chemistry

Peroxidases have also proven to be very useful to enzymologists since these enzymes form distinct, relatively long-lived reaction intermediates whose kinetics and biophysical properties can be studied in depth. The generalized peroxidase reaction occurs in three well-defined steps:

$$\mathrm{Fe(III)} + \mathrm{H_2O_2} \longrightarrow \underset{\text{Compound I}}{\mathrm{Fe(IV)}{=}\mathrm{O}\ \mathrm{R}^{\bullet}} + \mathrm{H_2O}$$

$$\mathrm{Fe(IV)}{=}\mathrm{O}\ \mathrm{R}^{\bullet} + \text{substrate} \longrightarrow \underset{\text{Compound II}}{\mathrm{Fe(IV)}{=}\mathrm{O}\ \mathrm{R}} + \text{oxidized substrate}$$

$$\mathrm{Fe(IV)}{=}\mathrm{O}\ \mathrm{R} + \text{substrate} \longrightarrow \mathrm{Fe(III)} + \mathrm{H_2O} + \text{oxidized substrate}$$

The first step in the reaction cycle is the oxidation of the enzyme by peroxide to compound I. Although no true Michaelis–Menten complex has been observed between peroxidase and peroxide, its existence has been inferred. The activation energy for compound I formation is near to that for the fluidity of water although the rate of compound I formation is much slower than that expected for a diffusion-controlled reaction [40]. These observations require that a prereaction complex forms. More recent experiments have supported this prediction. Baek and Van Wart [41] identified what they call peroxidase compound 0′ using rapid reaction low-temperature techniques. This is a peroxidase–peroxide complex which forms prior to compound I and, while the structure of this intermediate remains to be elucidated, it could be the prereaction complex that forms prior to cleavage of the O-O bond.

The nature of compound I has been the subject of considerable study and certain generalizations can now be made about its electronic structure. The two oxidizing equivalents of peroxide are stored by promoting iron from the +3 to +4 state and the second is stored as an organic radical. One peroxide-derived oxygen atom remains bound to the iron atom giving the so-called oxyferryl center. The existence of the iron–oxo bond has been demonstrated by both extended absorption line-structure spectroscopy (EXAFS) [42] and crystal structure studies [43] of compound I. In most peroxidases, the radical center is the heme itself giving a porphyrin π-cation radical [44]. CCP is unique since the radical is located on an amino acid side chain [45], mostly likely Trp191 [46]. Location of the CCP radical center will be considered in more detail in Sec. 4.1.

In the second step, compound I is reduced in a one-electron-transfer reaction to compound II. If the substrate is an aromatic molecule as with many of the plant peroxidase substrates, the substrate is oxidized to a radical which then reacts further, usually with a second substrate radical, in a nonenzymatic event. Compound II contains the oxyferryl center but no porphyrin radical. The situation is not so clear with CCP since it appears that the oxyferryl center is preferentially reduced by cytochrome c rather than the radical [47]. This leaves iron in the +3 state which then should be available for reaction with another peroxide molecule. Hence, under steady-state conditions, CCP may not form the traditional compound II. This additional complexity will be considered in more detail in Sec. 4.1.1.

The third and last step is the reduction of compound II back to the resting, ferric enzyme by a second substrate molecule.

2. PEROXIDASE CRYSTAL STRUCTURES

2.1. Cytochrome c Peroxidase and Lignin Peroxidase

2.1.1. *Introduction*

CCP was the first heme enzyme whose X-ray structure was determined [48] and as such has served as a paradigm for peroxidase structure and function. This has been especially true in understanding how peroxidases activate and store the oxidizing equivalents of peroxides. Although it is generally accepted that the reaction mechanism with peroxides to form compound I is the same in all fungal and plant heme peroxidases, the specificity of any particular peroxidase is in the types of molecules oxidized. CCP is unique among this family of peroxidases since its substrate is another macromolecule, cytochrome c, while most other peroxidases which have been studied operate on small molecules. This presents a limitation on using CCP as a paradigm for understanding the reduction of compound I. Fortunately, the 2.0 Å structure of a second peroxidase in this group, LiP, is now available [49,50]. The refined model ($R = 0.15$ for data between 8 and 2.0 Å and at least twice above background) exhibits excellent geometry (rms bond distances = 0.014 Å) and consists of all 343 residues, two calcium ions, and one glucosamine linked to Asn257. The availability of both refined models, CCP and LiP, now allows for some detailed comparisons to be made. Even before refinement of the LiP structure it was clear that despite the well-defined functional differences between CCP and LiP and the relatively low sequence homology between the two [51], the three-dimensional structures are remarkably similar. Hence, rather than describe each structure separately, we will discuss both structures together and emphasize important differences. The MPO structure will be described in a separate section since this represents a new class of peroxidase structure.

Before discussing the CCP and LiP structures, however, we should further elaborate on some of the functional differences between CCP and LiP. The most obvious are the substrates. CCP is designed to form a specific complex with cytochrome c and is not a very effective enzyme at oxidizing small aromatic molecules as are the plant peroxidases [52]. In contrast, LiP is able to oxidize aromatic and especially nonphenolic aromatic compounds [6–9]. One of these is veratryl alcohol (3,4-dimethoxybenzyl alcohol), a natural metabolite of *P. chrysosporium*. Veratryl alcohol also is relevant to another important difference between CCP and LiP. LiP readily converts to the oxy complex, usually called compound III or LiPIII [53,54], in the presence of excess peroxide and does not form a very stable compound I while CCP compound I is quite stable [45]. The presence of veratryl alcohol prevents LiPIII formation by reducing LiPI back to the resting state [53,54]. Thus, veratryl alcohol exhibits a protective effect by insuring that LiP is available for oxidation of LiP substrates, either lignin itself or degradation products of lignin. This unusual reactivity of LiP may be required for oxidation of those nonphenolic substrates with high redox potentials that cannot be oxidized by plant peroxidases.

One final difference between LiP and other peroxidases is the very low pH (near pH = 3) optimum for LiP reactions [55,56]. This is due to the reduction of compounds I and II and not in the reaction with peroxide [55]. As with CCP and plant peroxidases, the reaction of LiP with peroxide to form compound I is insensitive to pH over a wide range (pH 5–8) [55–57]. How pH controls the reduction of compounds I and II is an intriguing problem we consider in Sec. 3.1.

2.1.2. *Protein Fold*

CCP and LiP consist of two domains divided approximately equally between the N-terminal and C-terminal halves of the protein (Fig. 1). We refer to these domains as the distal (first half of sequence) and proximal (second half of sequence) domains. Eleven helical segments dominate the structure and compose about 50% of the molecule. In CCP the β structure is limited to the proximal domain and contains two antiparallel β pairs and one small β sheet. In LiP, the nonhelical region of the proximal domain corresponding to the β structure in CCP appears less regular.

LiP is larger than CCP (343 residues vs. 294 residues) and most of the extra protein is at the C-terminal region. At the point where the homology between LiP and CCP breaks down, near residue 295 of LiP, the polypeptide traverses over the surface of the protein and does not form any regular extended region of secondary structure. Indeed, this

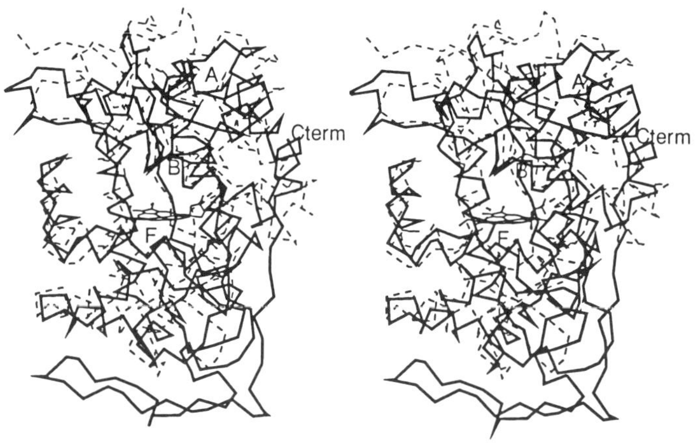

FIG. 1. Stereo α-carbon backbone models of LiP superimposed on CCP. Solid line, LiP; dashed lines, CCP.

C-terminal region unique to LiP is quite noisy in electron density maps with relatively high temperature factors indicating more disorder in the last 49 residues. LiP also differs from CCP in that LiP contains four disulfide bonds while CCP has none. Two of these occur at the N-terminal region involving the following pairs: Cys3-Cys14 and Cys15-Cys285. This produces a complicated twist at the N terminus not present in CCP. The remaining disulfides are between Cys34-Cys120 and Cys317-Cys349. The former could be readily accommodated in the CCP fold while the latter could not since Cys317 is part of the extra C-terminal tail not present in CCP. One other important difference is that LiP is a glycoprotein while CCP is not. The refined map at 2 Å clearly shows the presence of one N-acetyl-glucosamine attached to Asn257 as expected [58], with weaker density for a second carbohydrate. Extra electron density extending from Ser334 indicates at least one other site of carbohydrate attachment although the electron density is not strong enough to warrant building in additional carbohydrate units.

One final and probably important difference in the overall structure is access to the distal pocket where peroxides and other ligands bind. The heme in both enzymes is situated between the distal helix B and proximal helix F at the bottom of an open crevice. Figure 2 shows a comparison of the van der Waals surface around the access channel in CCP and LiP. It should be clear that the pocket of CCP is larger than the corresponding pocket in LiP. This difference is due primarily to side chain interactions in LiP not present in CCP. This important difference explains why the δ-meso heme position in CCP is available for chemical modification [52] while that of LiP is less reactive [59]. The compounds used to modify the heme (hydrazines) simply do not have as ready an access to the heme edge in LiP owing to the very small access channel. In contrast, the access channel in CCP is large enough to allow modification of the heme and also allows styrene to approach close enough to the iron-linked oxo atom in compound I for formation of styrene oxide [60]. In this regard, access to the ferryl center allows CCP to behave like a cytochrome P-450. As predicted [60], accessibility to the heme and iron center plays a large role in controlling activity.

LiP resembles horseradish peroxidase (HRP) [61] and differs from CCP in that LiP contains at least two calcium sites shown in Fig. 3. One site located in the distal domain is formed primarily by one continuous loop from residues 65–70. The side chains of Asp68, Ser70, Asp48, the peptide carbonyl of Gly65, and two water molecules provide the ligands.

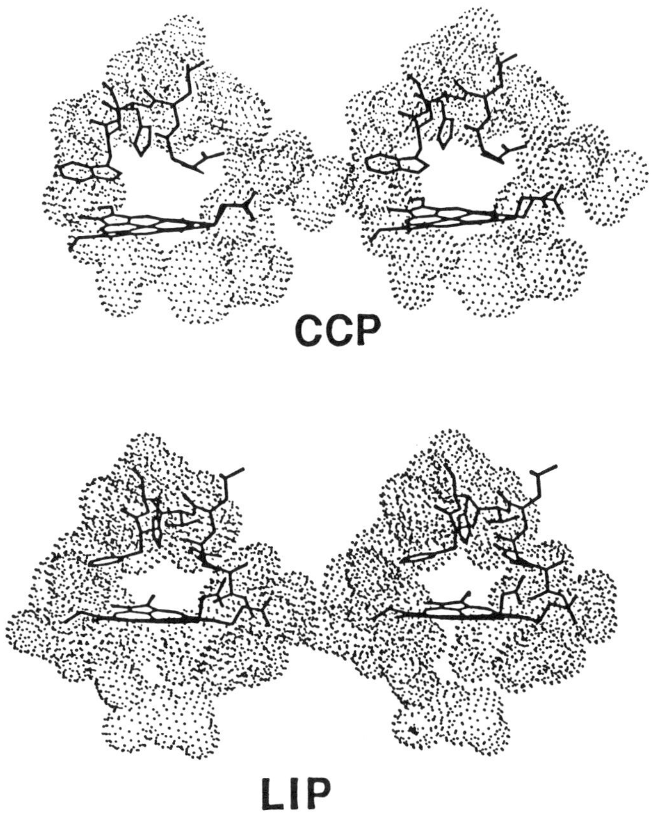

FIG. 2. Stereo views of the heme access channels in LiP and CCP. The dots represent van der Waals surfaces. Notice that this opening is much larger in CCP than LiP. The opening in LiP is approximately the correct size to accommodate an aromatic substrate and is the postulated substrate binding site.

One of the water ligands hydrogen-bonds with Glu78. The second site located in the proximal domain is more intricate with no observed solvent ligands. Here residues 196–201 provide ligands via the side chains of Thr196, Asp201, and the peptide carbonyl oxygen atoms of Pro196 and Ile199. The side chains for Ser177 and Asp194 provide the remaining ligands. The location of these ions is most interesting since each contains a ligand attached to one end of the distal and proximal helices. In the distal site, Asp48, which is highly conserved in plant peroxidases [51] and is near the C-terminal end of the distal helix and immediately follows the distal His47, is a calcium ligand. In the proxi-

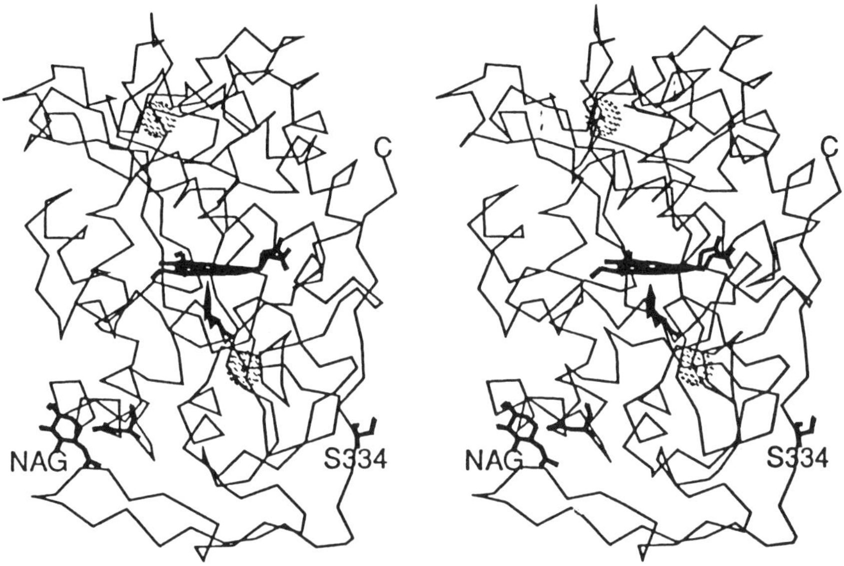

FIG. 3. Stereo α-carbon backbone model of LiP. Locations of glycosylation sites are indicated as NAG and S334. The small spheres represent the location of calcium ions.

mal site, Ser177, which is directly adjacent to the proximal ligand His176, is one ligand. Furthermore, Asp194 is a ligand and immediately follows Phe193 whose phenyl group lies parallel to and in contact with the proximal His ligand. It would therefore appear that divalent metal ions could be important for maintaining structural integrity of the heme pocket. Indeed, it has been shown in HRP that removal of calcium results in ≈40% loss of activity which can readily be restored by the addition of calcium [61]. As will be discussed in Sec. 2.2.2, MPO also contains a calcium ion analagous to the distal calcium site found in LiP.

It was anticipated that LiP would contain an aromatic binding site which enables the enzyme to form a Michaelis–Menten complex with aromatic substrates such as veratryl alcohol. Based on the work of Ortiz de Montellano and colleagues [52,59,62], those peroxidases that interact with aromatic substrates are expected to have a binding site near the heme edge. This is the case for plant peroxidases where direct oxidation at the heme edge by the porphyrin π-cation radical to give an aromatic radical is expected to occur. LiP, however, does not fit this pattern since the hydrazines used to probe the accessibility of the heme edge do not effect LiP [59]. As noted above, this is consistent with the crystal structure, which shows that the heme edge is much less accessible than in CCP (Fig. 2). Nevertheless, the opening in LiP that connects the surface of the enzyme to the distal heme pocket is just large enough to accommodate a molecule like veratryl alcohol [50]. This region also provides a relative close approach of the aromatic substrate to the heme edge, about 6–7 Å.

2.1.3. Active Sites

The homology between LiP and CCP extends to the active site (Fig. 4). The heme is sandwiched between the distal helix B and proximal helix F. The distal pocket is very similar in both proteins. Each contains a distal His and Arg which have been postulated to operate in the formation of CCP compound I. Both have ordered solvent in the distal pocket. One of these waters sits directly over the heme iron atom. The iron–water distance in CCP is about 2.4 Å [63] and in LiP approximately 2.5 Å [50]. Moreover, there is no strong electron density connecting the distal water to the iron in either CCP or LiP indicating that the water is not strongly coordinated to the iron if at all. This would be consistent with a pentacoordinate high-spin iron found with a variety of spectral probes [64–66].

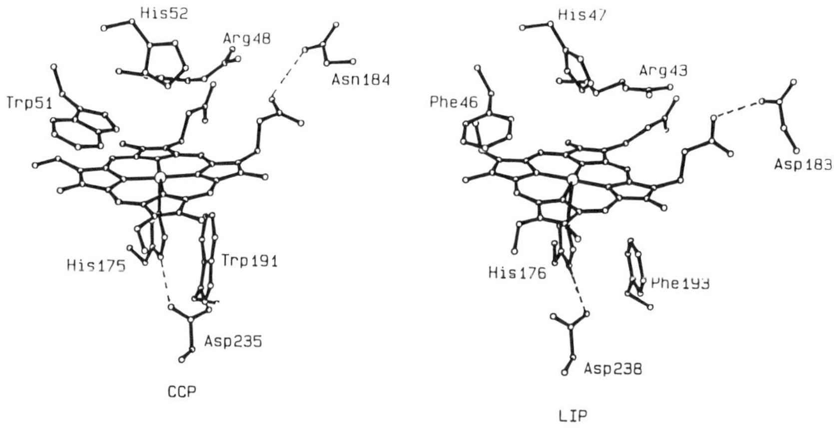

FIG. 4. Comparison of LiP and CCP active sites.

The proximal pocket also is very similar. In both LiP and CCP the proximal His ligand is hydrogen-bonded to a buried Asp side chain. Poulos and Finzel [67] postulated that this H bond imparts a greater anionic character to the His ligand than in other heme proteins like the globins, which provides additional stability to the ferryl (Fe^{4+}) iron in compound I [68]. Overall, these similarities support the view that the detailed mechanism of compound I formation proposed for CCP to be discussed in Sec. 3.1 operates in LiP as well. The most striking difference in the active site was anticipated from sequence comparisons [51]. That is, where CCP has Trp residues located on both the proximal and distal pockets, LiP has Phe (Fig. 4). Plant peroxidases resemble LiP in this regard since the proximal and distal Phe residues are conserved [69].

2.2. Myeloproxidase

2.2.1. Introduction

The X-ray crystal structure of canine myeloperoxidase was determined using crystals in space group $P4_12_12$ ($a = b = 133.0$ Å, $c = 203.6$ Å) grown by precipitation with 1.7 M ammonium sulfate in 50 mM pipes buffer at pH 6.5 [70]. The structure was initially solved by the multiple isomorphous replacement method. Both solvent flattening [71] and noncrystallographic symmetry averaging were used to improve the quality of the MIR phased electron density map to permit unambiguous tracing of the four polypeptide chains comprising an entire MPO molecule. Subsequent restrained least squares refinement using the method of Tronrud et al. [72] resulted in a final model with a crystallographic R factor of 0.257 and root mean square (rms) deviations from ideality for bond lengths and angles of 0.011 Å and 3.8° [73].

MPO differs from other mammalian peroxidases in being dimeric, consisting of two large polypeptides (466 residues) and two small polypeptides (108 residues). The two halves are related by a dyad axis and covalently linked by a single disulfide bridge between equivalent cysteines (C153) in the large polypeptide chains. Selective reduction of this disulfide yields hemi-MPO, which has catalytic and spectral properties indistinguishable from those of the native enzyme [74]. Each half molecule contains a covalently bound heme for which the exact chemical structure has not been determined, a single tightly bound calcium ion, and four potential sites for asparagine-linked glycosylation. Since the amino acid sequence of the canine enzyme has not been determined, the electron density map was interpreted with the aid of the sequence for the human enzyme. To facilitate comparisons with the sequences of other mammalian peroxidases belonging to the same gene family the first cysteine in the mature enzyme has been numbered residue 1. Different members of the family have varying numbers of residues prior to Cys1, two (Val, Thr) in the case of human MPO.

2.2.2. *Three-Dimensional Structure*

The polypeptide backbone structure for one half of the MPO molecule is illustrated in Figs. 5 and 6. The secondary structure is predominantly α-helical with 19 helices ranging in length from 5 to 29 residues (Table 1) making up 36% of the polypeptide backbone. A central core consisting of five α helices, four from the larger polypeptide (H5, H6, H8, and H12) and one from the small (H2), contains the covalently bound heme. Core helices H5 (24 residues) and H6 (29 residues) are connected by a five residue loop at the surface of the molecule and interact closely to form a very prominent inverted "V" structure with an interaxial angle of about 30°. These helices are amphipathic and their interaction involves complementary hydrophobic surfaces. The other three core helices are arranged at the open end of this "V" and their axes are all approximately perpendicular to the plane of the "V". This core is surrounded by regions of the large polypeptide that are folded into local domains. The amino terminal portion of the large polypeptide folds into two domains consisting of residues 113–164 and 165–243. Similarly, the carboxy terminal portion of the large polypeptide chain also forms two distinct domains consisting of residues 429–501 and 502–578. A fifth domain comprises residues 339–416 that separate the H8 and H12 core helices. There is no apparent similarity in amino acid sequence or three dimensional structure between any two of these domains. The large polypeptide is completed by a short stretch of 24 residues (302–325) that link core helices H6 and H8. The small polypeptide is folded into a largely extended configuration resembling a figure "6", with the bulk of the chain traversing the surface of the molecule and its carboxy terminal portion forming the core helix H2 that passes through the center of the molecule.

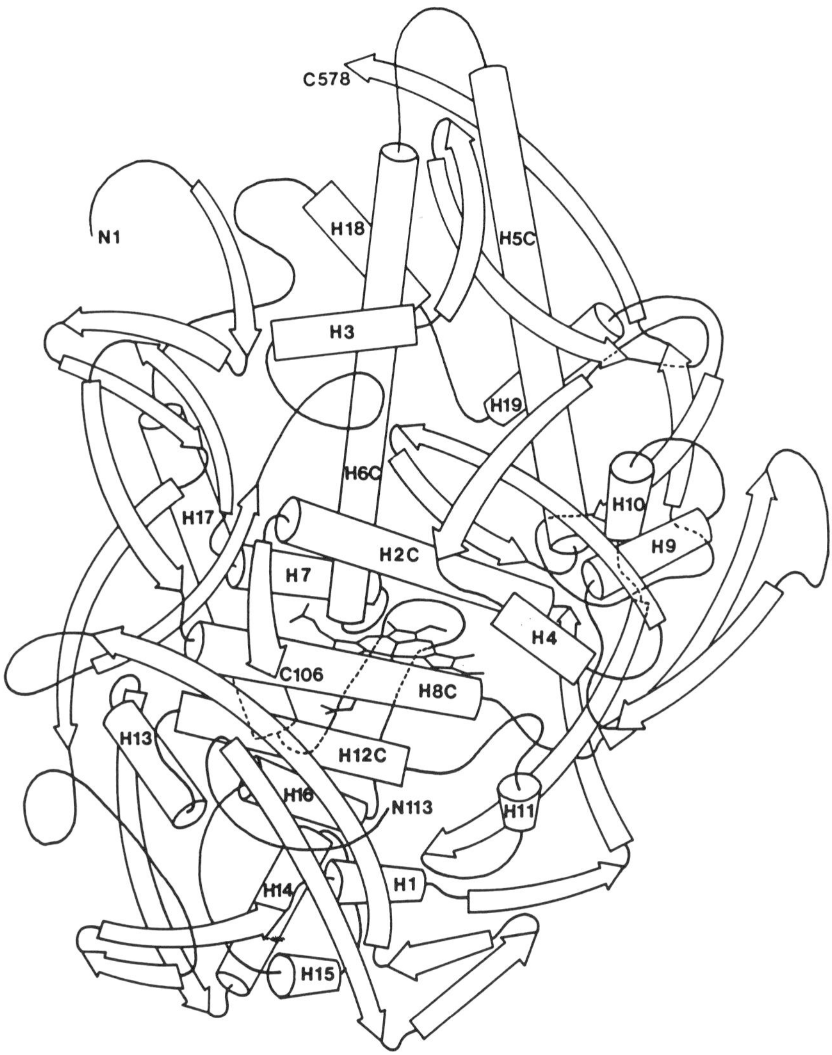

FIG. 5. Schematic drawing of one half of the myeloperoxidase molecule. Helices are shown as cylinders and extended chain as ribbons. The strands of extended chain do not necessarily form a β structure. Helices have been numbered according to their positions in the polypeptide chains defined in Table 1. The chain termini are identified as N1, C106 for the small polypeptide and N113, C578 for the large polypeptide. The heme is partially concealed by helix H8.

FIG. 6. Stereo diagram showing the backbone conformations of the polypeptides composing one half of the myeloperoxidase molecule. The location of the bound calcium ion is indicated by a cross. The amino terminus of the small polypeptide is at the upper left and the carboxyl terminus of the large polypeptide at the top. Chain termini generated by the excision of six residues resulting in the formation of separate large and small polypeptides can be seen just above and below the heme in the center of the molecule. The view is parallel with the molecular dyad, identical to that shown in Fig. 5.

There are six intrachain disulfide bridges in each half of the molecule as follows: Cys115–Cys125, Cys119–Cys143, Cys221–Cys232, Cys440–Cys497 and Cys538–Cys564 in the large polypeptide and between Cys1 and Cys14 in the small polypeptide. In addition to Cys153, which participates in the interchain disulfide between the two halves of the molecule, human MPO has one further cysteine at position 150. Since the high electron density expected for the presence of a sulfur atom was not observed at this position, the canine enzyme may lack this cysteine. The five disulfides in the large polypeptide all involve pairs of cysteines from within individual domains with no crosslinking between pairs of domains.

Both myeloperoxidase and lactoperoxidase have been shown to contain tightly bound calcium ion in 1:1 stoichiometry with the heme

TABLE 1

Location of Helices in Myeloperoxidase

Helix	Residues	Length	Helix	Residues	Length
H1	61–69	9	H11	400–405	6
H2	84–97	14	H12	417–428	12
H3	181–187	7	H13	433–439	7
H4	220–224	5	H14	448–455	8
H5	244–267	24	H15	458–467	10
H6	273–301	29	H16	476–482	7
H7	303–309	7	H17	493–509	17
H8	326–338	13	H18	522–529	8
H9	368–374	7	H19	533–540	8
H10	377–385	9			

Starting and ending residues for each helix were determined by the ϕ and ψ angles from the Ramachandran plots. Helices H2, H5, H6, H8, and H12 make up the central core.

iron [75]. The calcium binding site in MPO (Fig. 7) is composed largely of a single loop (168–174) in which calcium ligands include the side chain hydroxyl oxygen of Thr168, the peptide carbonyl oxygen of Phe170, a side chain carboxyl oxygen of Asp172, and the side chain hydroxyl oxygen of Ser174. An additional ligand is provided by a carboxyl oxygen of Asp96 which is adjacent to the distal histidine-95. This arrangement is very similar to that found in LiP where an aspartic acid residue adjacent to the distal histidine is also liganded to a bound calcium ion. The calcium-binding loop in MPO does not resemble the "EF-hand" type of structure seen in several calcium-binding proteins such as calmodulin and troponin c. The bound calcium ion seen in the structure of MPO is likely to be a common feature throughout the family of mammalian peroxidases. The amino acid sequences for residues 168–174 are almost identical in all four mammalian peroxidases for which sequences are available. The single exception is the structurally conservative substitution of leucine for valine at position 171 in both TPO and LPO. In all four peroxidases the distal histidine is followed by an aspartic acid at position 96.

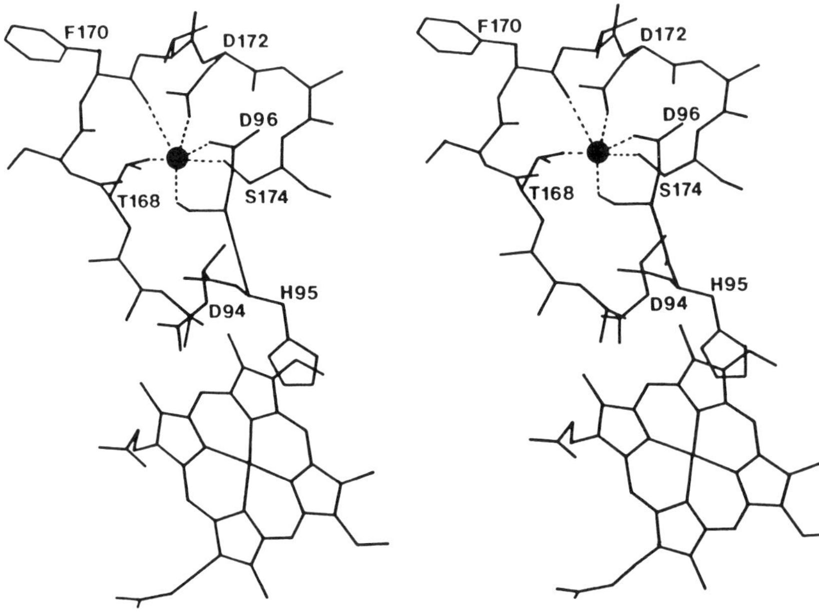

FIG. 7. Stereo drawing of the calcium binding loop, showing its close proximity to the heme. Possible calcium ligands include one carboxyl oxygen from each of the aspartates (D96 and D172), the hydroxyl oxygens of threonine (T168) and serine (S174), and the peptide carbonyl oxygens of phenylalanine (F170) and aspartate (D96) as indicated by the dashed lines.

Electron density consistent with the presence of asparagine-linked N-acetylglucosamine has been found at asparagines-189 and -225 in each half of the molecule. Both of these residues occur at the molecular surface where thermal motion can be expected to result in disordered electron density for subsequent sugar residues in the carbohydrate chains. Density for glycosylation at position 157 was not observed. While it is possible that this residue is not an asparagine in canine MPO, the poor quality of the electron density in this region does not rule out the possibility that asparagine-linked glycosylation also occurs in the canine enzyme at this location. In contrast, considerable well-defined electron density can be seen at the asparagine-317 glycosylation site that is located at the dimer interface. Noncovalent protein–protein interactions between the two halves of the molecule are minimized by the presence of bulky carbohydrate chains at this glycosylation site. Other than the disulfide bridge, the only protein–protein contact at the dimer interface involves residues 19–21 and 35–40 in the small polypeptides. Two pairs of hydrogen bonds are possible between the peptide carbonyl of alanine-35 and the hydroxyl of serine-19 and between the peptide carbonyl of glycine-40 and the hydroxyl group of threonine-21. Further interactions at the dimer interface occur through sugar–sugar and sugar–protein hydrogen bonding involving the carbohydrate chains at the asparagine-317 glycosylation site. Electron density was observed for two 1–4 linked N-acetylglucosamines followed by three mannose sugars. Additionally, electron density close to the first N-acetylglucosamine is consistent with the presence of a 1–6 linked fucose. While the orientations of N-acetylglucosamine sugars are well defined by the appearance of density for the bulky N-acetyl groups, the precise orientations of the mannose sugars could not be determined at the present resolution. Full characterization of the specific pattern of hydrogen bonding at the dimer interface must await higher resolution X-ray studies that will hopefully permit more accurate modeling of these sugar residues.

2.2.3. *Active Site Structure*

The heme crevice in MPO is about 15–20 Å in depth with access to the solvent via a channel on the distal side. The local heme environment is composed of amino acid side chains from within or closely flanking four of the core helices: H2, H5, H8, and H12. Since the true chemical structure of the heme in MPO remains uncertain, protoporphyrin IX was used to model the observed electron density. The azimuthal orientation of the porphyrin macrocycle is well defined by the appearance of electron density corresponding to the two propionic acid side chains. However, the relatively low resolution of the electron density map has not permitted distinction between two possible orientations related by the approximate twofold axis lying in the plane of the porphyrin macrocycle between pyrrole rings A,D and pyrrole rings B,C. This ambiguity has complicated the absolute assignment of specific interactions between the porphyrin macrocycle and amino acid side chains in the protein.

The proximal ligand to the heme iron atom is provided by the NE2 nitrogen of histidine-336 in helix H8, while the ND1 nitrogen of this residue is hydrogen-bonded to Asn421 in the adjacent helix H12 (Fig. 8). Note that in LiP and CCP the His ligand hydrogen bonds with an Asp residue and not an Asn residue. On the distal side, the imidazole group of histidine 95, located in the small polypeptide helix H2, is directed to the heme with its NE2 nitrogen approximately 5.7 Å from the iron atom. This histidine is flanked by Arg239 at the entrance to the distal cavity and by Glu91 at the back of the cavity (Fig. 9). The orientations of the distal histidine and the guanidinium group of Arg239 with respect to the heme are almost identical to the corresponding groups in CCP and LiP. It is therefore likely that these groups participate directly in the catalysis of hydrogen peroxide–mediated compound I formation

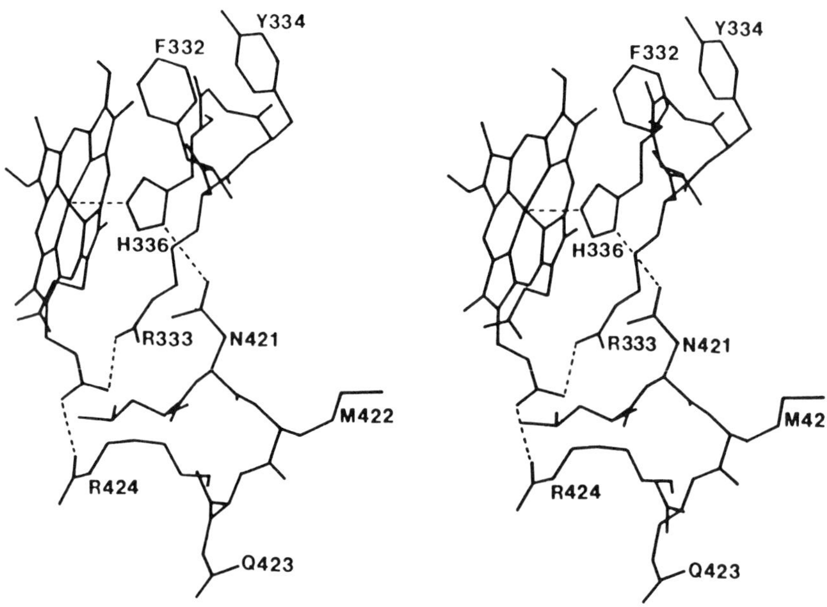

FIG. 8. Stereo drawing of the protein environment on the proximal side of the heme in myeloperoxidase. Residues 332–336 and 421–424 are shown. Dashed lines indicate the proximal ligand to the heme iron (H336); the hydrogen bond between this residue and asparagine (N421) and possible ionic interactions between the carboxyl group of one of the heme propionates and the guanidinium groups of two arginines: R333 and R424.

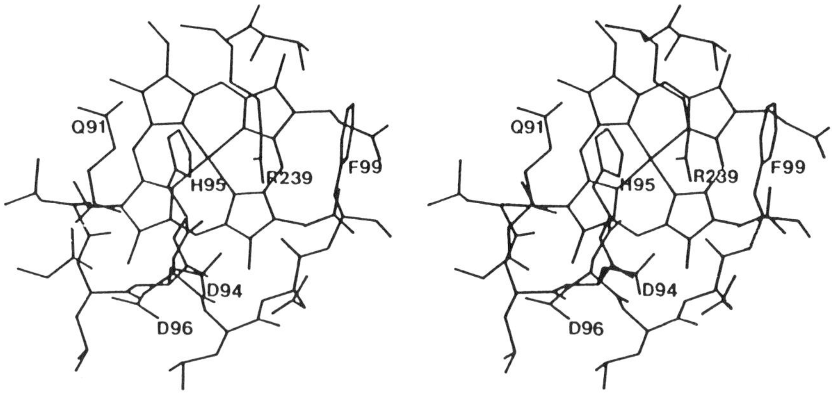

FIG. 9. Stereo drawing of the protein environment on the distal side of the heme in myeloperoxidase. Residues 91–99 in the H2 helix are shown together with residues 238 and 239 of the large polypeptide. H95 is the distal histidine and the adjacent R239 also is likely to be involved in the catalytic mechanism.

according to the mechanism proposed by Poulos and Kraut [76] and Finzel et al. [63] for the analogous reaction in CCP.

Unlike CCP and LiP, the heme in MPO, as well as other members of the mammalian peroxidase gene family, is covalently bound to the protein. The chemical nature of this linkage in MPO remains uncertain. In the electron density map of canine MPO continuous density between the side chain of Glu242 and the periphery of the heme macrocycle suggests that this residue is involved in the covalent linkage. Modification of pyrrole ring B is presumed to allow the formation of either an ester or a thioester bond with the side chain carboxyl groups of Glu242. The amino acid sequence of human MPO in the vicinity of Glu242 is consistent with the amino acid composition of a covalent heme peptide isolated by Schultz et al. [77] following digestion of the large polypeptide of canine MPO with trypsin and chymotrypsin. The present resolution of the X-ray crystal structure does not permit specific characterization of the pyrrole ring modification. Resonance Raman spectra of MPO have been interpreted by several groups to indicate that the heme group is an iron chlorin [78–80], in which case ring reduction would likely be accompanied by hydroxylation, providing a suitable group for ester formation with Glu242.

An additional close interaction between an amino acid side chain and the porphyrin macrocycle involves the side chain carboxyl of Asp94 which lies close to the presumed methyl group on pyrrole ring D. Wever et al. [81] suggested that the pH dependence of the red shift in the Soret band of the optical absorption spectrum of MPO could be explained by the presence of a negatively charged group in close proximity with the heme macrocycle. The presence of a formyl group in place of the methyl on pyrrole ring D would permit an energetically favorable hydrogen bond interaction with the side chain carboxyl of Asp 94. Some evidence for the presence of a formyl group in the porphyrin macrocycle has previously been presented [81,82], although Wever et al. were unable to detect a formyl group in the Raman spectrum of the native enzyme.

2.2.4. Mammalian Peroxidase Homology

Similarities between the amino acid sequences of human and mouse MPO, human EPO, bovine LPO, and the amino terminal portions of human, pig, and rat TPO suggest that these enzymes are also likely to have similar three-dimensional structures. All 12 cysteines involved in the six intrachain disulfides of MPO are conserved in all of these sequences. While there are no chemical data on the disulfide pairing in any of these proteins, if the pattern of disulfides seen in MPO is conserved this would undoubtedly constrain the three-dimensional conformations to be very similar. Conservation of amino acids in the immediate vicinity of the heme in MPO further suggests that the modes of heme binding and mechanism of catalysis are also likely to be very similar for other members of the mammalian peroxidase gene family. Residues that are believed to be essential for catalysis, including the proximal histidine and both the distal histidine and arginine, are conserved in all known sequences. Table 2 illustrates the high level of conservation of amino acids that compose the heme environment in MPO. Only two amino acids exhibit nonconservative variation between different peroxidases. Methionine-243 lies adjacent to Glu242, which participates in the putative covalent link between the heme and the polypeptide chain in MPO. This residue is replaced by threonine in EPO, glutamine in LPO, and valine in all three known sequences of TPO. Interspecies variation in the amino acid at position 98 suggests that this

TABLE 2

Conservation of the Heme Environment in Mammalian Peroxidases

Res.	HMPO	MMPO	HEPO	HTPO	PTPO	RTPO	BLPO	HLPO
87	M	*	*	*	*	P	*	
90	G	*	*	*	*	*	*	
91	Q	*	*	*	*	*	*	
94	D	*	*	*	*	*	*	
98	D	T	*	A	A	A	A	
99	F	L	*	*	*	L	*	
100	T	*	S	*	*	*	*	
146	F	*	*	*	*	*	*	
239	R	*	*	*	*	*	*	
242	E	*	*	*	*	*	*	
243	M	*	T	V	V	V	Q	
328	F	*	*	*	*	*	*	*
329	T	*	*	*	*	*	*	*
332	F	*	*	*	*	*	*	*
333	R	*	*	*	*	*	*	*
335	G	*	*	*	*	*	*	*
336	H	*	*	*	*	*	*	*
339	I	*	L	*	*	V	V	V
365	F	*	*	*	*	*	*	*
406	L	*	*	*	*	*	*	*
407	F	*	*	*	*	*	*	*
417	L	*	*	*	*	*	*	*
420	L	*	*	I	I	*	I	I
424	R	*	*	*	*	*	*	*

List of all residues in the model of MPO that have atoms within 4.5 Å of a heme atom. HMPO, human MPO; MMPO, mouse MPO; HEPO, human EPO; HTPO, human TPO; PTPO, pig TPO; RTPO, rat TPO; BLPO, bovine LPO; HLPO, human LPO. *indicates residues identical to those in human MPO. Data for human LPO up to residues 243 not available.

One-letter code of the amino acid residues: A, Ala; D, Asp; E, Glu; F, Phe; G, Gly; H, His; I, Ile; L, Leu; M, Met; Q, Gln; R, Arg; S, Ser; T, Thr; V, Val.

residue may be less significant in influencing the specific catalytic properties of different members of the peroxidase family. The calcium binding site in MPO is also likely to be a feature of other mammalian peroxidases. The sequence of the calcium binding loop 168–174 together with Asp96 is highly conserved in all known mammalian peroxidase sequences. While the presence of bound calcium has so far been demonstrated only in MPO and LPO, this site is likely to be present in both EPO and TPO.

3. PEROXIDASE CATALYTIC MECHANISMS

3.1. Compound I Formation

The similarity in the active site structures between the three peroxidase structures just described indicates a similar mechanism for compound I formation. Of particular importance is the proposed role of the distal His and Arg residues in promoting heterolytic fission of the peroxide O-O bond outlined below [76]:

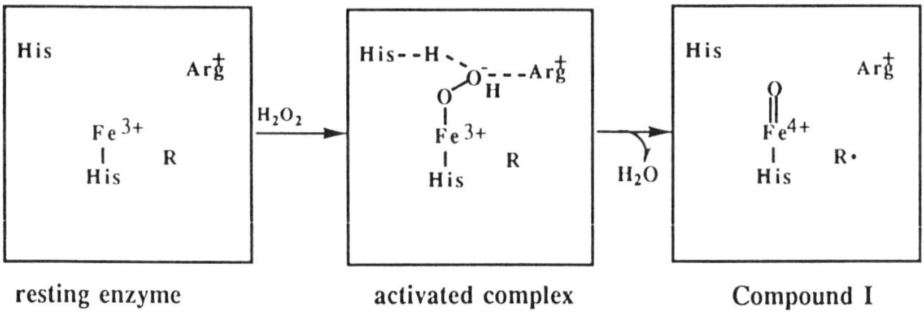

In this scheme, the distal His operates as an acid–base catalyst by abstracting a proton from the iron-linked oxygen atom and the arginine aids in stabilizing a developing negative charge on the leaving group. This catalytic machinery assures that the O-O bond cleaves heterolytically leaving behind an oxygen atom with only six valence electrons. This hypothetical iron–oxo intermediate is a powerful oxidizing center. It rapidly abstracts one electron from the iron to give Fe^{4+} and a second electron is removed from either the heme as in the plant peroxidases and MPO or a neighboring amino acid side chain as in CCP.

Another important feature in the proposed mechanism is the stabilization of compound I and the oxyferryl center. In both LiP and

CCP, the proximal His ligand is hydrogen bonded to a buried Asp residue (Fig. 4). The presence of this hydrogen bond imparts a greater anionic character to the His which in turn can stabilize the Fe^{4+} center. Nuclear magnetic resonance (NMR) spectroscopy has shown that this hydrogen bond is quite strong in peroxidases [83] relative to the weaker His peptide carbonyl oxygen hydrogen bond found in the globins [84]. MPO also has a hydrogen bond to the His ligand but in this case the hydrogen bond acceptor is an asparagine. This would suggest that in MPO the His ligand is not as effective in stabilizing the oxyferryl center.

The above scheme is supported by a wide range of biochemical and biophysical studies. The O-O bond is known to cleave heterolytically [85] and the protonated form of peroxides bind to peroxidases [86] as required by the above scheme. The hydrogen bonding pattern of the distal His in both CCP and LiP shown schematically below ensures that the imidazole nitrogen oriented toward the peroxide substrate is available to receive the peroxide proton. In both CCP and LiP, the distal His H bonds with the side chain of an Asn residue. Although there often is some ambiguity in the orientation of Asn and Gln residues, in both CCP and LiP the Asn must be oriented as shown since the side chain amide nitrogen H bonds with the peptide carbonyl oxygen atom of Glu76. Thus, the favored distal His tautomer will have the proton on ND1 of the imidazole ring leaving NE2 free to accept a proton from a peroxide molecule.

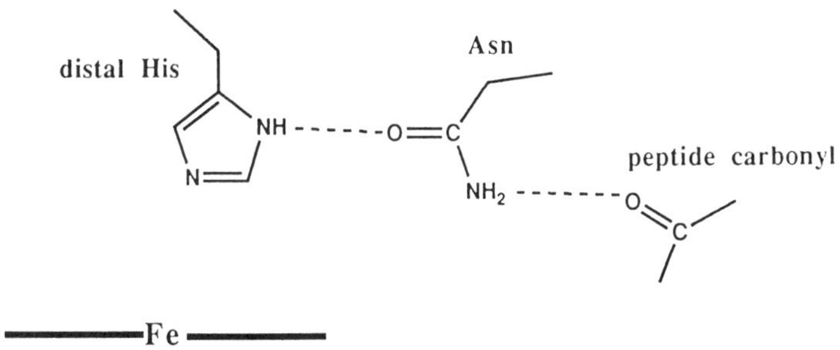

hydrogen bonding pattern of the distal His in CCP and LiP

The existence of the oxyferryl center and the presence of an iron-linked oxygen atom in compound I has been established by Mössbauer spectroscopy [87,88], magnetic susceptibility [89], EXAFS [42], and the CCP compound I crystal structure [43]. More recently, Baek and Van Wart [41] found a peroxide–enzyme intermediate that could be the

Michaelis complex also required by most formalisms for compound I formation. Moreover, recent molecular dynamics simulations of the CCP–peroxide complex support the proposed roles of the distal His and Arg and also indicate formation of a prereaction complex [90]. The nature of the radical site also is reasonably clear although we understand little about the energetics of what controls the location of the radical. With plant peroxidases, the radical is centered on the heme while in CCP it is an amino acid residue and very likely Trp191 [46].

If the distal His plays a crucial role as an acid–base catalyst in the rate-limiting step of compound I formation, then protonation of the distal His should effect the rate of reaction with peroxides. However, the formation of compound I is relatively insensitive to pH over a wide range for HRP [91], LiP [55], and CCP [57], and slows only at pH values lower than 5. With LiP, the rate of compound I formation is constant at pH 3–7 [92]. The distal His may have a depressed pK owing to its close proximity to the distal Arg [93], although it is not at all clear that the pK of the distal His could drop low enough (near pH 4) to explain the pH dependence of HRP compound I (HRPI) formation [91]. The overall importance of proton transfer in the rate-limiting step of compound I formation also is questionable since the rate of HRPI formation exhibits a relatively weak deuterium isotope effect of 1.6 [91]. Still we must contend with the fact that some group with a p$K \approx 4$ controls HRPI formation [91]. Unless this is due to some "nonmechanistic" and hence less interesting change such as a structural change, the observed p$K \approx 4$ indicates a carboxyl group [91]. Based on the available peroxidase structures, the most logical candidate(s) are the heme propionates. In both LiP and CCP, one heme propionate interacts with the distal Arg residue either directly or via a water molecule. We also know that the distal Arg is flexible and moves in response to the binding of various ligands [94]. Disruption of the intricate distal hydrogen bonding network by protonation of the heme propionate(s) could effect either proton abstraction from the incoming peroxide or delivery of the proton to the leaving OH^- group. As an interesting historical note, Jones and Suggett [95] predicted a propionate–Arg–peroxide hydrogen bonding network for the catalase mechanism based solely on kinetic and theoretical considerations. While this turned out not to be correct in detail for catalase [96], it was remarkably close to what the crystal structure of CCP revealed 12 years later.

3.2. Substrate Specificity

3.2.1. CCP–Cytochrome c Complex

The precise biological function of a peroxidase is not dictated by the reaction with peroxides to form compound I but rather by the types of molecules oxidized by compounds I and II. Of the three peroxidases considered in this article, CCP is the only one specific for another macromolecule. CCP forms a 1:1 complex with cytochrome c [97–102]. Knowledge of the CCP and cytochrome c crystal structures has prompted the use of computer graphics to model the CCP–cytochrome c complex [67,103]. The model shown in Fig. 10 was generated by visually optimizing the interaction of negative charges on CCP with positive charges on cytochrome c. The model predicts that the heme edges cannot approach closer than 18 Å and that the two hemes are nearly parallel. The primary value of this as well as other similar hypothetical models is as a basis for designing experiments. This has proven especially useful in

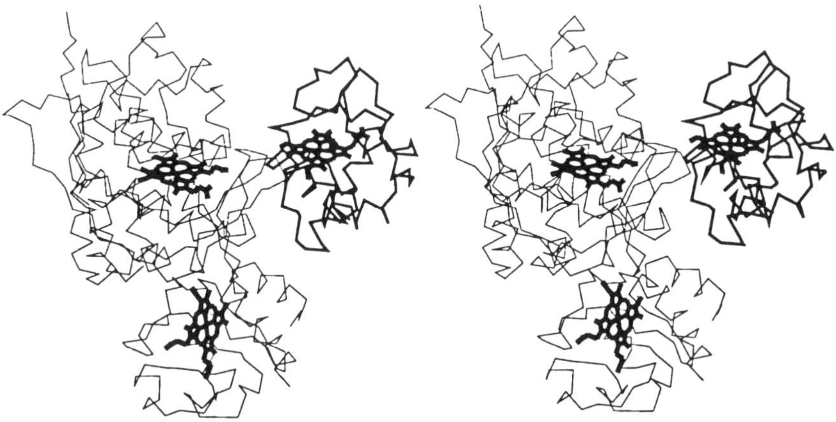

FIG. 10. Hypothetical and experimental models of the CCP–cytochrome c complex. The larger of the three molecules shown is CCP. The smaller molecule with thick bonds shows the location of cytochrome c based on the hypothetical model taken after Poulos and Finzel [67]. The location of cytochrome c based on the X-ray structure of the complex is shown as the smaller molecule with thin lines and is based on the structure described by Pelletier and Kraut [120]; it was generated using published diagrams and distances.

the CCP–cytochrome c system since the cDNA for both has been cloned and expressed, so that it has become possible to design, produce, and analyze mutants of each protein. While we consider mutagenesis work in more detail further on, the general conclusions from the protein-engineering approach are consistent with the hypothetical model. Earlier chemical modification studies [104,105], including crosslinking studies, also are consistent with the model [106–110]. Various spectroscopic probes also have been used including fluorescence [100,101,111] and NMR spectroscopy [112,113]. These, too, are generally in accord with the hypothetical model.

Nevertheless, it has become increasingly clear that the complex may not be static. Northrup et al. [114] used Brownian dynamics simulations to study the CCP–cytochrome c complex and while the dominant site of interaction corresponds to the hypothetical model, the theoretical approach predicts additional sites well removed from the primary site. McLendon and coworkers [115] also found that various techniques ranging from fluorescence and NMR spectroscopy to kinetics give very different association constants between CCP and cytochrome c with values ranging from 4×10^4 to 10^7 M^{-1}. Since each method probes a different property of the complex, McLendon concluded that the discrepancy must mean that the CCP–cytochrome c interaction is not stereospecifically unique but interconverts between many binding modes. Simple energetic considerations also indicate that a complex dominated by electrostatic forces and not nonbonded contacts would not be unique. Since the surfaces of CCP and cytochrome c are complementary with respect to charge and not shape, then electrostatic contacts and not nonbonded contacts are expected to control complex stability. Electrostatic energy falls off as $1/r$ where r is the distance between charges while nonbonded energy falls off as $1/r^6$. Hence, an electrostatic complex is much less sensitive to precise orientation.

Another interesting problem is the very fast rate of association between the two redox partners. Northrup and Erickson [116] showed that the fast rate of association between CCP and cytochrome c [117] need not invoke any special preorientation of the two proteins as suggested by others [118]. Rather, such rates can be simply explained by multiple collisions of the protein pair surrounded and trapped by water [116]. Therefore, two proteins could sample multiple orientations and still maintain very high rates of productive association. If, then, the

CCP–cytochrome c complex is dynamic or exhibits multiple binding modes, two problems remain. Is there only one orientation between the two that is optimal for electron transfer and, if so, does the protein provide a specific conduit for electron transfer?

In order to directly address these questions, attempts have been made to cocrystallize the CCP–cytochrome c complex. Poulos et al. [119] were the first to succeed in cocrystallizing CCP with horse heart cytochrome c and yeast iso-1-cytochrome c. The complexes were crystallized at low ionic strength and the crystals were found to have a 1:1 ratio of CCP to cytochrome c. Nevertheless, only CCP was visible in the electron density maps for the CCP–horse heart cytochrome c complex with large "empty" spaces between CCP molecules. Since an analysis of the crystals confirmed the presence of cytochrome c, the cytochrome c must be orientationally disordered in the crystal either by binding at multiple sites or by binding at a primary site but with considerable "wobble". More recently, Pelletier and Kraut [120] reported the crystal structures of CCP complexed with horse heart and yeast iso-1-cytochromes c. The horse heart cytochrome c–CCP crystals were isomorphous with the ones described in 1987 but were more ordered allowing for the location of cytochrome c. The yeast cytochrome c–CCP structure also was solved and showed a very similar orientation of cytochrome c relative to CCP. The orientation of the cytochrome c shown in Fig. 10 obtained from the crystal structures clearly is quite different than that proposed for the hypothetical complex. Pelletier and Kraut [120] argue that the complex observed in the crystals is the one unique complex important for electron transfer and have proposed a specific electron conduit involving Trp191 and its associated polypeptide chain.

Nevertheless, the problem of dynamic vs. a static complex and unique vs. multiple binding modes remains a problem. Neither the experimentally determined complex or the hypothetical model is fully consistent with solution studies. For example, Asakura and Yonetani [121] found that methylation of the CCP heme groups blocks electron transfer but not compound I formation. Moreover, Bosshard et al. [122] found that the cytochrome c prevents reaction with His181 in CCP. These two results are consistent with the hypothetical model. On the other hand, protection of CCP carboxylates in the C-terminal region by cytochrome c from chemical modification [104] is more consistent with the Pelletier and Kraut model. The Pelletier and Kraut model will undoubtedly receive close scrutiny to resolve some of these discrepancies.

The orientation problem is especially difficult to approach. To study this question, it is necessary to "freeze" the complex in some known orientation and then study electron transfer rates. This has been accomplished to some extent by crosslinking the two proteins [106,107, 109,110], yet the reagents used in these studies are relatively nonspecific and give heterogeneous products. The second question of selective pathways now can be approached by using mutagenesis methods. For example, in the hypothetical CCP–cytochrome c complex, His181 of CCP forms a continuous hydrogen bonding conduit from the surface of the enzyme to the distal pocket via a heme propionate and the distal Arg48. His181 sits at the center of the hypothetical CCP–cytochrome c complex which is supported by chemical modification studies [67]. Poulos and Finzel [67] proposed that His181 might provide part of the path for electron transfer. However, conversion of His181 to Gly did not alter activity [123], clearly showing that His181 is not essential for electron transfer. However, this study did not eliminate the possibility that solvent could substitute for His181 thus completing the hydrogen bonding network. The obvious experiment here is to replace His181 with a group like Leu, which would prevent any hydrogen bonding. Conversion of Trp191 to Phe, however, does eliminate electron transfer from cytochrome c to CCP [124]. To date, this is the most clear-cut mutagenesis study implicating a specific residue in electron transfer and supports the idea that electron transfer may prefer a specific path via Trp191.

3.2.2. Lignin Peroxidase

The question of substrate specificity in LiP is more problematic since the precise physiological substrate(s) are unknown. While it is becoming increasingly clear that LiP actively participates in lignin degradation [6,7,125], it is not known if LiP directly attacks lignin or first oxidizes a small molecule which then shuttles oxidizing equivalents to lignin. Manganese peroxidase, also secreted by *P. chrysosporium*, has an easier-to-rationalize modus operandi. Here the enzyme oxidizes Mn^{2+} to Mn^{3+} and in the presence of α-hydroxy acids, which stabilize Mn^{3+}, the Mn^{3+} complex oxidizes lignin-like compounds [126]. One well-known substrate of LiP which could serve a similar function is veratryl alcohol [53,54,127], a natural metabolite of *P. chrysosporium*.

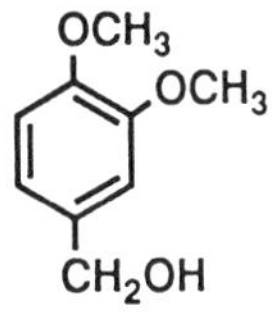

veratryl alcohol

Veratryl alcohol has two possible functions. It could serve the same role as manganese in manganese peroxidase by first being oxidized by LiP compound I to an aryl cationic radical and then the oxidized veratryl alcohol reacts with lignin in a nonenzymatic reaction. Alternatively, it is known that veratryl alcohol prevents LiP from inactivation in the presence of excess peroxide [53], presumably by feeding in electrons to LiP and thus preventing autoxidation and/or conversion to compound III, the oxy–LiP complex. In this case, veratryl alcohol operates as a protective agent.

Whatever the precise role of veratryl alcohol might be, it is assumed that the LiP possesses a site to which the alcohol binds. Poulos et al. [50] postulated that the opening to the heme crevice (Fig. 2) serves this purpose in LiP. A model of veratryl alcohol was manually docked to this site and the resulting protein–substrate complex energy minimized. It was possible to orient the substrate such that the substrate OH group forms an H bond with His82. His82 also forms an H bond with Glu146. The veratryl alcohol is then about 6–7 Å from the heme edge. This is a working model and will require verification by soaking crystals of LiP in solutions containing veratryl alcohol and solving the structure of the complex.

Despite these uncertainties, LiP is unusual among the peroxidases in its ability to oxidize nonphenolic aromatic molecules and the low pH optimum (optimum near pH 3.0) [128,129]. The structure of LiP reveals a possible link between structure and the low pH optimum in the oxidization of various substrates. In LiP, one of the heme propionates H-bonds with an Asp residue while the corresponding residue in CCP is an Asn (Fig. 4). While carboxylate–carboxylate H bonds are not common in proteins, they do occur and have been well documented [130]. Protonation of one or both of these acid groups could disrupt an important part of the structure required for activity. The disruption cannot be so large as to significantly alter the rate of compound I formation since this reaction is relatively pH-insensitive. However, disruption of an important heme–protein interaction could alter either the stability of compound I once it forms and/or its redox potential which would then have a significant effect on substrate oxidation. Alternatively, increases in pH could directly effect substrate binding. The hypothetical model of the veratryl alcohol–LiP complex suggests that His82 H-bonds with the substrate. Since His82 also H-bonds or ion-pairs with Glu146, an increase in pH would be expected to weaken the His82–Glu146 interaction, which might then result in a perturbation of the substrate pocket. One final possible explanation for the low pH optimum relates to regulation of the compound I redox potential. Compound I for both CCP [131] and horseradish peroxidase [132] exhibits higher redox potentials as the pH is lowered, which means that compound I becomes a better oxidizing agent at low pH. If the same holds true for LiP, then LiP may be designed to operate at low pH in order to achieve the redox potentials necessary for the oxidation of nonphenolic substrates.

4. PEROXIDASE ENGINEERING

While some preliminary data are available for horseradish peroxidase [133], CCP has served as the best system thus far for developing the field of peroxidase engineering. Indeed, the only other heme enzyme for which any detailed information exists for site-directed mutants including crystal structures is cytochrome P-450cam [134–136]. Part of the reason for this is the relative ease of producing active recombinant CCP in *E. coli*. For reasons not understood, CCP is produced in *E. coli* primarily as the apoenzyme with only a small fraction of the protein containing heme [137]. However, this has not presented a problem since the apoenzyme readily incorporates heme. The crystal structures and spectral properties of the recombinant and "natural" yeast CCP are indistinguishable [138,139]. Nevertheless, one puzzling and unexplained property of recombinant CCP is that the protein crystallizes with different cell dimensions and with the molecules oriented differently in the unit cell [138]. One of us (T.L.P.) has observed the same phenomenon with recombinant CCP although most of the crystals are exactly the same as those used for the initial structure determination. The reason for these differences is not clear but it is not due to detectable structural differences as judged by comparison of highly refined X-ray structures.

4.1. Mutagenesis and the CCP Radical Site

The earliest use of site-directed mutagenesis with CCP was directed at the longstanding question on the location of the amino acid radical in CCPI. When the crystal structure of CCP became available, Trp51 was found to lie parallel to and in contact with the distal heme surface directly adjacent to the peroxide binding site. Such close proximity of the indole ring to the iron-linked ferryl-oxo atom prompted the suggestion that Trp51 is the site of the free radical [76] and to the scheme shown below:

$$\mathrm{Fe(III)\ P\ R} \xrightarrow{H_2O_2} \begin{array}{c} \mathrm{Fe(IV)\ P^{\cdot +}\ R} \\ \downarrow\uparrow \\ \mathrm{Fe(IV)\ P\ R^{\cdot +}} \end{array}$$

P=porphyrin and R=side chain

In this scheme the first product formed in CCP is the porphyrin radical as with other peroxidases. The relative stability of the side chain vs. porphyrin radical controls the above equilibrium. Since plant peroxidases contain a Phe at the position corresponding to Trp51, these peroxidases favor porphyrin radicals since phenylalanine does not form as stable a radical as tryptophan. However, the heme radical has never been observed in CCP presumably because the equilibration to the tryptophan radical is too fast to be measured even by stopped-flow techniques. Nevertheless, a recent stopped-flow study of a mutant CCP with diminished activity showed formation of the green, heme cationic radical characteristic of the plant peroxidases [140]. Overall localization of the radical on Trp51 seemed reasonable based on the available evidence.

Nevertheless, various spectroscopic methods did not correlate well with a tryptophan radical. Using ENDOR spectroscopy, Hoffman and coworkers concluded that a tryptophan could not be the radical site. Instead, they postulated a methionine centered radical [141]. The CCP structure was consistent with this idea as well since the proximal pocket contains methionine residues close to the heme. Moreover, magnetic circular dichroism spectra of CCP compound I were interpreted as being incompatible with a tryptophan radical [142]. Hori and Yonetani [143] reexamined this question using single-crystal EPR spectroscopy and found evidence once again for a tryptophan radical.

The ability to generate site-specific CCP mutants has greatly aided in coming close to settling these conflicting views and conclusions. Neither converting Trp51 to phenylalanine nor removing a proximal methionine residue decreased enzyme activity or changed the characteristic shape of the compound I electron paramagnetic resonance (EPR) signal [137,144]. However, converting Trp191 to phenylalanine did dramatically decrease activity [124] and also significantly altered the stability and EPR spectral properties of compound I. These results point to Trp191 as a likely site for the free radical. Further support for a Trp191 radical stems from ENDOR studies of CCP containing deuterated tryptophan residues [46]. Incorporation of deuterated tryptophan caused a large change in the compound I spectrum indicating that tryptophan is a critical component in the CCP radical. The single-crystal EPR work of Hori and Yonetani [143] also is of interest here. They found that the axially symmetric $g_{\parallel}$ component of the EPR signal of CCP compound I is aligned along the heme normal and also found that the wavelength dependence of photochemical perturbation of the compound I radical closely resembles what one expects for a tryptophan radical. Then if tryptophan is the radical site the indole ring must be either parallel or perpendicular to the heme plane since $g_{\parallel}$ of the radical signal is aligned along the heme normal. Only Trp51 and 191 have the indole rings either parallel or perpendicular to the heme normal. Since the mutagenesis work has demonstrated that Trp51 is not essential, this leaves Trp191 as the only other reasonable candidate.

Recent crystallographic studies further point to Trp191 as an important site [94,145]. Based on 1.85 Å refined structures, it was found that nitric oxide (NO) coordinates with the CCP iron and specifically alters the Trp191 region of CCP by causing an increase in crystallographic B or thermal factors and leads to a slight movement of Trp191 away from the proximal ligand, His175. Other ligands, like CN^-, CO, and F^-, do not have this effect. Of these ligands, NO is unique since it is paramagnetic and the unpaired spin is transferred to the heme system [146]. Edwards and Poulos [94] postulated that NO imparts its effect on

Trp191 by delocalization of electron density via the proximal ligand, His175. Whatever the precise explanation of why NO perturbs Trp191, it is clear that Trp191 is uniquely sensitive to the electronic state of the heme which lends some support to conclusions derived from the mutagenesis work that Trp191 plays a special role in catalysis.

4.1.1. Possible Roles for Trp191

Certain interesting possibilities emerge regarding how CCP works under steady-state conditions when the Trp191 mutagenesis work is viewed in light of other kinetic studies. One of the more interesting recent observations regarding the role of Trp191 is that the radical center of CCP may not be the initial site to receive an electron from ferrocytochrome c. Hazzard and Tollin [47] found that in a single-turnover experiment, one ferrocytochrome c molecule is reduced simultaneously with the reduction of CCPI($Fe^{4+}Trp^{\bullet}$) to CCPII($Fe^{3+}Trp^{\bullet}$). Erman and Summers [147] came to the same conclusion using stopped-flow methods although the thermodynamically stable form of CCPII is Fe^{4+}Trp. Hazzard and Tollin [47] further concluded that there is no intramolecular electron transfer between the heme and Trp191 on the time scale of steady-state turnover as suggested by Ho et al. [148]. The implication of this assumption for the steady-state reaction is that the Trp radical is not reduced [149]. Most peroxidases have the radical site of compound I reduced first giving compound II. If we assume for the moment that it is the oxyferryl center which is reduced first in CCPI and not the radical site, then the $Fe^{3+}Trp^{\bullet}$ can form after the first encounter with cytochrome c. There is now a ferric iron ready to react with another peroxide molecule giving the following steady-state reaction scheme which is taken after Miller et al. [149].

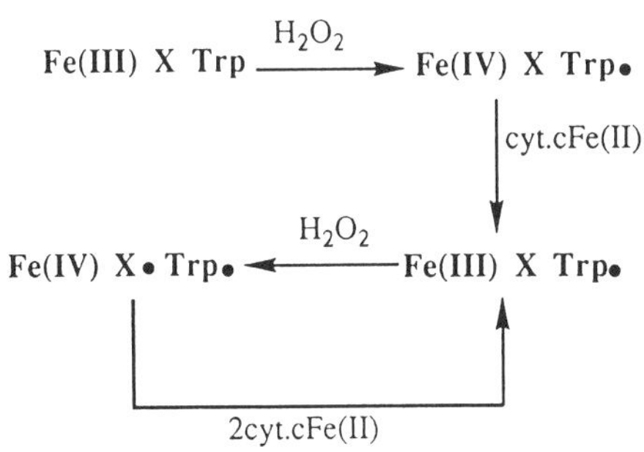

For $Fe^{3+}Trp^{\bullet}$ to build up, we must further assume that the rate of electron transfer to the Trp radical will be slow compared to the reaction of Fe^{3+} with peroxide. This scheme postulates a new intermediate where some group, X, is oxidized which could be the heme or one/some of the many oxidizable side chains in CCP. This additional complexity also could help to explain the multiphasic kinetics observed under conditions of excess peroxide and varying concentrations of cytochrome c [97,150]. These data were interpreted as representing multiple binding sites for cytochrome c on CCP [97], although there is no clear evidence for more than a 1:1 complex. With multiple routes for electron in to CCP (i.e., X in the above scheme and the oxyferryl center), biphasic kinetics might be expected.

In direct contrast to this view, Millet and coworkers [151,152] found that the Trp radical is reduced first. In this system, ruthenium selectively attached to either a lysine or a cysteine in cytochrome c is reduced by a light pulse. The reduced Ru transfers an electron to cytochrome c giving ferrocytochrome c and the ferrocytochrome c in turn transfers an electron to CCPI. At present there is no simple way of reconciling these conflicting views. Part of the problem is the close similarity in the absorption spectrum of $Fe^{4+}Trp^{\bullet}$ and $Fe^{4+}Trp$. However, it would be difficult to conclude that the Trp radical is not important in light of the mutagenesis work which clearly shows that Trp191 is essential for electron transfer in both steady-state and single-turnover experiments [124]. Although the question on the relevance of the Trp radical to the steady-state reaction may still be open to debate, it is clear that Trp191 is essential for electron transfer from ferryocytochrome c and provides support for the idea that electron transfer proceeds along specific polypeptide and side chain conduits.

4.2. Other Proximal Side Mutants

Two other key residues have been altered in CCP: Asp235 and the proximal ligand, His175. Asp235 is important since its side chain H-bonds with the indole nitrogen of Trp191 (Fig. 2). Since Trp191 is the radical site, then Asp235 may play a role in stabilizing a cationic Trp191 radical. Changing Asp235 to Asn decreases activity and the X-ray structure shows that the Trp191 indole ring rotates out of its native position in order to form an indole nitrogen–peptide carbonyl oxygen H bond with Leu177 [138]. The Asn235Asp mutant also has been studied by a variety of spectral techniques [153–156]. The NMR data show that Asp235 is important for imparting a greater anionic character to the proximal histidine ligand [155] as predicted [67]. The effect of this mutation on the resonance Raman of CCP and the CO adducts also has been studied [153] and these data are consistent with a strong H-bond between Asp235 and His175. The precise role of Asp235 is difficult to assess since a detailed analysis on the activity of the Asp235 mutants has not been published. Nevertheless, the Asp235–His175 hydrogen bond appears not to be essential in the formation or stability of the oxyferryl center in compound I [157]. That we find the same Asp–His interaction in LiP indicates that the Asp–proximal His H bond is an important property of fungal peroxidases and is probably required to partially deprotonate the proximal His. In CCP, Asp235 could have an additional function, which is to help anchor Trp191 in place so the indole ring can perform its electron transfer function [138]. Changing Asp235 to Glu [158] further shows that the EPR properties and hence the precise nature of the compound I radical is strongly influenced by Asp235.

The proximal ligand, His175, has been converted to Glu and Gln [139]. These residues were chosen since simple modeling studies indicate that Gln has about the same length as His and should be able to form a bond with the iron atom. Indeed, the crystal structures of both His175Gln and His175Glu mutants show that the new ligands do form a bond with the iron. The surprising result is that these mutations do not decrease enzyme activity under steady-state conditions. The Gln175 mutant has been most extensively studied and the turnover number is 1683 sec^{-1} for the mutant vs. 1463 sec^{-1} for the wild-type enzyme [139]. The proximal ligand mutants react with peroxide to give the characteristic compound I spectral changes but the stability of compound I is greatly diminished. Thus, it appears that the precise nature of the ligand is not important for activity but that the ligand is important in stabilizing the oxyferryl center.

The double mutant where His175 has been converted to Gln and Trp191 to Phe also has been partially characterized and the X-ray structure determined [159]. The double mutant exhibits about 20% wild-type activity yet the single Trp191Phe mutant exhibits a nearly undetectable (<0.05%) level of activity. Therefore, changing the proximal ligand from His to Gln partially reverses the deleterious effect of changing Trp191 to Phe. Since the rate-limiting step in the steady-state reaction is most likely electron transfer [160], it is most useful to refer to simple electron transfer theory [161] in seeking an explanation for this increase in activity. The activation energy of an electron transfer reaction, ΔG^*, depends on the driving force of the reaction, ΔG^0, and the reorganization energy, λ, according to the following equation:

$$\Delta G^* = \frac{(\Delta G^0 - \lambda)^2}{4\lambda}$$

From transition state theory, the rate of the reaction is proportional to $\exp(-\Delta G^*/RT)$. λ represents the energy required to reorganize protein–metal bond distances and geometry prior to electron transfer. In addition, the medium through which the electron must pass and the distance between redox centers will affect rates. The redox potential for the Fe^{3+}/Fe^{4+} couple in CCP is 1.08V [131]. Mutations cannot increase this value to more than that of H_2O_2 (1.5 V) so the overall driving force of the reaction cannot increase by more than 0.4 V as a result of mutations. A 0.4 V increase in ΔG^0 could come close to explaining a 500-fold increase in activity [162–164]. This would require that the reaction of the mutant with peroxide is 100% efficient leading to storage of the full oxidizing power of peroxide within the active site assuming all other factors, such as distance and λ, are equal.

While an increase in driving force may in part contribute to increased rates, Choudhury et al. [159] suggested that the difference in activity is related to the stability of compound I and how the oxidizing equivalents derived from H_2O_2 are stored. In the wild-type and single Trp191Phe mutant, the oxyferryl center of compound I is still quite stable with half-lives measured in hours [124]. In sharp contrast, both the single His175Gln and double His175Gln/Trp191Phe mutants form compound I but the decay of the typical compound I spectrum occurs within minutes rather than hours. Moreover, in wild-type CCP spontaneous decay of compound I leads to destruction of aromatic residues [45] demonstrating that the oxidizing equivalents can migrate throughout the protein. If this happens much faster in the glutamine mutants, which the instability of compound I suggests, then "sites" receiving electrons will no longer be confined to the heme pocket. This could significantly change both the distance the electron must traverse from cytochrome c and the medium of transfer. In other words, the rapid intramolecular migration of oxidizing equivalents in the glutamine mutants has opened up new and very likely nonspecific routes of electron transfer which bypass the requirement for a tryptophan at position 191.

4.3. Distal Side Mutants

To directly test the proposed role [67,103] of the distal His and Arg in compound I formation, both residues have been altered. Changing the distal Arg to Leu lowers the second-order rate of reaction with H_2O_2 from $3.2 \times 10^7\ M^{-1}\ s^{-1}$ to $1.5 \times 10^5\ M^{-1}\ s^{-1}$ [165]. When the distal His is converted to Leu, the rate of compound I formation decreases five orders of magnitude to $2.9 \times 10^2\ M^{-1}\ s^{-1}$. These results support the general idea that the distal His plays a critical role as an acid–base catalyst in the formation of compound I (see Sec. 3.1) but that the Arg is not so important. In light of these distal side mutant results, it would appear that the role of the distal Arg may be more in the stabilization of compound I once formed. It is known from crystal structure work that the distal Arg is capable of movement in order to optimize hydrogen bonding interactions with distal side iron ligands including the ferryl oxygen atom of compound I [43,94]. These mutagenesis results require a slight modification of the original stereochemical mechanism of compound I formation [103]. As predicted, the distal His is important but the role of the distal Arg may not be so critical in the cleavage of the peroxide O-O bond but rather serves a role in the stabilization of the ferryl center once formed. Taken together, mutagenesis of proximal and distal side residues have enabled a deeper and more detailed understanding on the precise role played by individual residues in the formation and stabilization of compound I.

4.4. Other Mutants

His181 has been changed to Gly [123] to test the proposed role of His181 in the electron transfer reaction [67]. His181 participates in a series of H bonds that connect the surface of CCP with active site via a heme propionate and Arg48. His181 also lies near the center of the CCP–cytochrome c interface in the hypothetical model, suggesting that perhaps His181 participates in an H-bonding conduit between cytochrome c and CCP [67]. Converting His81 to Gly does not affect activity nor is there a significant change in the crystal structure [123]. It would appear that His181 does not operate as proposed [67]. However, these results can be viewed to be at odds with earlier work of Asakura and Yonetani [121] who found that methylating the heme propionates blocks electron transfer from cytochrome c to CCP but does not effect the reaction with peroxides to give a normal compound I. Since one heme propionate H-bonds with His181, why would disruption of this H bond by esterification of the propionate block electron transfer while changing His181 to Gly, which disrupts the same H bond, not lead to a loss in activity? There are two possibilities. First, esterification presents a steric problem not present in the mutant. Second, replacing His181 with Gly does *not* disrupt the H-bonding network since water substitutes for His thus retaining the H-bonding network found in the native enzyme. These possibilities can be tested by producing other mutants at position 181 like Leu which should disrupt any propionate–solvent interactions.

4.5. Myeloperoxidase Mutant Studies

Human MPO cDNA has been overexpressed in cultured Chinese hamster ovary cells [166]. The secreted product is an 84 kD single polypeptide lacking the signal sequence but including the 116 residue amino terminal pro-sequence not found in the mature enzyme. Further differences occur in the level of processing of the asparagine-linked carbohydrate chains, which collectively have a higher mannose content than those of the native enzyme. Human recMPO was shown to be catalytically active and to display visible absorption spectra almost identical to those of the native enzyme in both oxidized and reduced forms [167]. The combination of a viable expression system for recMPO and knowledge of the three-dimensional structure of the native enzyme provides an excellent opportunity to further pursue structure–function relationships through application of the techniques of site-directed mutagenesis. Four mutant recMPOs have now been expressed, purified, and analyzed for spectral and catalytic properties in Alex Bollen's group at the Université Libre de Bruxelles, Belgium. The His336-to-Ala MPO mutant was found to have no visible absorption spectrum and no catalytic activity, consistent with the assignment of this residue as the proximal heme ligand in the X-ray crystal structure of canine MPO [168]. Conversion of His95 to Ala was also found to abolish catalytic activity, consistent with its assignment as the distal histidine [169]. In contrast, the MPO His250Ala mutant has a visible absorption spectrum and catalytic properties that are almost identical with those of the wild type. A fourth mutant, MPO Asp94Asn, was constructed to test the hypothesis that an acidic residue close to the heme macrocycle could influence the red shift of the Soret band in the absorption spectrum of MPO. This mutant exhibits similar catalytic activity to the native enzyme, indicating that this aspartic acid is not essential to the catalytic mechanism. However, significant differences were found in the visible absorption spectrum. The Soret peak for the mutant was blue-shifted to 414 nm compared with 428 nm for the native enzyme. This observation supports the hypothesis that the red-shifted Soret peak in MPO may be in part related to the presence of this negatively charged amino acid in close proximity to the heme macrocycle, although there are likely to be additional mechanisms by which the spectral properties of the chromophore are influenced by the immediate protein environment [169].

5. CONCLUSIONS

The first peroxidase crystal structure, CCP, was published in 1980 [48] and since then has served as a paradigm for peroxidase structure and function. However, CCP is limited as a model for all peroxidases for two reasons. First, one of the two oxidizing equivalents in compound I is stored as an amino acid free radical while in most other heme peroxidases the radical is centered on the porphyrin; second, CCP is highly specialized to operate on cytochrome c while most other well studied peroxidases oxidize small aromatic molecules or halides. Fortunately, the situation is rapidly changing since this past year has witnessed the solution of two new peroxidase crystal structures, myleoperoxidase [73] and lignin peroxidase [49,50]. Moreover, cDNAs and sequences are now available for many other peroxidases [69]. Recombinant systems are being developed for a number of peroxidases and crystals are available for others. The next few years should be exciting in peroxidase structure and function research since protein engineering studies will expand to other peroxidases besides CCP and new crystal structures will be forthcoming.

ABBREVIATIONS

Ala	alanine residue
Arg	arginine residue
Asn	asparagine residue
Asp	aspartic acid residue
CCP	cytochrome c peroxidase
Cys	cysteine residue
ENDOR	electron nuclear double-resonance spectroscopy
EPO	eosinophil peroxidase
EPR	electron paramagnetic resonance
EXAFS	extended absorption fine-structure spectroscopy
Gln	glutamine residue
Glu	glutamic acid residue
Gly	glycine residue
His	histidine residue
HRP	horseradish peroxidase
Ile	isoleucine residue
Leu	leucine residue
LiP	lignin peroxidase
LPO	lactoperoxidase
Met	methionine residue
MIR	multiple isomorphous replacement
MPO	myeloperoxidase
NMR	nuclear magnetic resonance
PGH	prostaglandin H
Phe	phenylalanine residue
Pro	proline residue
rms	root mean square

Ser	serine residue
Thr	threonine residue
TPO	thyroid peroxidase
Trp	tryptophan residue
Val	valine residue

REFERENCES

1. A. Campa, in *Peroxidases in Chemistry and Biology*, Vol. 2 (J. Everse, K. Everse, and M. B. Grisham, eds.), CRC Press, Boca Raton, 1991, p. 25.
2. H. Grisebach, in *The Biochemistry of Plants*, (P. K. Stumpf and E. E. Conn, eds.) Academic Press, New York, 1981, Chap. 15.
3. T. Higuchi, in *Encyclopedia at Plant Physiology*, Vol 13B (W. Tanner and F. A. Coeuvs, eds.), Springer-Verlag, Berlin, 1981, Chap. 9.
4. S. F. Yang, *Arch. Biochem. Biophys. 122*, 481 (1967).
5. K. V. Sarkanen and C. H. Ludwig, *Lignins: Occurrence, Formation, Structure, and Reactions*, Wiley-Interscience, New York, 1971.
6. M. H. Gold, H. Wariishi, and K. Valli, in *Biocatalysis in Agricultural Biotechnology*, (J. R. Whitaker and P. Sonnet, eds.), ACS Symp. Series 389, Am. Chem. Soc., Washington, D.C., 1989, p. 127.
7. T. K. Kirk and R. L. Farrell, *Ann. Review Microbiol., 41*, 465 (1987).
8. T. Higuchi, *Wood Sci. Technol., 24*, 23 (1990).
9. H. E. Shoemaker, *Recl. Trav. Chim. Pays-Bas Beg., 109*, 255 (1990).
10. K. E. Hammel, *Enzyme Microb. Technol., 11*, 776 (1989).
11. K. Valli and M. H. Gold, *J. Bacteriol., 173*, 345 (1991).
12. K. Valli and H. Wariishi, and M. H. Gold, *J. Bacteriol., 174*, 2131 (1992).
13. S. J. Klebanoff, in *Peroxidases in Chemistry and Biology*, Vol. 1 (J. Everse, K. E. Everse, and M. B. Grisham, eds.), CRC Press, Boca Raton, 1991, pp. 1–35.

14. R. A. Clark, P. J. Stone, A. El Hag, J. D. Calore, and C. J. Franzblau, *J. Biol. Chem., 256*, 3348 (1981).
15. N. R. Matheson, P. S. Wong, M. Schuyler, and J. Travis, *Biochemistry, 20*, 331 (1981).
16. O. Stendahl, B. I. Coble, C. Dahlgren, J. Hed, and L. Molin, *J. Clin. Invest., 73*, 366 (1984).
17. R. A. Clark and N. Borregaard, *Blood, 65*, 375 (1985).
18. B. Reiter, in *The Lactoperoxidase System: Chemistry and Biological Significance* (K. M. Pruitt and J. O. Tenovuo, eds.), Marcel Dekker, New York, 1985, p. 123.
19. R. P. Magnusson, in *Peroxidases in Chemistry and Biology*, Vol. 1 (S. Everse, K. E. Everse, and M. B. Grisham, eds.), CRC Press, Boca Raton, 1991, pp. 199–219.
20. K. R. Johnson, W. M. Nauseef, A. Care, M. J. Weelock, S. Shane, S. Hudson, H. P. Koeffler, M. Selsted, C. Miller, and G. Rovera, *Nucl. Acids Res., 15*, 2013–2028 (1987).
21. K. Morishita, N. Kubota, S. Asano, Y. Kaziro, and S. Nagata, *J. Biol. Chem., 262*, 9391 (1987).
22. D. Venturelli, S. Bittenbender, and G. Rovera, *Nucl. Acids Res., 17*, 7987 (1989).
23. K. Sakamaki, M. Tomonaga, K. Tsukui, and S. Nagata, *J. Biol. Chem., 264*, 16828 (1989).
24. R. M. Ten, L. R. Pease, D. J. McKean, M. P. Bell, and G. J. Gleich, *J. Exp. Med., 169*, 1757 (1989).
25. S. Kimura, T. Kotani, O. W. McBride, K. Umeki, K. Hirai, S. Nakayama, and S. Ohtaki, *Proc. Natl. Acad. Sci. USA, 84*, 5555 (1987).
26. F. Libert, J. Ruel, M. Ludgate, S. Swillens, N. Alexander, G. Vassart, and C. Dinsart, *EMBOJ, 6*, 4193 (1987).
27. R. P. Magnusson, J. Gestautas, A. Taurog, and B. Rapoport, *J. Biol. Chem., 262*, 13885 (1987).
28. M. Derwahl, P. Seto, and B. Rapoport, *Nucl. Acids Res., 17*, 8380 (1989).
29. T. J. Dull, C. Uyeda, A. D. Strosberg, G. Nedwin, and J. J. Seilhammer, *DNA Cell Biol., 9*, 499 (1990).
30. M. M. Cals, P. Maillart, G. Brignon, P. Anglade, and B. R. Dumas, *Eur. J. Biochem., 198*, 733 (1991).

31. J. P. Merlie, D. Fagan, J. Mudd, and P. Needleman, *J. Biol. Chem.*, *263*, 3550 (1988).
32. J. Kaput, S. Glotz, and G. Blobel, *J. Biol. Chem.*, *257*, 11186 (1982).
33. M. L. Maccecchini, Y. Rudin, G. Blobel, and G. Schatz, *Proc. Natl. Acad. Sci. USA*, *76*, 343 (1979).
34. P. G. Williams and P. R. Stewart, *Arch. Microbiol.*, *107*, 63 (1976).
35. M. Erecinska, N. Oshino, P. Loh, and E. Brockelhurst, *Biochim. Biophys. Acta*, *292*, 1 (1973).
36. M. Yamada and K. Kurashi, *J. Biol. Chem.*, *259*, 3021 (1984).
37. W. M. Nauseef, *Blood*, *70*, 1143 (1987).
38. K. Arnljots and I. Olsson, *J. Biol. Chem.*, *262*, 10430 (1987).
39. K. L. Taylor, G. S. Guzman, C. A. Burgess, and J. M. Kincade, Jr., *Biochemistry*, *29*, 1533 (1990).
40. P. Jones and H. B. Dunford, *J. Theor. Biol.*, *69*, 457 (1977).
41. H. K. Baek and H. E. Van Wart, *J. Am. Chem. Soc.*, *114*, 718 (1992).
42. B. Chance, L. Powers, Y. Ching, T. Poulos, G. R. Schonbaum, I. Yamazaki, and K. G. Paul, *Arch. Biochem. Biophys.*, *235*, 596 (1984).
43. S. L. Edwards, N-h. Xuong, R. C. Hamlin, and J. Kraut, *J. Biol. Chem.*, *26*, 1503 (1987).
44. D. Dolphin, A. Forman, D. C. Borg, J. Fajer, and R. H. Felton, *Proc. Natl. Acad. Sci. USA*, *68*, 614 (1971).
45. T. Yonetani, in *The Enzymes*, Vol. 13 (P. D. Boyer, ed.), Academic Press, Orlando, 1976, p. 345.
46. M. Sivaraja, D. B. Goodin, M. Smith, and B. M. Hoffman, *Science*, *245*, 738 (1989).
47. J. T. Hazzard and G. Tollin, *J. Am. Chem. Soc.*, *113*, 8956–8957 (1991).
48. T. L. Poulos, B. T. Freer, R. A. Alden, S. L. Edwards, U. Skoglund, K. Takio, B. Eriksson, N. Xuong, T. Yonetani, and J. Kraut, *J. Biol. Chem.*, *255*, 575 (1980).
49. S. L. Edwards, R. Raag, H. Wariishi, M. H. Gold, and T. L. Poulos, *Proc. Natl. Acad. Sci. USA*, *90*, 750 (1993).
50. T. L. Poulos, S. L. Edwards, H. Wariishi, and M. H. Gold, *J. Biol. Chem.*, *268*, 4429 (1993).

51. B. Henrissat, M. Saloheimo, S. Lavaitte, and J. K. C. Knowles, *Proteins*, *8*, 251 (1990).
52. G. D. DePillis, B. P. Sishta, A. G. Mauk, and P. R. Ortiz de Montellano, *J. Biol. Chem.*, *266*, 19334 (1991).
53. K. Valli, H. Wariishi and M. H. Gold, *Biochemistry*, *29*, 8535 (1990).
54. H. Wariishi and M. H. Gold, *J. Biol. Chem.*, *265*, 2070 (1990).
55. L. Marquez, H. Wariishi, H. B. Dunford, and M. H. Gold, *J. Biol. Chem.*, *263*, 1059 (1988).
56. A. Andrawis, K. A. Johnson, and M. Tien, *J. Biol. Chem.*, *263*, 1195 (1988).
57. C. Balny, H. Anni, and T. Yonetani, *FEBS Lett.*, *221*, 349 (1987).
58. T. G. Ritch, Jr., V. J. Nipper, L. Akileswaran, A. Smith, D. Prbnow, and M. H. Gold, *Gene*, *107*, 119 (1991).
59. G. D. DePillis, H. Wariishi, M. H. Gold, and P. R. Ortiz de Montellano, *Arch. Biochem. Biophys.*, *280*, 217 (1990).
60. V. P. Miller, G. D. DePillis, J. C. Ferrer, G. Mauk, and P. R. Ortiz de Montellano, *J. Biol. Chem.*, *267*, 8936 (1992).
61. R. H. Haschke and J. M. Friedhoff, *Biochem. Biophys. Res. Commun.*, *80*, 1039 (1978).
62. M. A. Ator, S. K. David, and P. R. Ortiz de Montellano, *J. Biol. Chem.*, *264*, 9250 (1989).
63. B. C. Finzel, T. L. Poulos, and J. Kraut, *J. Biol. Chem.*, *259*, 13027 (1984).
64. L. A. Andersson, V. R. Renganathan, A. A Chiu, T. M. Loehr, and M. H. Gold, *J. Biol. Chem.*, *260*, 6080 (1985).
65. L. A. Andersson, V. Renganathan, T. M. Loehr, and M. H. Gold, *Biochemistry*, *26*, 2258 (1987).
66. D. Kuila, M. Tien, J. A. Fee, and M. R. Ondrias, *Biochemistry*, *24*, 3394 (1985).
67. T. L. Poulos and B. C. Finzel, in *Peptide and Protein Reviews*, Vol 4 (M. T. Hearn and W. Hearn, eds.), Marcel Dekker, New York, 1984, p. 115.
68. T. L. Poulos, *Adv. Inorg. Biochem.*, *7*, 1 (1987).
69. K. G. Welinder, *Curr. Opinion Struct. Biol.*, *2*, 388 (1992).
70. J. Zeng and R. E. Fenna, *J. Mol. Biol.*, *210*, 681 (1989).

71. B. C. Wang, *Meth. Enzymol.*, *115*, 90 (1985).
72. D. E. Tronrud, L. F. Ten Eyck, and B. W. Matthews, *Acta Crystallogr. Sect. A*, *43*, 489 (1987).
73. J. Zeng and R. E. Fenna, *J. Mol. Biol.*, *226*, 185 (1992).
74. P. C. Andrews and N. I. Krinsky, *J. Biol. Chem.*, *256*, 411 (1981).
75. K. S. Booth, S. Kimura, H. C. Lee, M. Ikeda-Saito, and W. W. Caughey, *Biochem. Biophys. Res. Commun.*, *60*, 897 (1989).
76. T. L. Poulos and J. Kraut, *J. Biol. Chem.*, *255*, 8199 (1980).
77. J. Schultz, N. C. Wu, and R. Marti, *Biochemistry*, *22*, 14A (1983).
78. S. S. Sibbett and J. K. Hurst, *Biochemistry*, *23*, 3007 (1984).
79. M. Ikeda-Saito, P. V. Argade, and D. I. Rousseau, *FEBS Lett.*, *184*, 52–55 (1985).
80. G. T. Babcock, R. T. Ingle, W. A. Oertling, J. S. Davis, B. A. Averill, C. A. Hulse, D. J. Stuffkens, B. G. J. M. Bolscher, and R. Wever, *Biochim. Biophys. Acta*, *828*, 58 (1985).
81. R. Wever, W. A. Oertling, H. Hoogland, B. G. J. M. Bolscher, Y. Kim, and G. T. Babcock, *J. Biol. Chem.*, *266*, 24308 (1991).
82. M. Sono, A. M. Bracete, A. M. Huff, M. Ikeda-Saito, and J. Dawson, *Proc. Natl. Acad. Sci. USA*, *88*, 11148 (1991).
83. G. N. LaMar, J. S. De Ropp, V. P. Chako, J. D. Satterlee, and J. R. Erman, *Biochim. Biophys. Acta*, *708*, 317 (1982).
84. J. S. Valentine, R. P. Sheridan, L. C. Allen, and P. C. Kahn, *Proc. Natl. Acad. Sci. USA*, *76*, 1009 (1979).
85. G. R. Schonbaum and S. Lo, *J. Biol. Chem.*, *247*, 3353 (1972).
86. D. Job and P. Jones, *Eur. J. Biochem.*, *86*, 565 (1978).
87. G. Lang, K. Spartalian, and T. Yonetani, *Biochim. Biophys. Acta*, *451*, 250 (1976).
88. C. Schulz, R. Chiang, and P. G. Debrunner, *J. Phys., Coeloq. (Orsay, Fr.)*, *40*, C2–534 (1979).
89. T. Iizuka, M. Kotani, and T. Yonetani, *J. Biol. Chem.*, *246*, 4731 (1971).
90. J. R. Collins, P. Du, and G. H. Loew, *Biochemistry*, *31*, 11166 (1992).
91. H. H. Dunford, W. D. Hewson, and H. Steiner, *Can. J. Chem.*, *56*, 2844 (1977).
92. D. Cai and M. Tien, *J. Biol. Chem.*, *264*, 14464 (1991).

93. G. Smulevich, M. A. Miller, J. Kraut, and T. G. Spiro, *Biochemistry*, *30*, 9546 (1991).
94. S. L. Edwards and T. L. Poulos, *J. Biol. Chem.*, *265*, 2588 (1990).
95. P. Jones and A. Suggett, *Biochem. J.*, *110*, 621 (1968).
96. I. Fita and M. G. Rossmann, *J. Mol. Biol.*, *185*, 21 (1985).
97. C. H. Kang, S. Fergusson-Miller, and E. Margoliash, *J. Biol. Chem.*, *252*, 919 (1977).
98. J. E. Erman and L. B. Vitello, *J. Biol. Chem.*, *255*, 6224 (1980).
99. R. J. Dowe, L. B. Vitello, and J. E. Erman, *Arch. Biochem. Biophys.*, *232*, 566 (1984).
100. J. J. Leonard and T. Yonetani, *Biochemistry*, *13*, 1465 (1974).
101. J. A. Kornblatt and A. M. English, *Eur. J. Biochem.*, *155*, 505 (1986).
102. P. Nicholls and E. Mochan, *Nature New Biol.*, *230*, 276 (1971).
103. T. L. Poulos and J. Kraut, *J. Biol. Chem.*, *255*, 10322 (1980).
104. R. Bechtold and H. R. Bosshard, *J. Biol. Chem.*, *260*, 5191 (1985).
105. B. Waldmeyer, R. Bechtold, H. R. Bosshard, and T. L. Poulos, *J. Biol. Chem.*, *257*, 6073 (1982).
106. B. Waldmeyer and H. R. Bosshard, *J. Biol. Chem.*, *260*, 5184 (1985).
107. J. E. Erman, K. L. Kim, L. B. Vitello, S. J. Moench, and J. D. Satterlee, *Biochim. Biophys. Acta*, *911*, 1 (1987).
108. S. J. Moench, J. D. Satterlee, and J. E. Erman, *Biochemistry*, *26*, 3821 (1987).
109. R. Bisson and R. A. Capaldi, *J. Biol. Chem.*, *256*, 4362 (1981).
110. B. Waldmeyer, R. Bechtold, M. Zurrer, and H. R. Bosshard, *FEBS Lett.*, *119*, 349 (1980).
111. L. B. Vitello and J. E. Erman, *Arch. Biochem. Biophys.*, *258*, 621 (1987).
112. R. K. Gupta and T. Yonetani, *Biochim. Biophys. Acta*, *292*, 502 (1973).
113. J. D. Satterlee, S. J. Moench, and J. E. Erman, *Biochim. Biophys. Acta*, *912*, *87* (1987).
114. S. H. Northrup, J. O. Boles, and J. C. L. Reynolds, *Science*, *241*, 67 (1988).

115. G. L. McLendon, *Proc. of the Robert A. Welch Found. Conf. on Chemical Res. XXXII Valency* (Houston, Oct. 31–Nov. 2, 1988), p. 201.

116. S. H. Northrup and H. P. Erickson, *Proc. Natl. Acad. Sci. USA*, *89*, 3338 (1992).

117. C. H. Kang, D. I. Brautigan, N. Osheroff, and E. Margoliash, *J. Biol. Chem.*, *253*, 6502 (1978).

118. W. H. Koppenol, S. Fergusson-Miller, N. Osheroff, S. H. Speck, and E. Margoliash, in *Oxidases and Related Redox Systems* (T. E. King, H. S. Mason, and M. Morrison, eds.), Pergamon Press, Toronto, Sydney, Frankfurt, New York, 1982, p. 1037.

119. T. L. Poulos, S. Sheriff, and A. J. Howard, *J. Biol. Chem.*, *262*, 13881 (1987).

120. H. Pelletier and J. Kraut, *Science*, *258*, 1748 (1992).

121. T. Asakura and T. Yonetani, *J. Biol. Chem.*, *244*, 4573 (1969).

122. H. R. Bosshard, J. Banziger, T. Hasler, and T. L. Poulos, *J. Biol. Chem.*, *259*, 5683 (1984).

123. M. A. Miller, J. T. Hazzard, J. M. Mauro, S. L. Edwards, P. C. Simons, G. Tollin, and J. Kraut, *Biochemistry*, *27*, 9081 (1988).

124. J. M. Mauro, L. A. Fishel, J. T. Hazzard, T. E. Meyer, G. Tollin, M. A. Cusanovich, and J. Kraut, *Biochemistry*, *27*, 6243 (1988).

125. K. E. Hammel and M. A. Moen, *Enzyme Microb. Technol.*, *13*, 15 (1991).

126. H. Wariishi, K. Valli, and M. H. Gold, *Biochemistry*, *28*, 6017 (1989).

127. M. Tien, K. Kirk, C. Bull, and J. A. Fee, *J. Biol. Chem.*, *261*, 1687 (1986).

128. H. Wariishi, J. Huang, H. B. Dunford, and M. H. Gold, *J. Biol. Chem.*, *266*, 20694 (1991).

129. V. Renganathan, M. Miki, and M. H. Gold, *Biochemistry*, *26*, 5127 (1985).

130. L. Sawyer and M. James, *Nature*, *295*, 79 (1982).

131. W. L. Purcell and J. E. Erman, *J. Am. Chem. Soc.*, *98*, 7033 (1976).

132. Y. Hayashi and I. Yamazaki, *J. Biol. Chem.*, *254*, 9101 (1979).

133. A. T. Smith, S. A. Sanders, R. N. F. Thorneley, J. F. Burke, and R. C. Bray, *Eur. J. Biochem.*, *207*, 507 (1992).

134. S. A. Martinis, W. M. Atkins, P. S. Stayton, and S. Sligar, *J. Am. Chem. Soc.*, *111*, 9252 (1989).

135. M. Imai, H. Shimada, Y. Watanabe, Y. Matsushima-Hibiya, R. Makino, H. Koga, T. Horiuchi, and Y. Ishimura, *Proc. Natl. Acad. Sci. USA*, *86*, 7823 (1989).

136. R. Raag, S. A. Martinis, S. G. Sligar, and T. L. Poulos, *Biochemistry*, *30*, 11420 (1991).

137. L. A. Fishel, J. E. Villafranca, J. M. Mauro, and J. Kraut, *Biochemistry*, *26*, 351 (1987).

138. J. Wang, J. M. Mauro, S. L. Edwards, S. J. Oatley, L. A. Fishel, V. A. Ashford, N-h. Xuong, and J. Kraut, *Biochemistry*, *299*, 7160 (1990).

139. M. Sundaramoorthy, K. Choudhury, S. L. Edwards, and T. L. Poulos, *J. Am. Chem. Soc.*, *113*, 7755 (1991).

140. J. E. Erman, L. B. Vitello, J. M. Mauro, and J. Kraut, *Biochemistry*, *28*, 7992 (1989).

141. B. M. Hoffman, J. E. Roberts, T. G. Brown, C. H. Kang, and E. Margoliash, *Proc. Natl. Acad. Sci. USA*, *76*, 6132 (1979).

142. D. Myers and G. Palmer, *J. Biol. Chem.*, *260*, 3887 (1985).

143. H. Hori and T. Yonetani, *J. Biol. Chem.*, *260*, 349 (1985).

144. D. B. Goodin, A. G. Mauk, and M. Smith, *J. Biol. Chem.*, *262*, 7719 (1987).

145. S. L. Edwards, J. Kraut, and T. L. Poulos, *Biochemistry*, *27*, 8074 (1988).

146. T. Yonetani, H. Yamamoto, J. E. Erman, J. S. Leigh, and G. Reed, *J. Biol. Chem.*, *247*, 2447 (1972).

147. F. E. Summers and J. E. Erman. *J. Biol. Chem.*, *263*, 14267 (1988).

148. P. S. Ho, B. M. Hoffman, N. Solomon, C. H. Kang, and E. Margoliash, *Biochemistry*, *23*, 4122 (1984).

149. M. A. Miller, D. Bandyopadhyay, J. M. Mauro, T. G. Traylor, and J. Kraut, *Biochemistry*, *31*, 2789 (1992).

150. D. S. Kang and J. E. Erman, *J. Biol. Chem.*, *257*, 12775 (1982).

151. S. Hahm, B. Durham and F. Millet, *Biochemistry*, *31*, 3472 (1992).

152. L. Green, S. Hahm, B. Durham, and F. Millet, *Biochemistry*, *30*, 9450 (1991).

153. G. Smulevich, J. M. Mauro, L. A. Fishel, A. M. English, J. Kraut, and T. G. Spiro, *Biochemistry*, *27*, 5486 (1988).

154. G. Smulevich, A. R. Mantini, A. M. English, and J. M. Mauro, *Biochemistry*, *28*, 5058 (1989).

155. J. D. Satterlee, J. E. Erman, J. M. Mauro, and J. Kraut, *Biochemistry*, *29*, 8797 (1990).

156. L. A. Fishel, M. F. Farnum, J. M. Mauro, M. A. Miller, J. Kraut, Y. L. Xiao-ling, and C. P. Scholes, *Biochemistry*, *30*, 1986 (1991).

157. J. E. Erman, L. B. Vitello, M. A. Miller, J. M. Mauro, and J. Kraut, *Biochemistry*, *31*, 11524 (1992).

158. D. B. Goodin and D. E. McRee (in press).

159. K. Choudhury, M. Sundaramoorthy, J. M. Mauro, and T. L. Poulos, *J. Biol. Chem.*, *267*, 25656 (1992).

160. D. B. Goodin, M. G. Davidson, J. A. Roe, A. G. Mauk, and M. Smith, *Biochemistry*, *30*, 4953 (1991).

161. R. A. Marcus and N. Sutin, *Arch. Biochem. Biophys.*, *811*, 265 (1985).

162. K. T. Conklin and G. McLendon, *J. Am. Chem. Soc.*, *110*, 3345 (1988).

163. E. Cheung, K. Taylor, J. A. Kornblatt, A. M. English, and G. McLendon, *Proc. Natl. Acad. Sci. USA*, *83*, 1330 (1986).

164. G. McLendon and J. R. Miller, *J. Am. Chem. Soc.*, *105*, 4470 (1985).

165. J. E. Erman, L. B. Vitello, M. A. Miller, and J. Kraut, *J. Am. Chem. Soc.*, *114*, 6592 (1992).

166. N. Moguilevsky, L. Garcia-Quintana, A. Jacquet, C. Tournay, L. Fabry, L. Pierard, and A. Bollen, *Eur. J. Biochem.*, *197*, 605 (1991).

167. A. Jacquet, C. Deby, M. Mathy, N. Moguilevsky, G. Deby-Dupont, A. Thirion, E. Goormaghtigh, L. Garcia-Quintana, A. Bollen, and J. Pincemail, *Arch. Biochem. Biophys.*, *291*, 132–138 (1991).

168. A. Jacquet, V. Deleersnyder, L. Garcia-Quintana, A. Bollen, and N. Moguilevsky, *FEBS Lett.*, *302*, 189–191 (1992).

169. A. Jacquet, V. Deleersynder, L. Garcia-Quintana, R. E. Fenna, A. Bollen, and N. Moguilevsky. *Proceedings 1993 Miami Bio/Technology Winter Symposium*, IRL Press, Oxford, 1993, p. 44.

3
Photosystem II

Curtis W. Hoganson and Gerald T. Babcock

Department of Chemistry, Michigan State University, Chemistry Building, East Lansing, Michigan 48824-1322, USA

1. INTRODUCTION

The stunning diversity of living organisms on earth came about in response to the formation of an atmosphere rich in molecular oxygen. This gas is a strong oxidant, but, because of its unusual triplet electronic ground state, it reacts at only moderate rates at ambient temperatures. Thus, organic material based on reduced carbon and nitrogen compounds does not spontaneously combust and may function as energy stores. This oxygen atmosphere arose as primitive organisms (cyanobacteria?) acquired the ability to use the energy of sunlight to remove electrons from molecules of water.

The development of the biochemical ability to oxidize water is remarkable, given that the formation of water from its elements is so strongly favored. A strong oxidant must be created, and it must be stabilized against side reactions that would destroy its potential for multiple turnovers, so necessary for a biochemical catalyst. Because this ability did develop, the amount of organic material that could be produced was greatly increased. Furthermore, the availability of an oxygen atmosphere made new biochemical reactions possible, i.e., those involving molecular oxygen or peroxides as reactants, some of which are described in other chapters of this book. For these reasons, we think of oxygen as a very important *product* of photosynthesis.

This chemical reaction within the green membranes of plants and algae has been the subject of many investigations. Much is now known about how light energy is converted to chemical energy. The catalytic process that releases molecular oxygen, however, is still the subject of intense research activity [1,2].

The purpose of this chapter is to discuss the occurrence and function of amino acid radicals in the oxidation of water by Photosystem II (PSII) of cyanobacteria and higher plants. We will provide an overview of PSII and then focus on the properties and reactions of its two tyrosine radicals. Although these radical species do not react with water, they do engage in electron transfer reactions with the cluster of manganese atoms that does oxidize water. We will also discuss other amino acid radicals that were recently reported to occur in PSII.

1.1. Function of Photosystem II

Photosystem II converts light energy to chemical energy in two forms. Redox energy is stored when a strong oxidant and a strong reductant are created simultaneously. Additional energy is stored in the gradient of the proton chemical potential across a membrane. These are reviewed in general form in the following two subsections; in Sec. 1.2 we begin a detailed discussion of the redox chemistry and the cofactors that mediate it.

1.1.1. Redox Reaction

In PSII, water is oxidized to molecular oxygen and protons, and plastoquinone is reduced to the corresponding quinol. These reactions occur at separate sites in the protein and these two sites are connected by a series of redox cofactors that transfer electrons from one site to the other. A specialized chlorophyll molecule (or a dimer of chlorophyll) is responsible for initiating the movement of electrons: its excited state is both a strong reductant and a strong oxidant. The other redox components function to stabilize this energy of molecular excitation and to separate the intermediate reductants and oxidants formed. Plastoquinol is reoxidized by the cytochrome b_6f complex, delivering its electrons to the mobile electron donor, plastocyanin, and thence to Photosystem I. Photoexcitation and charge separation in this photosystem produces the reductant nicotinamide adenine dinucleotide phosphate, reduced form (NADPH) used in the Calvin cycle reactions that reduce carbon dioxide to carbohydrate. For a review of the overall electron transport scheme in photosynthesis, see [3].

1.1.2. Proton Gradient Creation

A proton gradient is created by the action of light on PSII because electron transport in the reaction center is vectorial and the sites of water oxidation and plastoquinone reduction lie on opposite sides of the membrane. Thus, the protons from the oxidation of water are released into the space bounded by the thylakoid membranes (the lumen), while the protons required to produce the neutral plastoquinol are taken up from the outer side of the thylakoid membrane (the stroma). For each electron transferred through PSII, one proton is taken up from the stroma and one is released into the lumen. The reactions of plastoquinone with the cytochrome b_6f complex [4] also contribute to the electrochemical potential gradients that develop during light-induced electron transfer. The energy of the proton chemical potential gradient across the thylakoid membrane allows the formation of ATP which is consumed, for example, in the reactions of the Calvin cycle.

1.2. Components of Photosystem II

Photosystem II contains chlorophylls, carotenoids, and several inorganic or organic molecules, some of which are redox-active. These species are bound to several polypeptides that combine to form an overall organizing superstructure. Most of the chlorophyll and all of the carotenoid molecules act by absorbing incident photons to produce excited states and then transferring the excitation energy to the reaction center. It is at the reaction center that the energy of molecular excitation is trapped and converted to chemical energy.

1.2.1. Electron Transfer Cofactors

Photosystem II contains at least a dozen redox-active cofactors that have been distinguished on the basis of spectroscopic measurements. These cofactors and the polypeptides of PSII are shown schematically in Fig. 1, which also indicates the normal flow of electrons during photosynthesis. The electron paramagnetic resonance (EPR) signals of these cofactors have been reviewed [5], and several more general reviews of PSII are available [2,6–8], as is one focusing on amino acid radicals [9].

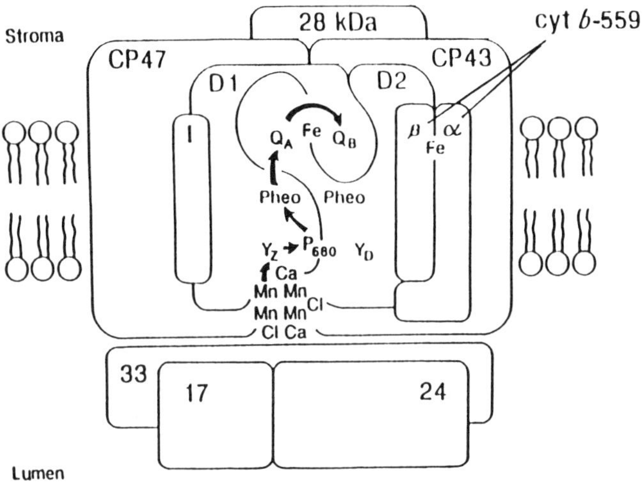

FIG. 1. Schematic diagram of Photosystem II showing the major polypeptides and the electron transfer cofactors. Solid arrows indicate the direction of electron flow. (Reprinted with permission, copyright 1992 by Elsevier Science Publishers B.V.)

Central to the scheme is the specialized chlorophyll a, P_{680}. Its excited state is the strongest reductant in PSII, about -0.7 V vs. normal hydrogen electrode (NHE). P^*_{680} reduces pheophytin in a few picoseconds, leaving the radical cation of P_{680}, which is the strongest oxidant of PSII, having a redox potential of about +1.17 V vs. NHE [10]. The pheophytin anion radical reduces a bound plastoquinone, Q_A, which transfers the electron further to a second plastoquinone, Q_B. Upon reduction by a second electron, Q_B protonates to form the neutral plastoquinol, leaves the Q_B binding site, and is replaced there by another plastoquinone molecule. (For a summary of the kinetics of these electron transfer reactions, see [11]). These reactions of the reducing side resemble strongly those of the reaction centers of the purple photosynthetic bacteria, of which the X-ray crystal structure has been determined [12,13]. These bacterial reaction centers contain bacteriochlorophyll and bacteriopheophytin instead of chlorophyll and pheophytin, and the bacteriochlorophyll cation radical produced by light is thermodynamically unable to oxidize H_2O.

The distinctive reactions of PSII are those of its oxidizing side. P^+_{680}

is reduced by a nearby tyrosine residue, Y_Z, of the D1 polypeptide, and the tyrosine radical is reduced to the diamagnetic state by the manganese complex. The complex of four manganese atoms goes through a cycle of five redox states, S_0–S_4, the most oxidized of which, S_4, initiates the reaction that releases molecular oxygen.

Current ideas on the structure and function of the manganese complex have been reviewed recently [2]. There is an emerging consensus that this complex consists of four manganese atoms strongly coupled to each other. A very probable structural element is the di-μ-oxo bridged manganese dimer, of which there may be two. The manganese is likely to be ligated by carboxylate residues and at least one histidine, which probably are part of the D1 polypeptide. The S_1 state is the most stable oxidation state in the dark and the oxidation states of its manganese atoms are probably +3, +3, +4, and +4.

Other redox-active cofactors occur in PSII, but they are not involved in the main pathway of electron transfer from water. These include the tyrosine residue, Y_D, cytochrome b_{559} (one or two copies per PSII), a nonheme iron atom ligated approximately midway between the Q_A and Q_B binding sites, and chlorophyll and carotenoid molecules that are liable to photooxidation at low temperature ($T < 200$ K).

1.2.2. *Other Inorganic Cofactors*

Calcium and chloride ions are required for oxygen evolution activity. They promote the proper turnover of the S states of the manganese ensemble (for an excellent recent review, see [2]). Chloride is a ligand to manganese [14] and may be required to prevent the formation and release of hydrogen peroxide [15,16]. Photosystem II also contains the nonheme iron atom noted above, which is analogous to that found in the reaction centers of purple bacteria in most respects, but which is also oxidizable at a redox potential of +0.4 V [17].

1.2.3. *Proteins*

Photosystem II is composed of many different protein subunits, having several different functions (reviewed in [18]). The core photochemical reactions occur on the cofactors bound to a central reaction center consisting of polypeptides called D1 and D2 (mol. wt. 34 and 32 kD, respectively). The reaction center as isolated by Triton X-100 extraction [19] also contains the two polypeptides of cytochrome b_{559} and one small polypeptide (5 kD). Larger complexes can be isolated by using milder detergent treatments and these retain the chlorophyll a binding proteins, CP47 and CP43 (mol. wt. 47 and 43 kD), as well as other intrinsic membrane proteins (mol. wt. 28 and 22 kD and smaller) with less obvious functions. Extrinsic membrane proteins (mol. wt. 33, 24, and 17 kD) are retained or not depending on the severity of treatment but are released by exposure to high ionic strength solutions. These extrinsic proteins facilitate manganese, calcium, and chloride binding and shield the manganese complex from exogenous reductants.

2. IDENTIFICATION AND SPECTROSCOPIC STUDIES OF THE TYROSINE RADICALS

The discovery of an organic free radical associated with PSII was made by using EPR [20], and this spectroscopic technique has been most useful in studying its reactions. UV absorption spectroscopy has become useful more recently after the kinetic parameters became known from EPR studies.

Illumination of chloroplasts or other photosynthetic material (oxygen-evolving or inhibited) produces an EPR signal (see Fig. 2) showing partially resolved hyperfine structure and an overall linewidth of about 2.0 mT and average *g* value of 2.0046, which was referred to in the early literature as signal II. Signal II disappears after several hours in the dark but can be reinduced by further illumination. Signal I comes from the oxidized form of the primary donor chlorophyll of PSI, P-700, which is induced in the light, having a linewidth of 0.7 mT and a *g* value of 2.0027. The EPR properties of P_{680}^{+} are similar to these. The differences in *g* value and linewidth allow the tyrosine radicals to be monitored in the presence of chlorophyll radicals, although spectral overlap between these species has and continues to create some difficulties.

When oxygen evolution is inhibited, illumination produces approximately a doubling of the EPR signal II. The light-induced increase decays in 1 sec or less. In oxygen-evolving material, flashes of light produce an increase of signal II that decays in about 1 msec. The stable

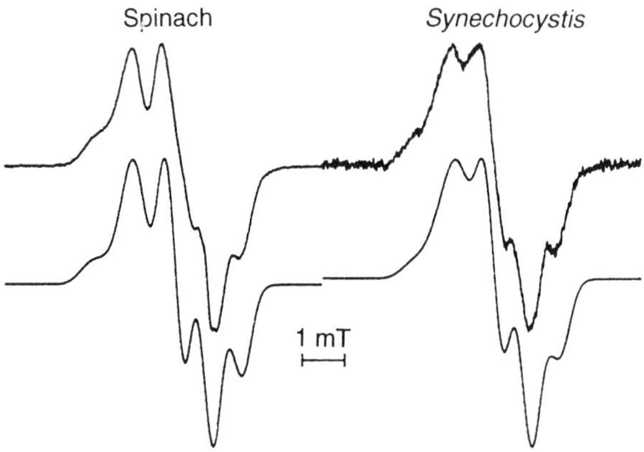

FIG. 2. Electron paramagnetic resonance spectra of $Y_D^{\bullet}$ (above) in PSII preparations from spinach and *Synechocystis*, together with computer-generated simulations (below) using the parameters in Table 1. The spectra were obtained at 130 K, using 0.2 mW microwave power at 9.59 GHz, and 100 kHz field modulation of 0.2 mT.

signal and these two rapidly decaying signals were referred to as signal II slow (s), fast (f), and very fast (vf), respectively. The radical giving rise to the stable signal is now called $Y_D^{\bullet}$ (or D^+), and that giving the rapid kinetic phases is called $Y_Z^{\bullet}$ (or Z^+). Quantitation shows that there is one Y_D and one Y_Z per PSII unit. Detection of $Y_D^{\bullet}$ and $Y_Z^{\bullet}$ requires a protein complex that includes D1, D2, cyt b559, and CP47 [21].

2.1. Identification by Isotopic Replacement

As is the case for most of the radicals discussed in this book, definitive assignment came as the result of experiments in which the radical precursor was supplied in an isotopically labeled form. Although amino acid auxotrophs are rare in cyanobacteria and higher plants, a functional auxotrophic condition in *Synechocystis* can be created for tyrosine in a medium containing 0.5 mM phenylalanine [22]. Under such conditions, the synthesis of all aromatic amino acids is inhibited, and externally supplied tyrosine is taken up by the organism and incorporated into protein. Perdeuterated tyrosine thus supplied produced a narrowing of the EPR spectrum, similar to that which occurs during growth in D_2O, and showed that the signal II is due to a tyrosine radical [22]. Parallel experiments in which deuterated methionine was supplied showed that plastoquinone was labeled with deuterium but that the EPR signal II was unchanged, and thereby put to rest the earlier notion that the EPR signal might derive from a plastoquinone species.

Further experiments in which tyrosine deuterated at specific ring positions was used showed that the photosynthetic radical has properties similar to those of the tyrosyl radical produced by UV illumination at low temperature [23], and showed that the tyrosine residues had not undergone posttranslational modification.

2.2. Site-Directed Mutagenesis

With the identification of the PSII organic radicals as tyrosyl radicals, their location within PSII became an addressable question. At that time, a D1–D2 reaction center preparation in which the primary photochemical reactions of P_{680} were intact had recently been developed [19]. The X-ray structure of the bacterial reaction center had been determined and shown to have an approximate, twofold rotation symmetry [12,13]. A strong sequence similarity had been demonstrated between D1, D2, and the bacterial reaction center L and M proteins [24]. These facts led two research groups to suspect that the tyrosine residues would reside on D1 and D2, where they would be near P_{680} and thus able to participate in the rapid electron transfer reactions that had been long known to occur. Iodination experiments suggested that Y_Z and Y_D might reside on D1 and D2, respectively [76]. A stretch of nine highly conserved residues was found to occur on D1 and D2 with conserved tyrosine residues as the fifth residue, and these became the targets for site-directed mutagenesis.

The first experiments replaced tyrosine-160 of the D2 polypeptide with phenylalanine and produced a photosynthetically competent mutant that lacked the stable tyrosine EPR signal [25,26]. Mutation of the neighboring residue methionine-159 to arginine did not alter signal II [26]. Loss of tyrosine-160 resulted in slower growth under photoautotrophic conditions, but no change in photoheterotrophic growth rate or in whole-cell oxygen evolution rates. These experiments established that tyrosine-160 of the D2 polypeptide is indeed Y_D, and they strongly suggested that the corresponding tyrosine residue on polypeptide D1 is Y_Z.

Further mutagenesis studies on D1 that replaced tyrosine-161 with phenylalanine produced a mutant that does not grow photoautotrophically or evolve oxygen [27,28]. The $Y_Z^{\bullet}$ signal was absent and fluorescence associated with a reduced Q_A was not observed, showing that electron transfer through PSII did not occur. These experiments demonstrated convincingly that Y161 of D1 is Y_Z.

2.3. Hyperfine Interactions from EPR and ENDOR

The distinctive EPR line shapes of $Y_D^{\bullet}$ and $Y_Z^{\bullet}$ result from the combined effects of an anisotropic g tensor [29] and anisotropic hyperfine interactions with primarily three protons [23]. Isotopic replacement of the hydrogen atoms by deuterium shows that significant spin density occurs at the positions ortho and para, but not meta, to the phenolic oxygen [23]. The ^{17}O substitution of tyrosyl radicals shows that significant spin density also occurs at the phenolic oxygen [30]. This spin density distribution is characteristic of an odd-alternant π system and is found in other phenoxyl radicals.

Because the tyrosyl radicals of PSII are fixed in the protein and prevented from rotating, their EPR spectra are broadened by the anisotropy of the g and hyperfine tensors, and determination of the hyperfine tensors from the spectra is not straightforward. Instead, a simulation approach can be used in which a spectrum is calculated from a set of parameters given as input to a computer program. This is feasible for $Y_D^{\bullet}$ because isotopically labeled spectra and spectra at different microwave frequencies are available. We used this approach with the tyrosine radicals in PSII and ribonucleotide reductase and the in vitro tyrosine radical to discover their hyperfine splittings [31]. In addition to noting the effect of the protein on determining the conformation of the methylene group and its proton hyperfine splittings, we concluded that the protein environment can alter the distribution of spin in the radicals. Electron nuclear double resonance (ENDOR) spectroscopy supports the hyperfine parameters obtained from simulations. The EPR spectra of

TABLE 1

Proton Hyperfine Parameters Used in the Simulation of $Y_D^{\bullet}$ EPR Spectra

Parameter	Spinach	*Synechocystis*
3,5: A_x, A_y, A_z	1.13, 0.27, 0.74	0.95, 0.28, 0.73
Methylene 1	1.0, 0.7, 0.7	0.95, 0.65, 0.65
Methylene 2	0.3, 0.1, 0.1	0.5, 0.16, 0.16

Hyperfine values are in mT. Simulations included an anisotropic g tensor (2.0074, 2.0044, 2.0023) [29], and splittings due to the 2,6 ring protons (0.2, 0.3, 0.1). The x and y axes of the 3,5 and 2,6 hyperfine tensors are oriented ± 23 and $\pm 10°$, respectively, away from the corresponding axis of the g tensor.

$Y_D^{\bullet}$ from spinach and *Synechocystis* show slight differences (Fig. 2) that can be accounted for by the small changes in the hyperfine splittings (Table 1) that were suggested by ENDOR [31].

2.4. Hydrogen Bonding of Y_D

The tyrosyl Y_D radical is deprotonated, judging by studies of in vitro tyrosyl and other phenoxyl radicals [22]. The pK_a of phenoxyl radicals is about -2, so that a deprotonated radical is expected on that account. Further, the g values of protonated phenoxyl radicals are distinctly lower (2.0033) than those of deprotonated phenoxyl radicals (2.0044), so the experimental g value of Y_D ($g_{iso} = 2.0046$) implies a deprotonated radical.

The diamagnetic phenol group has a pK_a of 10, however, and should be protonated. Presumably, a nearby basic chemical group accepts the proton released on oxidation. The strength of this base would have a strong influence on the redox potential of the tyrosine. The protonated base would also be able to hydrogen-bond to the radical. If it does, the hydrogen-bonded proton should be detectable by ENDOR spectroscopy. Resonances that slowly disappeared during D_2O exchange were assigned to a hydrogen-bonded proton [32,33], having an axially symmetric, traceless tensor with couplings of 3.5 and 7.1 MHz.

Speculation as to the identity of this basic group has centered on histidine-190 of the D2 polypeptide, which is highly conserved [25]. Modeling based on the bacterial reaction center supports this [34]. Mutagenesis of this residue leads to replacement of the usual spectrum of $Y_D^{\bullet}$ by a signal about 1.0 mT wide with little hyperfine structure [35]. A tyrosine deuteration experiment as described above shows narrowing of the electron spin resonance (ESR) line, indicating that the radical does derive from a tyrosine residue [36]. Mutagenesis of glutamine-164 alters the spectrum of $Y_D^{\bullet}$ slightly [35], suggesting that it is near Y_D but not directly hydrogen bonded to Y_D, as had been suggested [37]. The D1 polypeptide has a corresponding histidine, which is a logical candidate for hydrogen bonding to Y_Z. Mutation of that residue to aspartate or glutamate does not alter the $Y_Z^{\bullet}$ spectrum, but kinetics of P_{680}^{+} reduction are altered [36].

2.5. Environment Around the Tyrosine Radicals

Clearly, Y_D is buried inside the protein. Deuterium exchange detected by ENDOR occurs very slowly. Exogenous reductants reduce $Y_D^{\bullet}$, but more slowly than they reduce $Y_Z^{\bullet}$ in manganese-depleted preparations. Paramagnetic lanthanide ions alter the electron spin relaxation behaviour of $Y_D^{\bullet}$, and allow a distance of 2.6 nm between Y_D and the protein surface to be calculated [38], which locates it nearly in the middle of the membrane-spanning region of the D2 polypeptide. Molecular modeling based on the bacterial reaction center suggests that Y_D is surrounded by conserved hydrophobic amino acids: 3 or 4 phenylalanines, 1 or 2 leucines and 1 alanine [34].

Because the Y_Z radical is much less stable, ENDOR studies of this species have not been done until recently [39]. The much higher reactivity with exogenous reductants suggests a location closer to the protein surface, which is negatively charged [40,41]. When manganese and the water-soluble polypeptides are present, Y_Z is probably quite far from the aqueous phase. Modelling as above suggests that the amino acids surrounding Y_Z are more hydrophilic than those around Y_D, and may include glutamine, aspartate, glutamate, histidine and serine [34]. This difference in the surroundings may contribute to the different reduction potentials of the two radicals. Magnetic interaction between $Y_Z^{\bullet}$ and P_{680}^{+} has been used to estimate a center–center distance between the two cofactors of 1–1.5 nm [42], in reasonable agreement with modeling studies [34,37]. Although magnetic interaction with the oxygen-evolving complex (OEC) enhances spin relaxation of $Y_Z^{\bullet}$, manganese depletion does not alter its EPR line shape [43]; this would indicate a substantial distance between the two paramagnetic entities. A recent model puts the manganese complex about 1 nm from Y_Z [44].

3. ELECTRON TRANSFER REACTIONS OF THE TYROSINE RADICALS—BOTH O_2-EVOLVING AND INHIBITED PHOTOSYSTEM II

3.1. Reactions of Y_Z, the Tyrosine in the Z Scheme

The function of Y_Z in interfacing the photochemistry initiated by P_{680} to the water-splitting reaction catalyzed in the manganese ensemble was originally identified by EPR techniques. More recently, detection by optical techniques in the UV region has become possible and a detailed understanding of the dynamics of its electron transfer function is emerging.

3.1.1. Oxidation by P_{680}^{+}

After transfer of excitation energy to P_{680}, an electron is transferred to a pheophytin residue and to Q_A in less than 1 nsec. P_{680}^{+} thus produced is a strong oxidant (E_m = 1.17 V) and is subsequently reduced by Y_Z ($E_m \approx$ 1.0 V). The P_{680}^{+} Y_Z electron transfer reaction is monitored most conveniently by measuring the absorbance changes around 830 nm where P_{680}^{+} absorbs, but where ground state chlorophyll does not absorb. The oxidation of Y_Z causes an absorbance increase at 300 nm, but because light of this wavelength is capable of promoting PSII photochemistry, this measurement is more difficult. Nonetheless several groups have developed techniques whereby Y_Z optical changes can be detected directly and reliably.

Reduction of P_{680}^+ occurs in as little as 20 nsec in uninhibited PSII, but the rate is dependent on the S state of the OEC. In S_0 and S_1, the halftime is about 20 nsec, and in S_2 and S_3, the halftime is pH-dependent with phases of 50 and 200 nsec [45]. On all S-state transitions, Y_Z is the only detected intermediate [43,46,47]. The S-state dependence has been suggested to reflect the charge of the OEC and may indicate oxidation of the manganese on the S_1-to-S_2 transition without loss of a proton from the immediate surroundings. In this interpretation, the alteration in the electron transfer kinetics is attributed to an electrostatically induced change in the redox potential of Y_Z. Recent work on proton release accompanying S-state advance has challenged this view, however; conformational change of the protein coupled to the S-state transition as the origin of the S-state dependence of the P_{680}^+ Y_Z electron transfer cannot be ruled out [48].

In preparations depleted of manganese by washing either with concentrated Tris buffer or with NH_2OH, the oxidation of Y_Z proceeds more slowly. The reason for this is not yet clear, although we can eliminate gross structural changes near the tyrosine as an explanation since we have shown that the $Y_Z^{\bullet}$ line shape is not strongly affected by manganese depletion [43]. The electron transfer is pH-dependent, having a half-time of about 2 μsec at pH 8 and 20 μsec at pH 5. Detailed studies of the ionic strength dependence suggest that the rate depends on the protonation state of a basic group with an inherent pK_a of 6.5 [49]. Divalent transition metal ions also bind near Y_Z and slow the P_{680}^+/Y_Z electron transfer rate to about 30 μsec [109].

3.1.2. Reduction by the Oxygen-Evolving Complex

In intact PSII membranes or particles, oxidized Y_Z is reduced by the four manganese atoms of the OEC. These manganese atoms are close enough to interact magnetically with each other, as is clear from the multiline ESR signal of the S_2 state [51]. To produce molecular O_2, the OEC undergoes four sequential oxidations. Recent X-ray absorption data indicate that manganese or its direct ligands are oxidized on each transition up to S_3 [52]. The rate of the electron transfer from the OEC to Y_Z is dependent on the S state; half-times for the reaction at room temperature are 30, 100, 200–300, and 1300 μsec for the transitions from S_0, S_1, S_2, and S_3, respectively [7]. This change in reaction rate presumably reflects the increasing difficulty of extracting further electrons from an already oxidized center. The extent to which positive charge is also accumulated in the OEC is somewhat less certain, as proton release occurs on several, if not all, of the oxidation steps. A detailed study of the temperature dependence of the S_n $Y_Z^{\bullet}$ electron transfer reactions was recently made and was interpreted to indicate that oxidation of S_2 to S_3 is accompanied by a more extensive structural rearrangement than is observed for the other S-state transitions [53].

3.1.3. Charge Recombination in the Reaction Center

Following light-induced charge separation, charge recombination can occur causing a return of the reaction center to its initial chemical state. All light-induced electron transfer couples are prone to these reactions to some extent, but they are nonproductive in the photosynthetic context, as no permanent chemical reaction has occurred. The quantum yield for charge separation in intact PSII reaction centers approaches 1.0, so charge recombination is studied in inhibited preparations. Depletion of manganese by treatment with Tris (0.8 M, pH 8) or with hydroxylamine (0.1 mM) in the dark are commonly used methods. Calcium or chloride depletion are also used because these treatments can prevent normal S-state advancement in the OEC (without release of manganese) and allow $Y_Z^{\bullet}$ to accumulate in the light [112,113]. In manganese-depleted PSII, if Y_Z has been preoxidized by light, P_{680}^{+} can be reduced by Q_A^- in 100–200 μsec at room temperature. Multiphasic kinetics have been reported for the same reaction in cyanobacterial PSII. The electron transfer from Y_Z to P_{680}^{+} has an activation energy of 11 kcal/mol in manganese-depleted PSII, while the charge recombination between P_{680}^{+} and Q_A^- has an activation energy of only 4.5 kcal/mol [54]; thus, at temperatures below −40°C, reduction of P_{680}^{+} occurs mainly by charge recombination from Q_A^-. Hydroxylamine in the presence of light can prevent the oxidation of Y_Z by a mechanism that is not yet understood and this condition allows P_{680}^{+} to be reduced by Q_A^- in about 200 μsec [55].

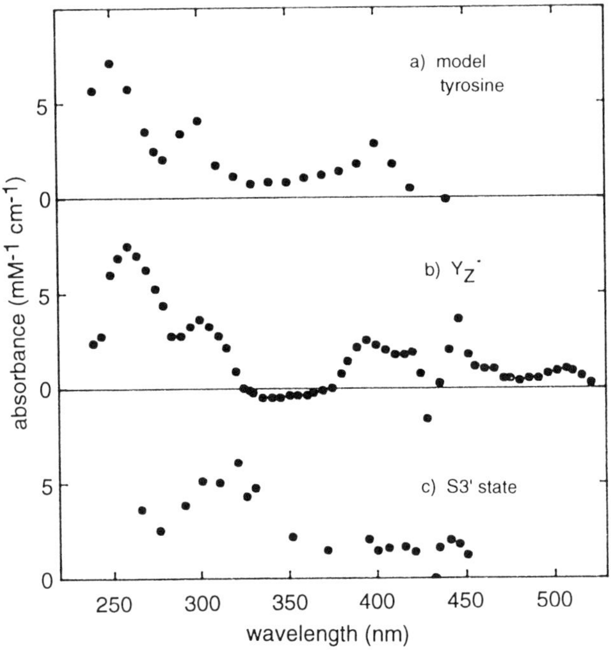

FIG. 3. Electronic absorption spectra of the tyrosine radical produced by 265-nm illumination [114] (a), $Y_Z^{\bullet}$ in manganese-depleted PSII membranes [56] (b), and the S_3' state in calcium-depleted PSII membranes [81] (c).

When Y_Z is photooxidized, charge recombination between it and Q_A^- occurs on the tens to hundreds of milliseconds time scale depending on conditions [56,57]. This relatively long lifetime facilitates measurement of the $Y_Z^{\bullet}$ UV spectrum (Fig. 3) and provides ample opportunity for reaction with exogenous reductants.

3.1.4. Reduction by Exogenous Reductants, Including Mn^{2+}

In the absence of an active OEC, $Y_Z^{\bullet}$ retains its strongly oxidizing character and reacts rapidly with many one-electron reductants. Reactions with the hydrophobic reductants benzidine, phenylenediamine, hydroquinone, and diphenylcarbazide were studied [58], as were the hydrophilic reductants ascorbate, ferrocyanide, iodide, and cyanotungstate [59]. From the ionic strength dependence of the rate of reaction with charged reductants, the surface charge density near Y_Z was estimated to be $-3.4\ \mu C/cm^2$ at pH 8 [40].

Mn^{2+} is also oxidized by $Y_Z^{\bullet}$ [60], and this reaction is likely to be involved in photoactivation (see below). Binding of Mn^{2+} precedes the electron transfer, which occurs in about 0.4 msec [60,61]. When the reaction is induced by flashes at low repetition rates, a pH-dependent dissociation constant for Mn^{2+} of a few micromolar is observed. The Mn^{3+} produced remains bound to the protein and under continuous illumination inhibits, with a K_I of 150 nM, the photooxidation of diphenylcarbazide [62]. Submicromolar K_M values for Mn^{2+} have been measured for the manganese-catalyzed photooxidation of H_2O_2 [63,64]. These reactions have been used together with modification by diethylpyrocarbonate or site-directed mutagenesis to show that the manganese binding site contains a histidine residue [65] and aspartate-170 of D1 [61].

3.1.5. *Involvement in Photoactivation*

Active PSII contains manganese in some oxidation states higher than Mn^{2+}. Chemical reductants can reduce the manganese, releasing Mn^{2+}. The process in which Mn^{2+} is reincorporated into PSII and in which oxygen evolution activity is restored requires light and is called photoactivation. The process occurs with a rather low quantum yield and requires the flow of electrons through PSII.

From experiments that used short flashes of light to induce photoactivation, a model for the photoactivation process has been suggested [66]. A single flash of light oxidizes one bound Mn^{2+} to Mn^{3+}, generating an intermediate that is unstable because of its tendency to be reduced. A second flash of light given before reduction of the Mn^{3+} can occur produces a stable intermediate containing a second bound Mn^{3+} ion. Incorporation of two more Mn^{2+} ions follows; this step does not require light [67]. Time-resolved studies of the Mn^{2+} photooxidation process have provided some support for this model and have shown directly that $Y_Z^{\bullet}$ is the endogenous oxidant of the bound metal [60].

3.2. Reactions of Y_D, the Stable Tyrosine Radical

In light, Y_D is oxidized and can remain in its radical state for hours.

Extensive dark adaptation usually returns Y_D to its reduced, diamagnetic state. As noted above, mutagenesis that replaces the Y_D tyrosine with phenylalanine slows growth but does not affect oxygen evolution [25,26], so the physiological importance of this radical is unknown. The highly conserved nature of the Y_D residue implies that the residue does have a function. $Y_D^{\bullet}$ is detected in PSII mutagenized so as not to contain Y_Z [27,28], which implies that oxidation of Y_D occurs by P_{680}^{+} in this case. Normally, electron transfer from the OEC to P_{680}^{+} outcompetes this reaction [68,69] as indicated by flash O_2 yields. In manganese-depleted PSII, Y_Z is oxidized by P_{680}^{+} and the $Y_Z^{\bullet}$ thus formed oxidizes Y_D [70,71]. The reduction potential of $Y_D^{\bullet}$ is estimated to be +720/+760 mV (vs. NHE) [50,72], which is significantly lower than that of Y_Z.

3.2.1. *Oxidation by the S_2 and S_3 States of the Oxygen-Evolving Complex*

When the OEC is intact, Y_D can be oxidized by both the S_2 and S_3 states, but not by S_0 or S_1 [68]. This was demonstrated both by EPR measurement of the signal from $Y_D^{\bullet}$ and by flash O_2 yield patterns. The half-time of these reactions was found to be about 2 sec. Further evidence of these reactions was found by delaying the times between one or two preflashes and the flash train given to elicit O_2 evolution [73], and by observing loss of multiline EPR signal of S_2 [74]. A detailed study of the pH dependence of the S_2 Y_D reaction has been made, and the redox potentials of the S_1/S_2 and S_2/S_3 couples have been estimated each to be about +920 mV [50].

3.2.2. *Reduction by the S_0 State of the Oxygen-Evolving Complex*

Flash O_2 yields have from the earliest measurements suggested that the distribution of S states attained after dark adaptation was approximately 75% S_1 and 25% S_0. Although light equalizes the populations of the four S states, reduction of the S_3 and S_2 states to S_1 occurs in the dark. This distribution gives maximal O_2 on the third flash with a smaller amount on the fourth. If the flash repetition rate used is low, however, one cannot use O_2 yields to distinguish between S_0 $Y_D^{\bullet}$ and S_1

Y_D [69] because of the S_2 Y_D and S_3 Y_D reactions. Therefore it was proposed that in the dark, all centers revert to S_1, and this was found to be the case [73]. $Y_D^{\bullet}$ oxidizes S_0 to S_1 in several tens of minutes [74]. From analysis of the pH dependence of the S_0 Y_D^+ reaction, the redox potential of the S_0/S_1 couple was estimated to be less than +700 mV [50].

The physiological significance of this reaction is not clear. Because the S_0 state is more reduced than S_1, it is more vulnerable to loss of manganese. $Y_D^{\bullet}$ may protect the OEC against reductive labilization of the manganese that can occur in darkness.

3.2.3. *Reduction by Exogenous Reductants*

$Y_D^{\bullet}$ is reduced by a number of reductants, such as hydroxylamine, carbonyl cyanide-m-chlorophenylhydrazone (CCCP) [75], reduced dichlorophenolindophenol (DCIP) [69], phenylenediamine [70], and iodide [76]. These observations are consistent with the estimated reduction potential of Y_D.

4. OTHER RADICALS IN PHOTOSYSTEM II

In addition to the tyrosine radicals, PSII has chlorophyll, pheophytin, and plastoquinone species that exist transiently as radicals during photosynthetic electron transport. During low-temperature illumination, both chlorophyll [71,77] and carotenoid radicals [78] have also been observed. Phenolic herbicides promote the photooxidation of carotenoid radicals in PSII [79]. This section, however, will deal with reports of other amino acid radicals in PSII.

4.1. 16-mT Radical

An EPR signal with $g = 2.0$ and having a linewidth of 9–16 mT has been observed in a number of inhibited PSII preparations that retain manganese. The chemical nature of the species giving rise to this signal is controversial. It has been suggested to be either a histidine radical or

$Y_Z^{\bullet}$. The following section describes the observations that have been made. Then, because there is no precedent for the occurrence of histidine radicals in enzymatic reactions, we will discuss the free radical chemistry of histidine and imidazole. Finally, we will discuss the proposal that the radical is $Y_Z^{\bullet}$.

4.1.1. Ca^{2+}-Depleted PSII/Histidine Hypothesis

The 16-mT EPR signal was first observed in PSII membranes depleted of the extrinsic 17- and 23-kD polypeptides and subsequently treated with either EGTA or NaCl in order to deplete them of Ca^{2+}. These membranes showed a modified multiline EPR signal that was stable in the dark [80] and was attributed to a formal S_2 state. Illumination of the membranes at 0° C caused the disappearance of the multiline signal and the appearance of the 16.4 mT EPR signal in what is considered a formal S_3 state.

The signal can be observed at liquid He temperatures and experiences microwave power saturation at powers considerably higher than do typical organic free radicals. These considerations suggest that the signal is due to an organic free radical, which accounts for the *g* value, that interacts magnetically with a paramagnetic transition metal ion, which accounts for the broadness of the signal and its saturation behavior. Because the signal is broadened by interaction with a transition metal ion, hyperfine splittings to protons or nitrogen atoms will be obscured and may not be useful in identifying the specific residue giving rise to the free radical.

On the basis of optical data and comparison of these with model compounds, however, the radical has been suggested to be derived from histidine. Measurements of the UV absorbance changes (Fig. 3) produced in parallel with the 16.4-mT signal show a peak around 315 nm [81] that is similar to the spectrum of a hydroxyl radical adduct of histidine [82]. Histidine had been proposed to function as a redox-active ligand to manganese in the photosynthetic water-splitting process [83], but the weakness of the magnetic interaction with manganese argues against direct liganding of the residue to manganese.

Other treatments that inhibit S-state advancement have led to the observation of similar EPR signals. These include Cl^- depletion or addition of F^- (16-mT-wide signal) [84], or Ca^{2+} depletion plus NH_3 (10 mT wide signal) [85]. The spectroscopic similarity of the radicals observed in these preparations suggests that all arise from the same species, although with some differences in the magnetic interaction that broadens the radical signal.

4.1.2. *Radical Chemistry of Histidine and Imidazole*

Oxidation of histidine has been measured at a graphite electrode [86] and the redox potential at pH 7 estimated to be +1.4 V (vs. NHE); the potential drops to +1.2 V (vs. NHE) at pH 11. The high pH potential, however, is relevant to histidine oxidation in PSII only if the histidine residue is deprotonated and negatively charged before oxidation, which is unlikely to be the case in a membrane environment. In general, the potentials reported for histidine oxidation are at least 400 mV more oxidizing than the potentials for the oxidation of tyrosine at comparable pHs. The redox potential of the imidazole side chain of histidine may be even higher: a derivative of histidine blocked to prevent reactions of the amino and carboxylic acid groups is oxidized at pH 7 at about +1.55 V vs. NHE [87,88]. Based on these data, it seems unlikely that histidine could be oxidized by $Y_Z^{\bullet}$ or by P_{680}^+, despite the well-known ability of the biological milieu to modulate redox potentials fairly substantially.

Chemical oxidation of histidine has been reported. Oxidation of histidine, arginine, and lysine by alkaline ferricyanide ($E = 0.46$ V) to the respective α-ketoacids proceeds by a radical intermediate formed by hydrogen abstraction from the α carbon [89]. Oxidation of histidine by Mn^{3+} in H_2SO_4 ($E_m = 1.5$ V) results in loss of CO_2, NH_4^+, and formation of an aldehyde at the α carbon [90]. In both cases, oxidation of the imidazole ring does not occur.

However, X-ray irradiation of crystalline histidine HCl at low temperature does produce several radicals, one of which was identified as the imidazole cation radical [91]. UV irradiation also produces a radical from 4-methylimidazole [92,93].

Histidine and other imidazole compounds may be oxidized by hydroxyl radicals generated by pulse radiolysis. The initial products are radical adducts [82], which may dehydrate at pH >10 to give an imidazole neutral radical [94]. The neutral imidazole radical is also generated by the strongly oxidizing $SO_4^{-\bullet}$ or $PO_4^{2-\bullet}$ radicals [95]. The reactions of histidine with $Br_2^{-\bullet}$ (E = 1.63 V), $(CNS)_2^{-\bullet}$ (E = 1.29 V), and $CO_3^{-\bullet}$ require high pH [96] and probably give the neutral imidazole radical of histidine as a major product. This radical has a broad, weak absorption at 400 nm and a stronger absorption below 280 nm, so that its spectrum is unlike that of the modified S_3 state.

The proposal of a hydroxyl radical adduct of histidine as a photosynthetically relevant species presents several problems. First, this adduct is formed by the reaction of the hydroxyl radical [82], making such a product unlikely in the absence of an extraordinarily strong oxidant. Furthermore, the hydroxyl radical adducts of imidazole compounds react with molecular oxygen in microseconds to form peroxy radicals, which have weak absorbances below 230 nm [82], so that a hydroxyl radical adduct is unlikely to persist in the PSII experiments. Finally, hydroxyl radical adducts of histidine and imidazoles are facile one-electron reductants of benzoquinone [97], menaquinone [98], and mercury electrodes [99]; thus, they are thermodynamically able to reduce the S_2 state or $Y_D^{\bullet}$.

4.1.3. Y_Z Hypothesis

Because the proposal of a histidine radical origin for the 16 mT EPR signal encounters the problems discussed above, alternate proposals must be considered. From a study of PSII depleted of Ca^{2+}, it was proposed that $Y_Z^{\bullet}$ was the radical interacting with the manganese cluster [100]. Although this work was criticized for having erred in its treatment of the microwave power saturation of the signals involved [101], the suggestion that the 10 to 16 mT radical is due to $Y_Z^{\bullet}$ has merit.

First, the UV absorption changes induced by formation of the modified S_3 state (a peak between 300 and 320 nm and a trough at 280 nm) [81] are not very different from the spectrum of $Y_Z^{\bullet}$ (peak at 305 nm, trough at 280 nm, peak at 260 nm) [56,102], while the absorbance changes in the region between 350 and 450 nm due to S_3 formation are similar to those of $Y_Z^{\bullet}$ [56] (Fig. 3). Second, the oxidation of Y_Z by P_{680}^{+} occurs with a high quantum yield at the temperatures used to trap S_3. After formation of the modified S_3 state, Y_Z could not be oxidized by P_{680}^{+} [103], perhaps because it was already oxidized. In manganese-depleted PSII, $Y_Z^{\bullet}$ can be trapped under conditions of temperature and illumination similar to those that produce the 16 mT signal [104]. Third, a thermoluminescence signal (the A_T band) has been interpreted to indicate histidine oxidation [105–108] but this can be equally well explained by oxidation of Y_Z. Indeed, the oxidation of Y_Z and the induction of the A_T band have very similar temperature dependencies between −90 and −20°C [108], strongly supporting this idea. Because Y_Z is probably hydrogen bonded to a histidine residue, histidine modification alters the oxidation kinetics of Y_Z, particularly at lower temperatures. We are aware of no compelling observation that precludes the radical from being $Y_Z^{\bullet}$.

4.2. Four-Line Radical in Mutant Photosystem II

A four-line EPR signal has been observed in highly resolved, manganese-depleted preparations from PSII mutants. In the Y_Z to phenylalanine mutant, an EPR signal spanning 3.5 mT (g = 2.004) was observed during illumination [110]. In the Y_D to either phenylalanine or tryptophan mutants, the four-line signal was observed during illumination, but $Y_D^{\bullet}$ and $Y_Z^{\bullet}$ were not [111]. Other workers do observe $Y_Z^{\bullet}$ in this mutant, however [36]. The four-line signal has been suggested to be due to a tyrosine radical. The physiological significance of the radical giving the four-line signal in PSII is unclear; it may be involved in photodegradation of PSII reaction center proteins.

5. CONCLUDING REMARKS

The basic polypeptide framework and electron transfer pathways in PSII have been defined over the past decade. The importance of redox-active amino acids in the water-splitting process has been recognized and the full extent of side chain–mediated redox chemistry in O_2 evolution is being actively explored. The manganese ensemble, the functional heart of the water oxidation reaction, is spectroscopically accessible, at least in some of its intermediate states, and its location and organization are emerging, but slowly. The mechanistic details of the interaction between the manganese complex and substrate water remain obscure—the mode of H_2O binding, the chemical structure of key intermediate species, and even the step in which the oxygen–oxygen bond is formed are unknown. These issues undoubtedly will fuel research in the next few years as we continue to elucidate the chemistry that underlies one of the most vital reactions in biology.

ACKNOWLEDGMENTS

This work was supported by the USDA Competitive Grants Office and NIH grant GM37300.

ABBREVIATIONS

ATP	adenosine 5′-triphosphate
CCCP	carbonyl cyanide-m-chlorophenylhydrazone
DCIP	2,6-dichlorophenolindophenol
EGTA	ethylene glycolbis(β-aminoethyl ether)-N,N,N′,N′-tetra-acetic acid
ENDOR	electron nuclear double resonance
EPR	electron paramagnetic resonance
ESR	electron spin resonance
NADPH	nicotinamide adenine dinucleotide phosphate, reduced form
NHE	normal hydrogen electrode
OEC	oxygen-evolving complex
PSII	Photosystem II
UV	ultraviolet

REFERENCES

1. N. Murata (ed.), *Research in Photosynthesis: Proceedings of the IXth International Congress on Photosynthesis* (Nagoya, Japan, August 30–September 4, 1992), Vol. 2, Kluwer Academic, Dordrecht, 1992.

2. R. J. Debus, *Biochim. Biophys. Acta*, *1102*, 269–362 (1992).

3. L. E. Andréasson and T. Vänngård, *Ann. Rev. Plant Physiol. Plant Mol. Biol.*, *39*, 379–411 (1988).

4. W. A. Cramer, M. T. Black, M. R. Widger, and M. E. Girvin, in *The Light Reactions* (J. Barber, ed.), Elsevier, Amsterdam, 1987, pp. 447–493.

5. A.-F. Miller and G. W. Brudvig, *Biochim. Biophys. Acta*, *1056*, 1–18 (1991).

6. G. T. Babcock, in *New Comprehensive Biochemistry* (J. Amesz, ed.), Elsevier, Amsterdam, 1987, pp. 125–158.

7. G. T. Babcock, B. A. Barry, R. J. Debus, C. W. Hoganson, M. Atamian, L. McIntosh, I. Sithole, and C. F. Yocum, *Biochemistry*, *28*, 9557–9565 (1989).

8. Ö. Hansson and T. Wydrzynski, *Photosynth. Res.*, *23*, 131–162 (1990).

9. B. A. Barry, *Photochem. Photobiol.*, *57*, 179–188 (1993).

10. V. V. Klimov, S. I. Allakhverdiev, S. Demeter, and A. A. Krasnovsky, *Dokl. Akad. Nauk. USSR*, *249*, 227–230 (1979).

11. A. W. Rutherford, *Tibs*, *14*, 227–232 (1989).

12. J. Deisenhofer, O. Epp, K. Miki, R. Huber, and H. Michel, *Nature*, *318*, 618–624 (1985).

13. J. Deisenhofer and H. Michel, *Science*, 245, 1463–1473 (1989).

14. P. O. Sandusky and C. F. Yocum, *Biochim. Biophys. Acta*, *849*, 85–93 (1986).

15. P. L. Fine and W. D. Frasch, *Biochemistry*, *31*, 12204–12210 (1992).

16. T. Wydrzynski, J. Ångström, and T. Vänngård, *Biochim. Biophys. Acta*, *973*, 23–28 (1989).

17. B. A. Diner and V. Petrouleas, *Biochim. Biophys. Acta*, *895*, 107–125 (1987).

18. D. F. Ghanotakis and C. F. Yocum, *Ann. Rev. Plant Physiol. Plant Mol. Biol.*, *41*, 255–276 (1990).

19. O. Nanba and K. Satoh, *Proc. Natl. Acad. Sci. USA*, *84*, 109–112 (1987).

20. B. Commoner, J. J. Heise, and J. Townsend, *Proc. Natl. Acad. Sci. USA*, *42*, 710–718 (1956).
21. J. Petersen, J. P. Dekker, N. R. Bowlby, D. F. Ghanotakis, C. F. Yocum, and G. T. Babcock, *Biochemistry*, *29*, 3226–3231 (1990).
22. B. A. Barry and G. T. Babcock, *Proc. Natl. Acad. Sci. USA*, *84*, 7099–7103 (1987).
23. B. A. Barry, M. K. El-Deeb, P. O. Sandusky, and G. T. Babcock, *J. Biol. Chem.*, *265*, 20139–20143 (1990).
24. A. Trebst, *Z. Naturforsch.*, *41c*, 240–245 (1985).
25. R. J. Debus, B. A. Barry, G. T. Babcock, and L. McIntosh, *Proc. Natl. Acad. Sci. USA*, *85*, 427–430 (1988).
26. W. F. J. Vermaas, A. W. Rutherford, and Ö. Hansson, *Proc. Natl. Acad. Sci. USA*, *85*, 8477–8481 (1988).
27. R. J. Debus, B. A. Barry, I. Sithole, G. T. Babcock, and L. McIntosh, *Biochemistry*, *27*, 9071–9074 (1988).
28. J. G. Metz, P. J. Nixon, M. Rögner, G. W. Brudvig, and B. A. Diner, *Biochemistry*, *28*, 6960–6969 (1989).
29. M. Brok, F. C. R. Ebskamp, and A. J. Hoff, *Biochim. Biophys. Acta*, *809*, 421–428 (1985).
30. C. W. Hoganson and G. T. Babcock, unpublished results (1993).
31. C. W. Hoganson and G. T. Babcock, *Biochemistry*, *31*, 11874–11880 (1992).
32. T. K. Chandrashekar, P. J. O'Malley, I. Rodriguez, and G. T. Babcock, *Photosynth. Res.*, *10*, 423–429 (1986).
33. I. D. Rodriguez, T. K. Chandrashekar, and G. T. Babcock, in *Prog. Photosynth. Res., Proc. 7th Int. Congr. Photosynth.*, Vol. 1 (J. Biggins, ed.) (1986), Nijhoff, Dordrecht, 1987, pp. 471–474.
34. B. Svensson, I. Vass, E. Cedergren, and S. Styring, *EMBO J.*, *9*, 2051–2059 (1990).
35. S. Styring, L. Davidsson, C. A. Tommos, W. Vermaas, and B. Svensson, in *Research in Photosynthesis*, Vol. 2 (N. Murata, ed.), Kluwer Academic, Dordrecht, 1992, pp. 251–254.
36. B. Diner, personal communication, (1993).
37. S. V. Ruffle, D. Donnelly, T. L. Blundell, and J. H. A. Nugent, *Photosynth. Res.*, *34*, 287–300 (1992).

38. J. B. Innes and G. W. Brudwig, *Biochemistry*, *28*, 1116–1125 (1989).

39. H. Mino, Y. Kodera, and A. Kawamori, in *Research in Photosynthesis*, Vol. 2 (N. Murata, ed.), Kluwer Academic, Dordrecht, 1992, pp. 61–64.

40. C. T. Yerkes and G. T. Babcock, *Biochim. Biophys. Acta*, *634*, 19–29 (1981).

41. D. F. Ghanotakis, G. T. Babcock, and C. T. Yerkes, *Arch. Biochem. Biophys.*, *225*, 248–255 (1983).

42. C. W. Hoganson and G. T. Babcock, *Biochemistry*, *28*, 1448–1454 (1989).

43. C. W. Hoganson and G. T. Babcock, *Biochemistry*, *27*, 5848–5855 (1988).

44. S. V. Ruffle and J. H. A. Nugent, in *Research in Photosynthesis*, Vol. 2 (N. Murata, ed.), Kluwer Academic, Dordrecht, 1992, pp. 191–194.

45. E. Schlodder and B. Meyer, *Biochim. Biophys. Acta*, *890*, 23–31 (1987).

46. R. E. Blankenship, G. T. Babcock, J. T. Warden, and K. Sauer, *FEBS Lett.*, *51*, 287–293 (1975).

47. S. Gerken, K. Brettel, E. Schlodder, and H. T. Witt, *FEBS Lett.*, *232*, 376–380 (1987).

48. J. Lavergne, in *Research in Photosynthesis*, Vol. 2 (N. Murata, ed.), Kluwer Academic, Dordrecht, 1992, pp. 273–280.

49. H. Conjeaud and P. Mathis, *Biophys. J.*, *49*, 1215–1221 (1986).

50. I. Vass and S. Styring, *Biochemistry*, *30*, 830–839 (1991).

51. J. Bonvoisin, G. Blondin, J.-J. Girerd, and J.-L. Zimmermann, *Biophys. J.*, *61*, 1076–1086 (1992).

52. M. Kusunoki, T. Ono, M. Suzuki, T. Noguchi, A. Uehara, T. Matsushita, H. Oyanagi, and Y. Inoue, in *Research in Photosynthesis*, Vol. 2 (N. Murata, ed.), Kluwer Academic, Dordrecht, 1992, pp. 293–296.

53. G. Renger and B. Hanssum, *FEBS Lett.*, *299*, 28–32 (1992).

54. S. Reinman and P. Mathis, *Biochim. Biophys. Acta*, *635*, 249–258 (1981).

55. D. F. Ghanotakis and G. T. Babcock, *FEBS Lett.*, *153*, 231–234 (1983).
56. J. P. Dekker, H. J. van Gorkom, M. Brok, and L. Ouwehand, *Biochim. Biophys. Acta*, *764*, 301–309 (1984).
57. C. T. Yerkes, G. T. Babcock, and A. R. Crofts, *FEBS Lett.*, *158*, 359–363 (1983).
58. G. T. Babcock and K. Sauer, *Biochim. Biophys. Acta*, *396*, 48–62 (1975).
59. C. T. Yerkes and G. T. Babcock, *Biochim. Biophys. Acta*, *590*, 360–372 (1980).
60. C. W. Hoganson, D. F. Ghanotakis, G. T. Babcock, and C. F. Yocum, *Photosynth. Res.*, *22*, 285–293 (1989).
61. B. A. Diner and P. J. Nixon, *Biochim. Biophys. Acta*, *1101*, 134–138 (1992).
62. B.-D. Hsu, J.-Y. Lee, and R.-L. Pan, *Biochim. Biophys. Acta*, *890*, 89–96 (1987).
63. H. Inoue, H. Akahori, and M. Noguchi, *Plant Cell Physiol.*, *28*, 1339–1343 (1987).
64. H. Inoue and T. Wada, *Plant Cell Physiol.*, *28*, 767–773 (1987).
65. N. Tamura, M. Ikeuchi, and Y. Inoue, *Biochim. Biophys. Acta*, *973*, 281–289 (1989).
66. N. Tamura and G. Cheniae, *Biochim. Biophys. Acta*, *890*, 179–194 (1987).
67. M. Miyao-Tokutomi and Y. Inoue, *Biochemistry*, *31*, 526–532 (1992).
68. G. T. Babcock and K. Sauer, *Biochim. Biophys. Acta*, *325*, 483–503 (1973).
69. B. R. Velthuys and J. W. M. Visser, *FEBS Lett.*, *55*, 109–112 (1975).
70. A. Boussac and A.-L. Etienne, *Biochem. Biophys. Res. Commun.*, *109*, 1200–1205 (1982).
71. C. A. Buser, L. K. Thompson, B. A. Diner, and G. W. Brudvig, *Biochemistry*, *29*, 8977–8985 (1990).
72. A. Boussac and A. L. Etienne, *Biochim. Biophys. Acta*, *766*, 576–581 (1984).
73. W. F. J. Vermaas, G. Renger, and G. Dohnt, *Biochim. Biophys. Acta*, *764*, 194–202 (1984).

74. S. Styring and A. W. Rutherford, *Biochemistry*, *26*, 2401–2405 (1987).

75. R. H. Lozier and W. L. Butler, *Photochem. Photobiol.*, *17*, 133–137 (1973).

76. Y. Takahashi, M. Takahashi, and K. Satoh, *FEBS Lett.*, *208*, 347–351 (1986).

77. J. C. de Paula, J. B. Innes, and G. W. Brudvig, *Biochemistry*, *24*, 8114–8120 (1985).

78. C. C. Schenck, B. Diner, P. Mathis, and K. Satoh, *Biochim. Biophys. Acta*, *680*, 216–227 (1982).

79. P. Mathis and A. W. Rutherford, *Biochim. Biophys. Acta*, *767*, 217–222 (1984).

80. A. Boussac, J.-L. Zimmermann, and A. W. Rutherford, *Biochemistry*, *28*, 8984–8989 (1989).

81. A. Boussac, J.-L. Zimmermann, A. W. Rutherford, and J. Lavergne, *Nature*, *347*, 303–306 (1990).

82. P. S. Rao, M. Simic, and E. Hayon, *J. Chem. Phys.*, *79*, 1260–1263 (1975).

83. S. Padhye, T. Kambara, D. N. Hendrickson, and Govindjee, *Photosynth. Res.*, *9*, 103–112 (1986).

84. M. Baumgarten, J. S. Philo, and G. C. Dismukes, *Biochemistry*, *29*, 10814–10822 (1990).

85. L.-E. Andréasson and K. Lindberg, *Biochim. Biophys. Acta*, *1100*, 177–183 (1992).

86. V. Brabec and V. Mornstein, *Biophys. Chem.*, *12*, 159–165 (1980).

87. R. M. Mobashar, Ph.D. thesis, Rutgers University, 1989.

88. S. S. Isied, in *Metal Ions in Biological Systems, Vol. 27*, Electron Transfer Reactions in Metalloproteins (H. Sigel and A. Sigel, eds.), Marcel Dekker, New York, 1991, pp. 1–56.

89. D. Laloo and M. K. Mahanti, *J. Chem. Soc. Dalton Trans.*, 311–313 (1990).

90. I. Pinto, B. S. Sherigara, and H. V. K. Udupa, *Bull. Chem. Soc. Jap.*, *63*, 3625–3631 (1990).

91. F. Q. Ngo, E. E. Budzinski, and H. C. Box, *J. Chem. Phys.*, *60*, 3373–3377 (1974).

92. A. W. Rutherford and A. Boussac, in *Research in Photosynthesis*, Vol. 2 (N. Murata, ed.), Kluwer Academic, Dordrecht, 1992, pp. 21–27.

93. A. Boussac, J.-L. Zimmermann, and A. W. Rutherford, in *Research in Photosynthesis*, Vol. 2 (N. Murata, ed.), Kluwer Academic, Dordrecht, 1992, pp. 235–238.

94. A. Samuni and P. Neta, *J. Phys. Chem.*, *77*, 1629–1635 (1973).

95. I. Dogan, S. Steenken, D. Schulte-Frohlinde, and S. Icli, *J. Phys. Chem.*, *94*, 1887–1894 (1990).

96. G. E. Adams, J. E. Aldrich, R. H. Bisby, R. B. Cundall, J. L. Redpath, and R. L. Willson, *Radiat. Res.*, *49*, 278–289 (1972).

97. M. Simic and E. Hayon, *Biochem. Biophys. Res. Commun.*, *50*, 364 (1973).

98. P. S. Rao and E. Hayon, *Biochim. Biophys. Acta*, *292*, 516 (1973).

99. K. M. Bansal and R. M. Sellers, *J. Phys. Chem.*, *79*, 1775–1780 (1975).

100. B. J. Hallahan, J. H. A. Nugent, J. T. Warden, and M. C. W. Evans, *Biochemistry*, *31*, 4562–4573 (1992).

101. A. Boussac and A. W. Rutherford, *Biochemistry*, *31*, 7441–7445 (1992).

102. S. Gerken, J. P. Dekker, E. Schlodder, and H. T. Witt, *Biochim. Biophys. Acta*, *977*, 52–61 (1989).

103. A. Boussac, P. Setif, and A. W. Rutherford, *Biochemistry*, *31*, 1224–1234 (1992).

104. Y. Kodera, K. Takura, H. Mino, and A. Kawamori, in *Research in Photosynthesis*, Vol. 2 (N. Murata, ed.), Kluwer Academic, Dordrecht, 1992, pp. 57–60.

105. T. Ono and Y. Inoue, *Biochemistry*, *30*, 6183–6188 (1991).

106. T. Ono and Y. Inoue, *FEBS Lett.*, *278*, 183–186 (1991).

107. T. Ono and Y. Inoue, *Biochim. Biophys. Acta*, *1099*, 185–192 (1992).

108. T. Ono, in *Research in Photosynthesis*, Vol. 2 (N. Murata, ed.), Kluwer Academic, Dordrecht, 1992, pp. 265–272.

109. C. W. Hoganson, P. A. Casey, and Ö. Hansson, *Biochim. Biophys. Acta*, *1057*, 399–406 (1991).

110. G. H. Noren and B. A. Barry, *Biochemistry*, *31*, 3335–3342 (1992).

111. R. J. Boerner, K. A. Bixby, A. P. Nguyen, G. H. Noren, R. J. Debus, and B. A. Barry, *J. Biol. Chem.*, *268*, 1817–1823 (1993).

112. D. F. Ghanotakis, G. T. Babcock, and C. F. Yocum, *FEBS Lett.*, *167*, 127–130 (1984).

113. T. Ono and Y. Inoue, *Biochim. Biophys. Acta*, *973*, 443–449 (1989).

114. D. V. Bent and E. Hayon, *J. Am. Chem. Soc.*, *97*, 2599–2606 (1975).

4
Ribonucleotide Reductase in Mammalian Systems

Lars Thelander and Astrid Gräslund

Department of Medical Biochemistry and Biophysics, University of Umeå, S-90187 Umeå, Sweden

1. INTRODUCTION

The best characterized ribonucleotide reductase up to now is the aerobic *E. coli* enzyme and there are a number of excellent recent reviews mainly focusing on this enzyme [1–3]. However, after the cDNA cloning and successful expression in bacteria, the active mammalian ribonucleotide reductase is as easily studied as the *E. coli* enzyme [4]. Although the mammalian reductase in many respects grossly resembles the bacterial enzyme, more detailed studies have revealed significant differences in the allosteric regulation, the subunit interaction, and the properties of the binuclear iron-tyrosyl free radical center. In this chapter, we will concentrate on these aspects. Furthermore, the gene structure and the control of the expression of the two nonidentical R1 and R2 subunits during the cell cycle, when cells go from quiescence to proliferation, and during the induction of DNA repair are unique to each species and we will describe what is known about this for the mammalian enzyme.

2. MAMMALIAN RIBONUCLEOTIDE REDUCTASE

2.1. Enzyme Structure

Mammalian ribonucleotide reductase consists of the two nonidentical subunits proteins R1 and R2. Protein R1 has a monomer molecular weight of 90000 and was first purified to homogeneity from calf thymus [5]. The protein contains redox-active disulfides participating as electron donors during catalysis, and it also binds the nucleoside triphosphate effectors and the ribonucleoside diphosphate substrates [6,7]. Depending on which effector is present, the R1 monomer forms a dimer or an inactive tetramer [7]. The complete amino acid sequence of 792 residues for the mouse [8] and human [9] R1 proteins was deduced from the corresponding cDNAs and showed 97.7% homology between the two species.

Mammalian R2 is a homodimer of molecular weight 2×45000 and it was purified to homogeneity from hydroxyurea-resistant, R2-overproducing mouse TA3 cells [10]. Each polypeptide chain contains a binuclear nonheme iron center and a tyrosyl free radical essential for activity [4]. The complete amino acid sequences for the mouse [11], human [12], and hamster [13] R2 proteins were deduced from the corresponding cDNA sequences and again showed very high homology (human to mouse 91.6% and mouse to hamster 94%) (see Fig. 1 in Sec. 4). Interestingly, the homology is lower for the N-terminal amino acid regions which together with activity measurements of N-terminal proteolytically truncated mouse R2 protein [4] indicate that this part of the molecule is not necessary for activity.

Each subunit is inactive by itself but together they form the active ribonucleotide reductase enzyme complex which has the structure $\alpha_2\beta_2$ [7]. Recombinant soluble, enzymatically active mouse R1 and R2 proteins can easily be purified to homogeneity in large amounts from *E. coli* bacteria containing T7 RNA polymerase pET expression vectors [4,14]. Recent data show that the recombinant mouse R1 protein contains an iron sulfur center giving a characteristic light absorption spectrum. The R2 protein is mainly obtained as the apoprotein but the iron radical center can be fully regenerated in vitro by the addition of ferrous iron and oxygen under controlled conditions [4,15]. The three-dimensional structure of the *E. coli* R2 protein has been determined at 2.2 Å resolution [16]. This protein contains about 70% α-helical structure while the mouse R2 protein only contains about 50% α helices as shown by circular dichroism (CD) measurements [4]. In this context it should be remembered that the amino acid sequence homology between the mouse and *E. coli* R2 proteins is only 24%.

2.2. Regulation of Enzyme Activity

Since ribonucleotide reductase catalyzes the first unique step leading to DNA synthesis, its activity is regulated at many different levels. First, on the enzyme level, both substrate specificity and overall activity are controlled by binding of nucleoside triphosphate effectors to the R1 protein [17]. Early equilibrium dialysis experiments on the calf thymus enzyme clearly identified two different classes of effector binding sites [7] although the exact stoichiometry within each site probably should be redetermined using recombinant mouse R1. One class of sites was specific for ATP and dATP binding and controlled overall activity. The other class in addition bound 2′-deoxythymidine 5′-triphosphate (dTTP) and 2′-deoxyguanosine 5′-triphosphate (dGTP) and determined the substrate specificity. The presence and nature of these allosteric effector binding sites on mammalian R1 was confirmed in an elegant way by isolating mutant R1 proteins with altered allosteric regulation from mutagenized cells using somatic cell genetics [18].

The general pattern of allosteric regulation is similar for the *E. coli* and mammalian enzymes but the effector requirements are more distinct with the mammalian enzyme. For example, dTTP only stimulates guanosine 5′-diphosphate (GDP) reduction for the mammalian enzyme [17] while with the *E. coli* enzyme dTTP stimulates both purine and pyrimidine reduction [19]. Furthermore, the mammalian enzyme is absolutely dependent on the presence of positive effectors for activity while the bacterial enzyme also shows activity in the absence of effectors.

Ribonucleotide reductase activity is tightly coupled to DNA synthesis with maximal values during the S phase [20]. This is controlled by the amount of enzyme and not by allosteric regulation. The R1 protein is proliferation-specific and it cannot be detected in resting or terminally differentiated cells [21,22]. The R1 levels are constant and in excess during the cell cycle [23]. In contrast, the levels of the R2 protein are strictly S-phase–correlated and this is regulated by de novo synthesis and breakdown of R2 protein and not by enzyme activation [24]. Therefore, holoenzyme activity during the cell cycle is limited and determined by the levels of the R2 protein. The half-life of the R1 protein is more than 20 hr [22] while the R2 protein has a half-life of 3 hr [24].

The R2 promoter of mouse ribonucleotide reductase is activated by DNA damage caused by UV irradiation [25]. This indicates that mammalian cells, like yeast cells, can activate ribonucleotide reductase expression in response to DNA damage even in resting noncycling cells [26].

2.3. Gene Structure and Localization

The genes for the R1 and R2 subunits have been localized to different chromosomes in the human, mouse, and hamster genomes. In the human genome the R1 gene was assigned to chromosome 11p15 [27] and the active R2 gene to chromosome 2p24-p25 [28]. In addition, there were R2 pseudogenes on both arms of chromosome 1 and on the X chromosome. In the mouse genome, the only R1-related sequences were localized to chromosome 7 [27] and the active R2 gene to chromosome 12 while R2 pseudogenes were identified on chromosomes 4, 7, and 13 [28]. Finally, in hamster the R1 gene was assigned to chromosome 3 and the R2 gene to chromosome 7 [29].

The functional R2 gene from mouse has been cloned and sequenced [30]. The transcribed sequence contains 10 exons distributed over 5770 base pairs (bp). The promoter has a TTTAAA homology and footprinting experiments identified four different protected regions, at −93 to −56, −584 to −527, −623 to −597, and −672 to −637 bp upstream from the transcription start. The proximal footprint was analyzed in detail by gel electrophoresis mobility shift assays and UV crosslinking which showed that three different protein complexes bound to this area. The transcription factor Sp1 was identified as forming one of the complexes [31].

The structure of the genomic mouse R1 gene was recently compiled from a number of overlapping λ clones isolated from a mouse genomic library [32]. The gene consists of 19 exons and covers around 35 kb of DNA. All exon/intron boundaries, the promoter region, and the 3′-flanking region were sequenced and the transcription start site was determined by S_1-nuclease protection experiments. Intron 7 starts with the uncommon variant GC instead of GT. The promoter lacks a TATA homology but footprinting experiments identified two nearly identical 23-bp-long protein binding regions. Three protein complexes binding to one of the 23 mers were characterized by gel electrophoresis mobility shift assays and UV crosslinking experiments. One of the proteins showed S-phase–specific binding and another was identified as Sp1.

2.4. Control of Gene Expression

The levels of R1 and R2 mRNA transcripts are very low in normal cells. Therefore a very sensitive RNase protection, solution hybridization assay was developed using R1 and R2 antisense RNA probes. The assay allowed measurements of these transcripts during the cell cycle in centrifugally elutriated cells and in cells synchronized by isoleucine or serum starvation [33]. The levels of both transcripts were very low or undetectable in G_0/G_1 cells but increased dramatically in parallel slightly before the rise in S-phase cells to reach the same levels and finally declined when cells progressed into $G_2 + M$. The earlier observed lack of cell cycle dependent variation in R1 protein levels may therefore be a consequence of the long half-life of the R1 protein.

Using mouse R2 promoter–luciferase reporter gene constructs, an unexpected early activation of the R2 promoter was observed when cells passed from quiescence to proliferation. These results were confirmed by in vitro nuclear run-on experiments again showing a very early activation of the R2 promoter, long before any R2 transcripts were observed in the earlier mRNA measurements. However, due to a transcriptional block, the early promoter activation only results in very short R2 transcripts until cells enter the S phase, when full-length R2 transcripts start to appear. The position of the transcriptional block was localized to a DNA sequence about 87 bp downstream from the first exon/intron boundary [34]. These results identify S-phase specific release from a transcriptional block as a mechanism to control cell cycle regulated gene expression.

The early activation of the R2 promoter was observed in reporter gene constructs containing about 1.0 kb of DNA upstream from the transcription start, i.e., containing all four footprinting regions. Constructs containing only the proximal footprint (−93 to −56) did not show any early activation when cells passed from quiescence to proliferation [31].

The mouse R1 promoter is quite different from the R2 promoter and so far much less is known about the S-phase specific expression of the R1 transcript. Interestingly, an S-phase specific protein was shown to bind to the 23-mer footprint in the R1 promoter and therefore R1 mRNA expression may be regulated mainly at the promoter level [32].

3. HERPES SIMPLEX VIRUS RIBONUCLEOTIDE REDUCTASE

Herpes simplex virus (HSV) encodes a ribonucleotide reductase consisting of a 136-kD polypeptide chain and a 38-kD polypeptide chain which together form an active enzyme complex of the type $\alpha_2\beta_2$ [35]. The two mRNAs encoding the HSV-1 large and small subunits are 5 and 1.2 kb, respectively. They are colinear and share the same 3′ end but the coding region of the large subunit stops 238 bp before the start codon of the small subunit [36]. The amino acid sequence homology between the mouse and herpes simplex virus R2 proteins is only about 30% but also the viral R2 protein contains a tyrosyl free radical and a binuclear iron center.

The viral ribonucleotide reductase is unique in the sense that it seems to completely lack allosteric regulation [37]. Another unique feature of the herpes simplex virus R1 protein is an N-terminal extension of some 320-amino-acid residues that is missing from the R1 protein isolated from other species. This extension is dispensable for ribonucleotide reductase activity [38] but instead possesses a protein kinase activity both in the case of HSV-1 and HSV-2 [39,40]. The kinase activity was proposed to have a function during the immediate–early phase of virus replication.

Both the large and small subunits of herpes simplex virus ribonucleotide reductase can now be obtained as recombinant proteins [4,41] and especially the small subunit can be obtained in large amounts in pure form from *E. coli* bacteria using the T7 RNA polymerase-based pET vector system.

As first shown for the herpes simplex virus ribonucleotide reductase [42,43] and later extended to the *E. coli* [44] and mammalian enzymes [45,46], short peptides corresponding to the C-terminal sequences of the small subunits inhibit enzyme activity. They inhibit by binding to the large subunit thereby preventing the association of the two subunits. This observation is being used to design antiviral drugs since the C-terminal sequence is different in the small subunits of different species such as *E. coli*, HSV, and mouse/homo.

4. COMPARISON OF AMINO ACID SEQUENCES AND SPECTROSCOPIC PROPERTIES FOR DIFFERENT SPECIES

Amino acid sequences of iron-containing ribonucleotide reductases from a variety of species ranging from bacteria to mammals and mammalian viruses have been reported and aligned [1]. Figure 1 shows alignments of protein R2 sequences from various species, including the human sequence. As indicated earlier, human, mouse, and hamster sequences are very homologous. Inspection of the other aligned sequences reveals that true amino acid conservation throughout all species is quite unusual. Out of about 700 amino acids in the large subunit, less than 40 are conserved (data not shown), and in a similar proportion, of about 400 amino acids in the small subunit, less than 20 are conserved. Yet all species seem to have similar operating principles regarding their redox functions. In the large subunit, the cysteines expected to be involved in the redox reactions are among those conserved. In the small subunit, iron ligands and residues close to the iron center are among the conserved ones, as is of course the tyrosyl residue carrying the stable free radical.

The tyrosyl radical of the small subunit was an early observed, clearly distinguishable feature of this class of ribonucleotide reductases. Electron paramagnetic resonance (EPR) spectra at low temperature of a few representative species are shown in Fig. 2. A common denominator is that the overall doublet nature of the EPR spectrum is evidence that the tyrosyl ring is in a locked configuration with a very similar geometry relative to the peptide backbone in these systems [47,48].

A significant variation between species occurs for another EPR observable property of the radical, namely, its relaxation behavior when exposed to increasing microwave power. The relaxation depends on the strength of magnetic interaction between the iron center and the

```
                     1         10         20         30         40         50         60         70         80         90        100
Human      .......... MLSLRVPLAP ITDPQQLQLS PLKGLSLVDK ENTPPALSGT RVLASKTARR IFQEPTEPKT KAAA-PGVED EPLLRENPRR FVIFPIEYHD IWQMYKKAEA
Hamster    .......... MFSVRVPLAT ITDQQQLQVS PLKALSLADK ENTPPSLSST PVLASKVARR ILQDVAEPES KVSTNPSVED EPLLRENPRR FVVFPIEYHD IWKMYKKAEA
Mouse      .......... MLSVRTPLAT IADQQQLQLS PLKRLTLADK ENTPPTLSST RVLASKAARR IFQDSAELES KAPTNPSVED EPLLRENPRR FVVFPIEYHD IWQMYKKAEA
Yeast      .....MPKET PSKAAADALS DLEIKDSKSN LNKELETLRE ENRVKSDMLK EKLSKDAENH KAYLKSHQVH RHKLKEMEKE EPLLNEDKER TVLFPIKYEH IWQAYKRAEA
E. coli    .......... .................................................. .....AYTTF SQTKNDQLKE PMFFGQPVNV ARYDQQKYDI FEKLIEKQLS
HSV1       .......... ...........................................MDSAA PA LSPALTAHTG HSATADLAIQ IPKCPDPERY FYTSQ.CPDI N.HLRSLSIL
Consensus  ---------- ---------- ---------- ---------- ---------- ---------- ---------- ---------- ---------- ---------- ----------

           101     110        120        130        140        150        160        170        180        190        200        210
Human      SFWTAEEVDL SKDIQHWES.L KPEERYFISH VLAFFAASDG IVNENLVERF SQEVQITEAR CFYGFQIAME NIHSEMYSLL IDTYIKDPKE REFLFNAIET MPCVKKKADW
Hamster    SFWTAEEVDL SKDIQHWEA.L KPDERHFISH VLAFFAASDG IVNENLVERF SQEVQVTEAR CFYGFQIAME NIHSEMYSLL IDTYIKDSKE REYLFNAIET MPCVKKKADW
Mouse      SFWTAEEVDL SKDIQHWEA.L KPDERHFISH VLAFFAASDG IVNENLVERF SQEVQVTEAR CFYGFQIAME NIHSEMYSLL IDTYIKDPKE REYLFNAIET MPCVKKKADW
Yeast      SFWTAEEIDL SKDIHDWNNRM NENERFFISR VLAFFAASDG IVNENLVENF STEVQIPEAK SFYGFQIMIE NIHSETYSLL IDTYIKDPKE SEFLFNAIHT IPEIGEKAEW
E.coli     FFWRPEEVDV SRDRIDYQA.L PEHEKHIFIS NLKYQTLLDS IQGRSPNVAL LPLISIPELE TWVETWAFSE TIHSRSYTHI IRNIVNDPSV VFDDIVTNEQ IQKRAEGISS
HSV1       NRWLETELVF VGDEEDVSK.L SEGELSFYRF LFAFLSAADD LVTENLGG.L SGLFEQKDIL HYYVEQECIE VAHSRVYNII QLVLFHNNDQ ARREYVAGTI NHPAIRAKVD
Consensus  --W---e--- --D-------l ---e--f--- ---f----d- -v--n----- ---------- --y--q---E --Hs--Y--- ---------- ---------- ----------
                                                                                     ↑

           211                                   220        230        240        250        260        270                280
Human      ALRWIG.... .......... .......... ...DKEA TYGERVVAFA AVEGIFFSGS FASIFWLKKR GLMPG.LTFSN ELISRDEGLH CDFACLMFK..........H
Hamster    ALRWIG.... .......... .......... ...DKEA TYGERVVAFA AVEGIFFSGS FASIFWLKKR GLMPG.LTFSN ELISRDEGLH CDFACLMFK..........H
Mouse      ALRWIG.... .......... .......... ...DKEA TYGERVVAFA AVEGIFFSGS FASIFWLKKR GLMPG.LTFSN ELISRDEGLH CDFACLMFK..........H
Yeast      ARLWIQ.... .......... .......... ...DADA LFGERLVAFA SIEGVFFSGS FASIFWLKKR GMMPG.LTFSN ELICRDEGLH TDFACLLFA..........H
E. coli    YYDELIEMTS YW.HLLGEGT HTVNGKTVTV SLRELKK KLYLCLMSVN ALEAIRFYVS FACSFAFAER ELMEG.NAKII RLIARDEALH LTGTQHMLNLLRSGADDPEMA
HSV1       WLEARVR... .......... .......... ....ECA SVPEKFILMI LIEGIFFAAS FAAIAYPLRT NNLLRVTCQSN DLISRDEAVH TTASCYIYN.........YLG
Consensus  ---------- ---------- ---------- ------- ---------- --EgifF--S Fa-i------ --m-g-----n -lI-RDE-lH --------- ----------

           281        290        300        310        320        330        340        350        360        370        380
Human      LVHKPSEERV REIIINAVRI EQEFLTE.ALP VKLIGMN..CTL MKQYIEFVAD RIMLELGFSK IFRV...ENPFDF MENISLEG..KT NFFEKRVGEY QRMGVMSSPT ENS...FTLDADF
Hamster    LVHKPSEQRV REIITNAVRI EQEFLTE.ALP VKLIGMN..CTL MKQYIEFVAD RIMLELGFNK IFKV...ENPFDF MENISLEG..KT NFFEKRVGEY QRMGVMSNST ENS...FTLDADF
Mouse      LVHKPAEQRV REIITNAVRI EQEFLTE.ALP VKLIGMN..CTL MKQYIEFVAD RIMLELGFNK IFRV...ENPFDF MENISLEG..KT NFFEKRVGEY QRMGVMSNST ENS...FTLDADF
Yeast      LKNKPDPAIV EKIVTEAVEI EQRYFLD.ALP VALLGMN..ADL MNQYVEFVAD RLLVAFGNKK YYKV...ENPFDF MENISLAG..KT NFFEKRVSDY QKAGVMSKST KQEAGAFTFNEDF
E. coli    EIAEECKQEC YDLFVQAAQQ EKDWADYLFRD GSMIGLN..KDI LCQYVEYITN IRMQAVGLDL PFQTRSNPIPWIN TWLVSDNV..QV APQEVEVSSY LVGQIDSEVD TDDLSNFQL....
HSV1       GHAKPPPDRV YGLFRQAVEI EIGFIRSQAPT DSHILSPAALAA IENYVRFSAD RLLGLIHMPP LFSA...PPPDAS FPLSLMSTDKHT NFFECRSTSY AGAVVNDL.. .............
Consensus  ---------- ------av-i e---------- ----g-n----- ---y----ad r--------- ---------P--- ----s------t nffE-----Y ---------- -------------
```

FIG. 1. Alignment of protein R2 amino acid sequences from several species. The amino acid sequences used are the following: human [12], hamster [13], mouse [11], yeast, *Saccharomyces cerevisiae* [71], *E. coli* [72], and HSV1, herpes simplex virus type 1 [73]. The conserved tyrosyl free radical residue is indicated by an arrow.

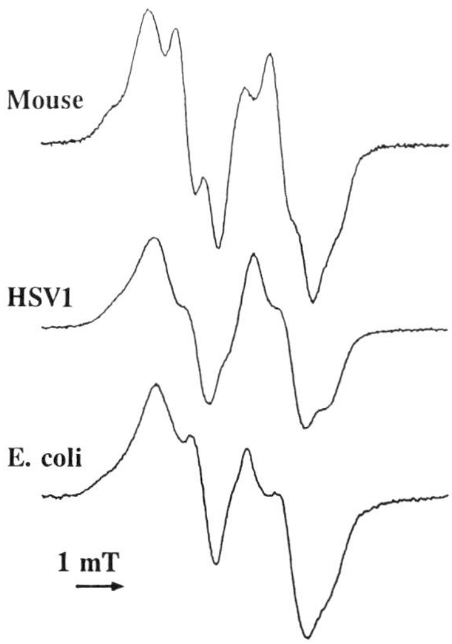

FIG. 2. EPR spectra of the tyrosyl free radical of ribonucleotide reductase from (top) mouse at 20 K [4]; (middle) herpes simplex virus type 1 at 32 K [74]; (bottom) *E. coli* at 30 K [49].

radical, and this property is highly variable among different species [49]. The differences appear to be at least partly determined by the distance between the radical and the iron center [4]. The rather extreme behavior of the mouse radical, with no apparent microwave saturation at 77 K, even at the highest powers available, is an important feature. It allows EPR detection and studies of this free radical with unusually high sensitivity, and was actually a decisive factor that allowed the first characterization of the mouse enzyme in hydroxyurea-resistant cells as an iron containing ribonucleotide reductase [50].

Optical absorption spectra in the near-UV/visible range show characteristic contributions from the iron center and the free radical [4,51] (see Fig. 3). The 410 to 415 nm line associated with the tyrosyl radical spectrum (phenoxy radical–like) is clearly visible in the spectra from mouse, HSV-1, and *E. coli*. Also in the other wavelength regions the spectra are quite similar, pointing to the similar nature of the iron/ radical center in all species investigated. A small variation in linewidth of the radical spectrum can be noted between mouse, HSV-1, and *E. coli*. The reason for this is not known but could be related to the different

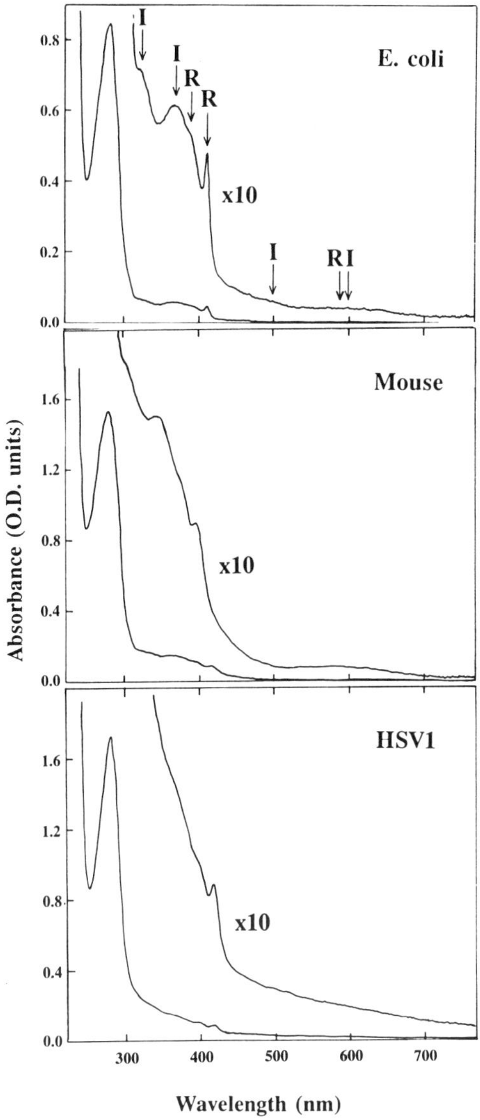

FIG. 3. Light absorption spectra of protein R2 of ribonucleotide reductase from (top) *E. coli*, 42 μM; (middle) mouse, 105 μM; (bottom) herpes simplex virus type 1, 95 μM. The concentrations refer to polypeptides. The spectra were recorded in 2-mm cuvettes at 10°C. I and R denote iron and radical light absorption bands, respectively. (Modified after [4].)

strengths of magnetic interaction between the radical and the iron center.

Paramagnetically shifted ^{1}H-NMR resonances from iron center ligands have been observed for the mouse, HSV-1, and *E. coli* proteins [4], and their similar nature again shows that the iron/free radical site in these three enzymes is spectroscopically very similar. Taken together, the similarities show that despite evolutionary divergence, the important features of one working center of this enzyme are strongly conserved, and we might expect the essential principles of the enzymatic reaction also to be retained.

5. FORMATION, PROPERTIES, AND STABILITY OF THE Fe-TYROSYL RADICAL CENTER

Active ribonucleotide reductase requires the presence of the iron/radical center in protein R2. Although it has never been directly demonstrated that this site participates in the reduction of the substrate, it has been postulated to be a redox partner in the enzyme reaction. For the reaction between *E. coli* ribonucleotide reductase and the suicidal substrate analog 2′-deoxy-2′-azido-CDP (azidoCDP), it has been shown that a transient free radical residing on the substrate analog appears concomitantly with the disappearance of the enzyme tyrosyl radical [52,53]. An almost identical reaction takes place between the mouse enzyme and azidoCDP, again leading to a transient radical with approximately the same characteristics as observed for the *E. coli* enzyme [54]. The transient radical is formed very slowly, but quantitatively as the normal tyrosyl radical disappears, and is the only radical transient so far observed in any reaction with ribonucleotide reductase. Its hyperfine structure arises from a nitrogen (of the azido group) and an additional doublet splitting, presumably from a proton, although its exact location has not been established [55].

In vitro, active protein R2 is formed from apoprotein, ferrous iron, and oxygen [51]. The formation of the iron/radical center involves at least three electrons: the two ferrous ions and the tyrosyl residue must lose an electron each. Since oxygen is used as oxidant, the reaction should involve four electrons, but the source of the fourth electron was not known. Quantitative studies of the radical regeneration reaction for mouse protein R2 with varying amounts of ferrous iron and oxygen established that the formation of one tyrosyl radical and one binuclear iron center in apoprotein R2 requires three Fe^{2+} and one molecule of O_2 [4,15]:

$$3Fe(II) + P + O_2 + Tyr\text{-}OH + H^+$$
$$\rightarrow Fe(III)O^{2-}Fe(III)\text{-}P + H_2O + Tyr\text{-}O^{\bullet} + Fe(III)$$

where P is the iron binding site of R2. Therefore, at least in vitro, the source of the fourth electron can be Fe^{2+}.

A similar stoichiometry was later also found for tyrosyl radical formation in the *E. coli* R2 protein [56]. For the bacterial protein, the reaction was shown to involve several transient species, the nature of which are not yet completely understood [56]. Obviously the iron–oxygen reaction with the protein is quite complex, involving not only the iron ligands and the radical-to-be tyrosyl residue, but also possibly other redox-active residues in the protein, which may or may not take part also in the enzymatic reaction. One such residue is the conserved mouse R2 Trp103, which is known to be required for enzyme activity. Mouse R2 containing Phe/Tyr instead of Trp103 is catalytically inactive but still binds to the R1 protein as shown in competition experiments. Recent experiments have indicated that Trp103 is necessary not only for catalysis but also for the proper formation of the iron/radical site [57].

6. INHIBITORS; RADICAL SCAVENGERS; IRON CHELATORS; SPECIFIC PEPTIDES

Specific inhibitors of ribonucleotide reductase can be found among compounds of at least three different classes: (1) radical scavengers, e.g., hydroxyurea, which specifically reduce (and destroy) the tyrosyl radical in the small subunit; (2) iron chelators, e.g., desferrioxamine, which prevent posttranslational formation of the iron center in the small subunit; (3) specific oligopeptides, with sequences corresponding to the carboxyl terminal sequence of the small subunit, which prevent the subunit interaction necessary for enzyme activity by binding to the large subunit.

Hydroxyurea reacts with protein R2 of all investigated iron-containing ribonucleotide reductases and can even be considered as a diagnostic tool for the iron-containing reductases [58]. The tyrosyl radical is destroyed in a one-electron redox reaction. A recent study has shown that one-electron oxidized hydroxyurea radicals appear as transients in the reaction [59]. In this reaction, the redox properties of hydroxyurea as well as the accessibility for the drug to the protein site of reduction are important. It has been shown with a series of hydroxamic acids with side chains of increasing bulk that on the whole the mouse protein R2 is easier to access than the *E. coli* protein, but that the bulkier side chains considerably slow down the reaction rate of hydroxyurea with the two proteins [60]. In the *E. coli* enzyme only the tyrosyl radical is destroyed in the reaction with hydroxyurea, whereas in the mouse enzyme both the radical and the iron center are lost [61]. The loss of the iron center is related to reduction of the ferric ions by hydroxyurea. Experiments using the ferrous iron chelator bathophenanthroline have shown that the iron is lost as ferrous ions. This makes hydroxyurea a competitive inhibitor to the mouse enzyme because it leaves a way of easy reactivation in the presence of iron and oxygen, whereas in the case of *E. coli* hydroxyurea can be considered as an irreversible inhibitor.

Two interesting questions can be raised about the hydroxyurea reaction: To what extent does protein dynamics mediate the reaction? Does hydroxyurea have to penetrate all the way into the free radical site, or is there an electron transfer route in operation? A study of the *E. coli* enzyme system has shown that the reaction rate is dependent on the regulatory state of the enzyme, with a considerably higher reaction rate for the complete enzyme system, than for pure protein R2 [62].

Another specific very potent inhibitor of mouse ribonucleotide reductase, presumably also acting through a reduction mechanism, is the iron (Fe^{2+}) chelate of 1-formylisoquinoline thiosemicarbazone. This drug is one of the most potent inhibitors known for mammalian ribonucleotide reductase (active in submicromolar concentrations [63]). It belongs to a class of α(N)-heterocyclic carboxaldehyde thiosemicarbazones which possess both antitumor and antiviral activities but unfortunately are too toxic for clinical applications. It has been shown that the active form is the iron chelate and that the tyrosyl radical of the small subunit is destroyed in a reaction that requires oxygen [64].

A related compound is 2,2′-bipyridyl-6-carbothiamide, which also has been shown to destroy the tyrosyl radical of mouse ribonucleotide reductase. Again the ferrous iron chelate appears to be the active species [65].

Cellular proliferation is known to be sensitive to iron deprivation due to selective inhibition of DNA synthesis. Since ribonucleotide reductase is the only ferroprotein in that metabolic pathway it has been assumed to be the target. Treatment of cells in tissue culture with the potent iron chelators parabactin or desferrioxamine results in inhibition of cell growth. This is paralleled by loss of protein R2 tyrosyl free radical and accumulation of apoR2 protein as seen by EPR measurements [66].

In vitro studies using pure recombinant mouse R2 protein showed unexpectedly that the iron center in this protein is quite labile and the protein loses half of its iron after 30 min incubation at 37°C [66]. The loss is temperature-dependent but relatively independent of the presence of chelators such as EDTA, parabactin, or desferrioxamine, or the presence of dithiothreitol or protein R1 plus a complete assay mixture. Therefore, mouse protein R2 needs a continuous supply of ferrous iron and oxygen to maintain its radical–iron center. This agrees with earlier data showing that assaying the mammalian enzyme under anaerobic conditions or in the presence of EDTA caused a rapid decline in enzyme activity [67]. In fact, oxygen deprivation gives a rapid loss of protein R2–specific tyrosyl radical in mammalian cells [68]. Therefore, the inhibitory effects of iron chelators on the mammalian R2 protein is due to a passive removal of iron from the medium, which prevents regeneration of the iron–radical center rather than an active removal of iron from the iron center of the protein. In this respect the *E. coli* enzyme is quite different since it does not easily lose its iron.

From a medical point of view, it is interesting to note that all three of these classes of compounds have the potential of providing an independent attack on the ribonucleotide reductase activity needed for cell division. It is known that adaptation to hydroxyurea by a gene amplification process leads to formation of more apoprotein R2 [69]. However, more apoprotein requires more iron to become active, and the activity would be strongly hindered by the presence of an iron chelator. This explains why hydroxyurea-resistant, R2-overproducing mouse cells are as sensitive to inhibition by parabactin or desferrioxamine as normal mouse cells [66]. Finally, addition of specific peptides (or peptide mimetics) would hinder enzyme activity even in the presence of the two active subunits. In this third step, the selectivity for a particular species (e.g., bacterial or viral) could be exploited most advantageously [46].

7. SUBUNIT INTERACTION AND ELECTRON TRANSFER

One key property to the understanding of the activity of ribonucleotide reductase is its subunit interaction. It has been suggested that one can regard the large subunit as the proper enzyme and the small subunit as a large cofactor. For activity the proper complex is needed. The stability constant of the complex of the *E. coli* enzyme has been reported to be about 0.3 μM under favorable conditions. One important observation is that subunit interaction involves only the last amino acids in the carboxyl terminal tail of the small subunit. This is the basis for understanding the inhibitory activity of short peptides with sequences corresponding to this tail.

The direct involvement of the carboxyl terminal tail for subunit interaction and enzyme activity of the HSV-1 ribonucleotide reductase has been shown in studies of a truncated mutant of HSV protein R2 lacking the last seven amino acid residues in its carboxyl terminus. The iron center and tyrosyl radical could readily be induced in this mutant, but it did not bind to the R1 subunit and was completely devoid of activity [70]. A similar observation was made for a truncated form of the mouse R2 protein lacking the C-terminal seven amino acid residues. A curious observation was that the EPR microwave saturation, reflecting the relaxation properties and thus the iron center–radical interaction, was significantly changed in the HSV R2 mutant.

The last 30 amino acid residues in the carboxyl terminal were not traced in the electron density map of the small subunit of *E. coli* protein R2, when its three-dimensional structure was reported [16]. The most likely explanation would be a too high flexibility and thus disordered structure of this part of the polypeptide chain. That this is actually the case has been shown in high-resolution NMR studies [75].

ACKNOWLEDGMENT

This work was supported from the Swedish Natural Science Research Council. We are grateful to H. Eklund for the preparation of Figure 1.

ABBREVIATIONS

A	adenine base in DNA
ATP	adenosine 5′-triphosphate
dATP	2′-deoxyadenosine 5′-triphosphate
azidoCDP	2′-deoxy-2′-azidocytidine 5′-diphosphate
bp	base pair
C	cytosine base in DNA
CD	circular dichroism
EDTA	ethylenediamine-N,N,N′,N′-tetraacetate
EPR	electron paramagnetic resonance
G	guanine base in DNA
GDP	guanosine 5′-diphosphate
dGTP	2′-deoxyguanosine 5′-triphosphate
HSV	herpes simplex virus
NMR	nuclear magnetic resonance
pET	plasmid for expression by T7 RNA polymerase
Phe	phenylalanine
T	thymine base in DNA
Trp	tryptophan
dTTP	2′-deoxythymidine 5′-triphosphate
Tyr	tyrosine
UV	ultraviolet

REFERENCES

1. S. Eriksson and B.-M. Sjöberg, in *Allosteric Enzymes* (G. Hervé, ed.), CRC Press, Boca Raton, 1989, p. 189 ff.
2. J. Stubbe, *Adv. Enzymol.*, *63*, 349 (1990).

3. M. Fontecave, P. Nordlund, H. Eklund, and P. Reichard, *Adv. Enzymol.*, *65*, 147 (1992).
4. G. J. Mann, A. Gräslund, E.-I. Ochiai, R. Ingemarson, and L. Thelander, *Biochemistry*, *30*, 1939 (1991).
5. Y. Engström, S. Eriksson, L. Thelander, and M. Åkerman, *Biochemistry*, *18*, 2941 (1979).
6. M. Luthman, S. Eriksson, A. Holmgren, and L. Thelander, *Proc. Natl. Acad. Sci. USA*, *76*, 2158 (1979).
7. L. Thelander, S. Eriksson, and M. Åkerman, *J. Biol. Chem.*, *255*, 7426 (1980).
8. I. W. Caras, B. B. Levinson, M. Fabry, S. R. Williams, and D. W. Martin Jr., *J. Biol. Chem.*, *260*, 7015 (1985).
9. N. J. Parker, C. G. Begley, and R. M. Fox, *Nucl. Acid Res.*, *19*, 3741 (1991).
10. M. Thelander, A. Gräslund, and L. Thelander, *J. Biol. Chem.*, *260*, 2737 (1985).
11. L. Thelander and P. Berg, *Mol. Cell. Biol.*, *6*, 3433 (1986).
12. N. Pavloff, D. Rivard, S. Masson, S.-H. Shen, and A.-M. Mes-Masson, *DNA-Seq.*, *2*, 227 (1992).
13. M. M. Chaudhuri, P. N. Tonin, and P. R. Srinivasan, *Biochim. Biophys. Acta*, *1171*, 117 (1992).
14. R. Davis, M. Thelander, and L. Thelander (manuscript in preparation).
15. E.-I. Ochiai, G. J. Mann, A. Gräslund, and L. Thelander, *J. Biol. Chem.*, *265*, 15758 (1990).
16. P. Nordlund, B.-M. Sjöberg, and H. Eklund, *Nature*, *345*, 593 (1990).
17. S. Eriksson, L. Thelander, and M. Åkerman, *Biochemistry*, *18*, 2948 (1979).
18. S. Eriksson, L. J. Gudas, S. M. Clift, I. W. Caras, B. Ullman, and D. W. Martin Jr., *J. Biol. Chem.*, *256*, 10193 (1981).
19. N. C. Brown and P. Reichard, *J. Mol. Biol.*, *46*, 39 (1969).
20. M. K. Turner, R. Abrams, and I. Lieberman, *J. Biol. Chem.*, *243*, 3725 (1968).
21. Y. Engström, B. Rozell, H.-A. Hansson, S. Stemme, and L. Thelander, *EMBO J.*, *3*, 863 (1984).

22. G. J. Mann, E. A. Musgrove, R. M. Fox, and L. Thelander, *Cancer Res.*, *48*, 5151 (1988).
23. Y. Engström, S. Eriksson, I. Jildevik, S. Skog, L. Thelander, and B. Tribukait, *J. Biol. Chem.*, *260*, 9114 (1985).
24. S. Eriksson, A. Gräslund, S. Skog, L. Thelander, and B. Tribukait, *J. Biol. Chem.*, *259*, 11695 (1984).
25. D. Filatov, S. Björklund, and L. Thelander (submitted) (1993).
26. S. J. Elledge and R. W. Davis, *Mol. Cell. Biol.*, *9*, 4932 (1989).
27. J. E. Brissenden, I. Caras, L. Thelander, and U. Francke, *Exp. Cell Res.*, *174*, 302 (1988).
28. T. L. Yang-Feng, D. E. Barton, L. Thelander, W. H. Lewis, P. R. Srinivasan, and U. Francke, *Genomics*, *1*, 77 (1987).
29. P. N. Tonin, R. L. Stallings, M. D. Carman, J. R. Bertino, J. A. Wright, P. R. Srinivasan, and W. H. Lewis, *Cytogenet. Cell Genet.*, *45*, 102 (1987).
30. M. Thelander and L. Thelander, *EMBO J.*, *8*, 2475 (1989).
31. S. Björklund, E. Skogman, and L. Thelander (submitted) (1993).
32. S. Björklund, K. Hjortsberg, E. Johansson, and L. Thelander, *Proc. Natl. Acad. Sci. USA* (1993).
33. S. Björklund, S. Skog, B. Tribukait, and L. Thelander, *Biochemistry*, *29*, 5452 (1990).
34. S. Björklund, E. Skogman, and L. Thelander, *EMBO J.*, *11*, 4953 (1992).
35. R. Ingemarson and H. Lankinen, *Virology*, *156*, 417 (1987).
36. J. McLauchlan and J. B. Clements, *J. Gen. Virol.*, *64*, 997 (1983).
37. D. R. Averett, P. A. Furman, and T. Spector, *J. Virol.*, *52*, 981 (1984).
38. J. Conner, J. Macfarlane, M. Lankinen, and H. Marsden, *J. Gen. Virol.*, *73*, 103 (1992).
39. T. D. Chung, J. P. Wymer, C. C. Smith, M. Kulka, and L. Aurelian, *J. Virol.*, *63*, 3389 (1989).
40. J. Conner, J. Cooper, J. Furlong, and J. B. Clements, *J. Virol.*, *66*, 7511 (1992).
41. J. Furlong, J. Conner, J. McLauchlan, H. Lankinen, C. Galt, H. S. Marsden, and J. B. Clements, *Virology*, *182*, 846 (1991).
42. B. M. Dutia, M. C. Frame, J. H. Subak-Sharpe, W. N. Clark, and H. S. Marsden, *Nature*, *321*, 439 (1986).

43. E. A. Cohen, P. Gaudreau, P. Brazeau, and Y. Langelier, *Nature*, *321*, 441 (1986).

44. I. Climent, B.-M. Sjöberg, and C.-Y. Huang, *Biochemistry*, *30*, 5164 (1991).

45. F.-D. Yang, R. A. Spanevello, I. Celiker, R. Hirschmann, H. Rubin, and B. S. Cooperman, *FEBS Lett.*, *272*, 61 (1990).

46. G. Cosentino, P. Lavallée, S. Rakhit, R. Plante, Y. Gaudette, C. Lawetz, P. W. Whitehead, J.-S. Duceppe, C. Lépine-Frenette, N. Dansereau, C. Guilbault, P. Gaudreau, Y. Langelier, L. Thelander, and Y. Guindon, *Biochem. Cell Biol.*, *69*, 79 (1991).

47. M. Sahlin, A. Gräslund, A. Ehrenberg, and B.-M. Sjöberg, *J. Biol. Chem.*, *257*, 366 (1982).

48. B. Barry, M. El-Deeb, P. Sandusky, and G. Babcock, *J. Biol. Chem.*, *265*, 20139 (1990).

49. M. Sahlin, A. Ehrenberg, A. Gräslund, L. Petersson, B.-M. Sjöberg, and L. Thelander, *Biochemistry*, *26*, 5541 (1987).

50. L. Åkerblom, A. Ehrenberg, A. Gräslund, H. Lankinen, P. Reichard, and L. Thelander, *Proc. Natl. Acad. Sci. USA*, *78*, 2159 (1981).

51. L. Petersson, A. Gräslund, A. Ehrenberg, B.-M. Sjöberg, and P. Reichard, *J. Biol. Chem.*, *255*, 6706 (1980).

52. B.-M. Sjöberg, A. Gräslund, and F. Eckstein, *J. Biol. Chem.*, *258*, 8060 (1983).

53. M. Ator, S. Salowe, J. Stubbe, M. Emptage, and M. Robins, *J. Am. Chem. Soc.*, *106*, 1886 (1984).

54. G. Behravan, S. Sen, L. Thelander, F. Eckstein, and A. Gräslund (manuscript in preparation).

55. J. Stubbe, *Ann. Rev. Biol.*, *58*, 257 (1989).

56. J. M. Bollinger, D. E. Edmondson, B. H. Huynh, J. Filley, J. R. Norton, and J. Stubbe, *Science*, *253*, 292 (1991).

57. U. Björn, K. Goodtzova, G. Behravan, A. Gräslund, and L. Thelander (manuscript in preparation).

58. M. Lammers and H. Follmann, *Struc. Bonding (Berlin)*, *54*, 27 (1983).

59. G. Lassmann, L. Thelander, and A. Gräslund, *Biochem. Biophys. Res. Commun.*, *188*, 879 (1992).

60. I. Kjöller-Larsen, B.-M. Sjöberg, and L. Thelander, *Eur. J. Biochem.*, *125*, 75 (1982).

61. S. Nyholm, L. Thelander, and A. Gräslund, *Biochemistry* (in press) (1993).
62. M. Karlsson, M. Sahlin, and B.-M. Sjöberg, *J. Biol. Chem.*, *267*, 12622 (1992).
63. E. C. Moore, M. S. Zedeck, K. C. Agrawal, and A. C. Sartorelli, *Biochemistry*, *9*, 4492 (1970).
64. L. Thelander and A. Gräslund, *J. Biol. Chem.*, *258*, 4063 (1983).
65. G. Nocentini, F. Federici, P. Franchetti, and A. Barzi, *Cancer Res.*, *53*, 19 (1993).
66. S. Nyholm, G. Mann, A. Johansson, R. Bergeron, A. Gräslund, and L. Thelander, *J. Biol. Chem.* (in press) (1993).
67. L. Thelander, A. Gräslund, and M. Thelander, *Biochem. Biophys. Res. Commun.*, *110*, 859 (1983).
68. H. Probst, H. Schiffer, V. Gekeler, and K. Scheffler, *Biochem. Biophys. Res. Commun.*, *163*, 334 (1989).
69. L. Thelander, in *Gene Amplification in Mammalian Cells* (R. E. Kellems, ed.), Marcel Dekker, New York, 1993, p. 265 ff.
70. D. Filatov, R. Ingemarson, A. Gräslund, and L. Thelander, *J. Biol. Chem.*, *267*, 15816 (1992).
71. S. J. Elledge and R. W. Davis, *Mol. Cell. Biol.*, *7*, 2783 (1987).
72. J. Carlsson, J. A. Fuchs, and J. Messing, *Proc. Natl. Acad. Sci. USA*, *81*, 4294 (1984).
73. D. J. McGeoch, M. A. Dalrymple, A. J. Davison, A. Dolan, M. C. Frame, D. McNab, L. J. Perry, J. E. Scott, and P. Taylor, *J. Gen. Virol.*, *69*, 1531 (1988).
74. R. Ingemarson, A. Gräslund, A. Darling, and L. Thelander, *J. Virol.*, *63*, 3769 (1989).
75. P. O. Lycksell, R. Ingemarson, R. Davis, A. Gräslund, and L. Thelander (submitted) (1993).

5

Manganese-Dependent Ribonucleotide Reduction and Overproduction of Nucleotides in Coryneform Bacteria

Georg Auling[1] and Hartmut Follmann[2]

[1]Institut für Mikrobiologie, Universität Hannover, D-30167 Hannover, FRG

[2]Biochemie, Fachbereich Biologie-Chemie, Universität Kassel, Heinrich-Plett-Strasse 40, D-34109 Kassel-Oberzwehren, FRG

1. INTRODUCTION: THE DIVERSITY OF RIBONUCLEOTIDE REDUCTASE SYSTEMS

The reductive formation of 2′-deoxyribonucleotides, which are required for DNA and cell replication, and the enzymes which catalyze the process are remarkable for many reasons.

They are restricted to the S phase of a cell cycle and not normally found in resting cells or tissues.

Ribonucleotide reductases are allosterically regulated by five (or more) different effector nucleotides in such a complex way that concise kinetic models have not even been attempted.

The reaction requires enzyme-generated organic free radicals, not too frequently encountered in biochemistry.

The universal pathway of de novo DNA precursor biosynthesis, seemingly identical in all living organisms, is nevertheless associated with a group of enzymes which differ grossly in different organisms.

It was the latter property that led us to propose and to prove the existence of yet another form of ribonucleotide reductase when certain unexplained effects of manganese ions on DNA synthesis in coryneform bacteria were analyzed several years ago [1].

Ribonucleotide reductases (EC 1.17.4) [2,3] catalyze the strongly exergonic ($\Delta G^{\circ\prime} \approx -70$ kJ/mol), irreversible deoxygenation of ribonucleotide 5′-di- or triphosphates to the corresponding 2′-deoxyribonucleotides concomitant with the oxidation of a pair of sulfhydryl groups to a disulfide (Scheme 1). The reaction mechanism, not described in detail here, has aptly been characterized "amazing and confusing" [4]. Key events, derived from our own early studies with ^{18}O-labeled

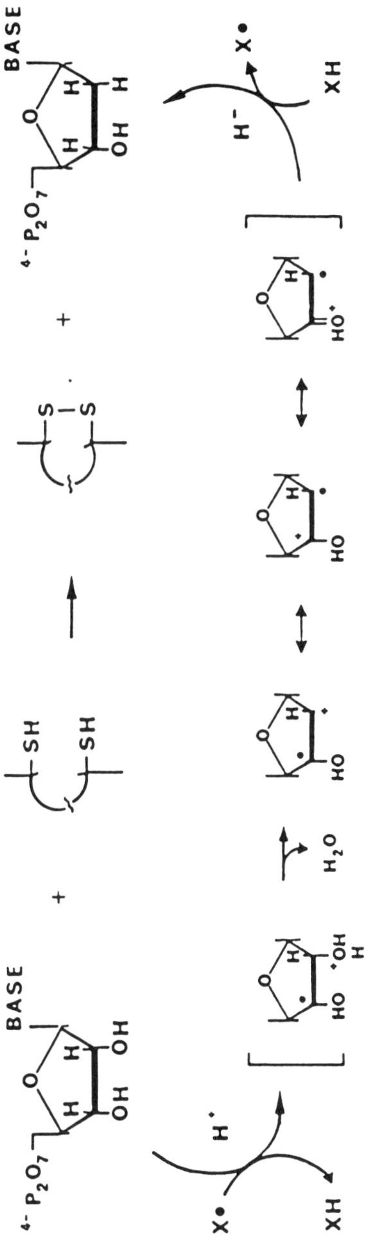

SCHEME 1.

substrates [5] and the work of Stubbe et al. using tritiated nucleotides [6], are the concerted abstraction of a hydrogen atom from the 3′ carbon and the protonated hydroxyl group from the 2′ carbon; however, the inferred nucleotide radical intermediate has not yet been intercepted.

The two originally known classes of ribonucleotide reductase are monomeric or dimeric enzymes requiring 5′-deoxyadenosylcobalamin (CoB_{12}) for activity [7], and more complex heterodimeric proteins containing dinuclear iron centers plus tyrosyl radicals [8]. Both enzymes are discussed in other chapters of this volume. Besides the obvious differences in structure and cofactor or metal requirement they clearly differ in distribution. The B_{12}-dependent ribonucleotide reductases, typified by the triphosphate-specific, monomeric enzyme of *Lactobacillus leichmannii* (M_r = 76000), are bacterial proteins; its presence in a eucaryotic alga, *Euglena gracilis*, is an exception. The cobalt requirement of an organism for DNA synthesis and growth can be directly linked to the availability and functioning of the cobalt(III)–corrinoid coenzyme in DNA precursor formation (e.g., in *Rhizobium* species [9]). The coenzyme is also the site of catalytic radical generation through homolytic cleavage of the cobalt–carbon bond in deoxyadenosylcobalamin, producing a short-lived cobalt(II)–deoxyadenosyl radical pair.

Knowledge of the structure and mechanisms of ribonucleotide reductases is most advanced in the enzyme systems isolated from *Escherichia coli* and from mammalian cells (cf. Chap. 4 in this volume). They consist of two nonidentical subunits. Protein R1, composed of two polypeptide chains ($M_r \approx 2 \times 85000$), contains all the substrate and effector nucleotide binding sites and several redox-active cysteine residues. Protein R2, also dimeric ($M_r \approx 2 \times 45000$), harbors two μ-oxo-bridged binuclear Fe(III) centers and tyrosyl free radicals, easily recognized by electron spin resonance (ESR) doublet signals at $g = 2.0047$. Although subunit R2 is a large protein it has long been recognized to interact with R1 and function like a dissociable cofactor required for catalysis of substrate reduction by the "apoenzyme" R1 [3]. Molecular oxygen is necessary for generation of the long-lived phenoxyl radicals from tyrosine residues in R2. DNA precursor and DNA synthesis are thus O_2-dependent processes in all eukaryotes (animals, plants, yeasts) expressing this enzyme system. A characteristic inhibitor of the Fe-tyrosyl ribonucleotide reductases is the radical scavenger hydroxyurea.

Our isolation and characterization of a manganese-containing ribonucleoside diphosphate reductase from *Brevibacterium ammoniagenes* and related aerobic bacteria [1,10,11] provided the third class of enzyme. Although similarities exist with the enzyme of *E. coli* and in the coordination chemistry and biochemistry of iron and manganese in general, the Fe and Mn ribonucleotide reductases differ substantially and do not appear to be mere variants of the same type of catalyst.

Finally, a fourth category of ribonucleotide reduction was predicted and has indeed been uncovered in strictly anaerobic microorganisms which, unlike other anaerobes, do *not* utilize coenzyme B_{12} for DNA precursor biosynthesis. Such novel systems have been demonstrated in the methanogenic archaebacteria *Methanobacterium thermoautotrophicum* [12] and as an alternative pathway in an oxygen-sensitive mutant of *E. coli* (oxyS) [13]. These enzymes are *in*activated by air. Organic free radicals have been detected by ESR spectroscopy but

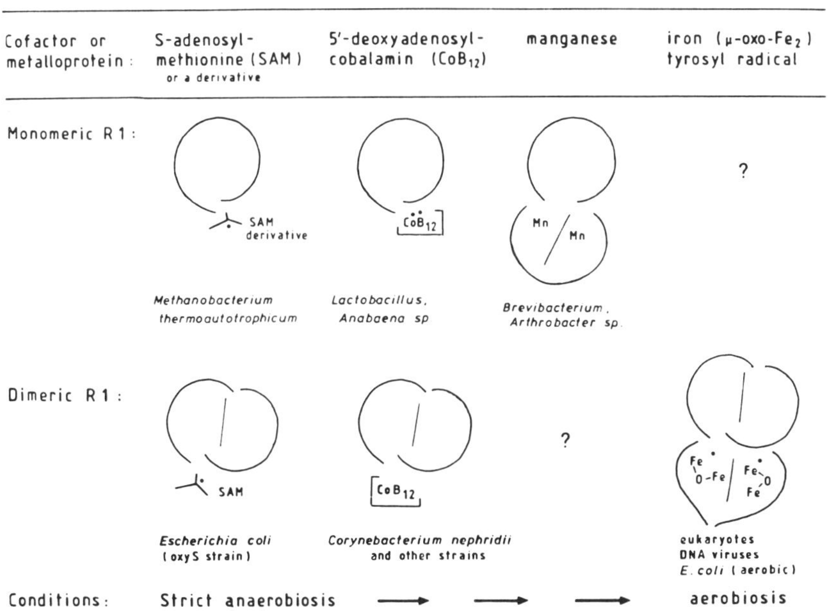

FIG. 1. Schematic view of the currently known types of ribonucleotide reductase and their distribution. ?, Structure not yet observed in nature.

transition metals are apparently not involved. A specialty of both enzymes is their interaction with S-adenosylmethionine (SAM) which activates the anaerobic reductase of *E. coli* but strongly inhibits the methanobacterial enzyme [14]. No such system was observed in anoxic cultures of yeast (*Saccharomyces cerevisiae*); eucaryotic cells rather cease DNA replication when the O_2-activated ribonucleotide reductase cannot function [15].

Our present understanding of the different enzymes, their radical cofactors or subunits, and the links to physiological circumstances of procaryotic and eucaryotic organisms are assembled in Fig. 1. Although some details remain speculative due to lack of more systematic studies, one notes an obvious increase in the complexity of ribonucleotide reductases accompanying the transgression from anaerobic to aerobic life, and the development from primitive to highly differentiated, de novo plus salvage nucleotide metabolism. The bacterial manganese enzymes evidently assume an interesting intermediate position in this scheme.

2. EFFECTS OF MANGANESE DEPLETION IN CORYNEFORM BACTERIA

2.1. Cellular and Metabolic Phenomena

It should be clear from the Introduction that any impairment of ribonucleotide reduction forces a cell to curtail DNA formation and thereby to block cell division. The transcription of ribonucleotide reductase genes has also been shown to be cell cycle-controlled in *E. coli* [16]. Microbiologists comprise perturbations in the coordination between macromolecular synthesis and cell division in the term "unbalanced growth". Lethal unbalanced growth is induced in *Brevibacterium ammoniagenes* ATCC 6872 and related bacteria when the external Mn^{2+} concentration remains below a threshold value of 1 μM. It is important for the industrial exploitation of this imbalance, described below, that the growth yield is *not* impaired by manganese depletion whereas viability is (Fig. 2). The specific manganese requirement for balanced growth cannot be compensated by other divalent cations (magnesium, iron, zinc, cobalt, calcium, or strontium) even when added in 10-fold or higher excess. In vivo labeling with radioactive precursors indicated that under conditions of manganese deprivation DNA formation is

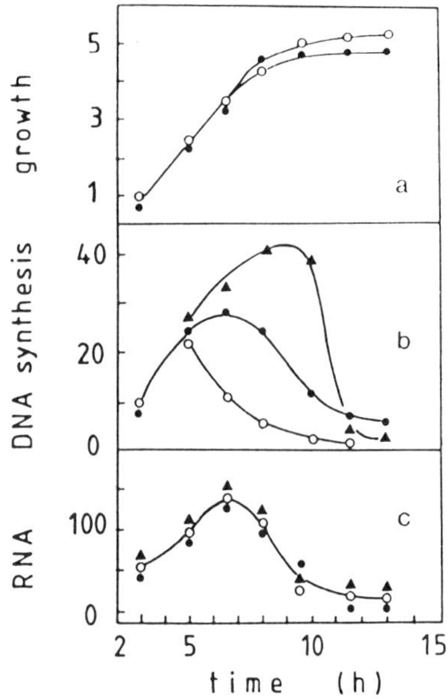

FIG. 2. Effect of manganese, manganese depletion, and resupplementation in coryneform bacteria (*Arthrobacter globiformis* [19]). (●) Regular manganese supply (2 μM); (○) manganese-free culture; (▲) 10 μM $MnCl_2$ added after manganese depletion. (a) Growth yield determined as dry weight ($g \cdot L^{-1}$). (b) DNA formation. (c) RNA formation measured by incorporation of radioactive precursors ($pmol \cdot hr^{-1} \cdot L^{-1}$).

selectively inhibited while RNA, protein, and peptidoglycan (cell wall) syntheses continue [17–19]. Consequently, unbalanced growth generates filamentous cells with reduced DNA content which can no longer proliferate. However, they remain metabolically active and sustain a membrane potential similar to that of cells grown in the presence of excess manganese [20].

Changes in the fatty acid, phospholipid, and protein composition of the bacterial membrane have been reported and are considered late events of manganese deprivation [20–22]. Further examination of highly purified membranes from *B. ammoniagenes* revealed membrane domains bound to ribosomes [23], which may be necessary for coupling the synthesis and secretion of proteins involved in extracellular nucleotide production (see below). However, a manganese deficiency–like protein pattern of the membrane could also be induced in the presence of 10 μM Mn^{2+} by inhibition of DNA precursor biosynthesis via ribonucleotide reduction with 20 mM hydroxyurea, or inhibition of DNA replication with 30 μM mitomycin C [20]. When 25 mM hydroxyurea (HU) is administered to exponentially growing cultures supplied with 10 μM Mn^{2+} [18], the cells display filamentous growth, release proteins into the culture medium, and convert added nicotinamide and adenine to NAD. Taken together, these observations contradict previous opinion that the extracellular nucleotide production of *B. ammoniagenes* is due to a defective cell envelope [21,24,25].

Manganese-deprived cells, deregulated by lethal unbalanced growth, are exploited in the Japanese fermentation industry for bulk production of ribonucleotides as taste-enhancing substances or nucleic acid–related pharmaceuticals (Table 1) [26]. Only such manganese-deprived cells accumulate phosphoribosyl pyrophosphate and proteins

TABLE 1

Ribonucleotides Produced by Fermentation with *Brevibacterium ammoniagenes* and Other Coryneform Bacteria Under Conditions of Manganese Starvation and Inhibition of Ribonucleotide Reduction

Flavor-enhancing food additives		
Guanosine 5′-monophosphate		
Inosine 5′-monophosphate		
Xanthosine 5′-monophosphate		
Modified nucleotides for biomedical application		
5-Fluorouridine 5′-monophosphate		
5-Hydroxyuridine 5′-monophosphate		
2-Thiouridine 5′-monophosphate		
6-Azauridine 5′-monophosphate		
2-Methylinosine 5′-monophosphate		
6-Thiopurine nucleotides		
8-Azapurine nucleotides		
Coenzymes		
NAD	ATP	CDP-choline
FAD	Coenzyme A	

extracellularly. Some of these proteins display salvage activity for the conversion of nucleobases to nucleotides (cf. Fig. 4). Based on the substrate specificity of phosphoribosyl transferases and other salvage pathway enzymes the accumulated, or added, bases are converted to the corresponding 5′-ribonucleotides. These reactions even allow microbiological production of unphysiological, semisynthetic nucleotides for medical applications.

2.2. Manganese Transport Systems

Although it is widely accepted that energy-dependent influx across the membrane is the first step of bacterial trace element nutrition [27], the question remains as to how *B. ammoniagenes* acquires its essential manganese ions from growth media which usually contain much higher amounts of magnesium ions. Isotope studies demonstrated that mid-log cells of *B. ammoniagenes* accumulate ^{54}Mn intracellularly 4000-fold over the initial environmental concentration, and that this accumulation obeys saturation kinetics characteristic of active transport. The kinetic data are in the range of other bacterial high-affinity uptake systems for manganese [28]. The high specificity of the Mn^{2+} transport system in *B. ammoniagenes* follows from the observation that even a 20000-fold excess of Mg^{2+} does not inhibit Mn^{2+} uptake [29] whereas Cd^{2+} and Zn^{2+} ions are competitive inhibitors with an inhibition constant in the range of the apparent K_m of Mn^{2+} transport. Nucleotide (NAD)-producing cells also show very active uptake of $^{54}Mn^{2+}$ [30]. This result again contradicts the opinion of a leaky membrane in nucleotide-producing cells [21,24] and argues for an induction of the high-affinity Mn^{2+} transport due to manganese deprivation. Three membrane proteins of 24, 27, and 49 kD have been suggested as potential carriers of high-affinity transport, and further transport systems for the energy-dependent uptake of Mn^{2+} by *B. ammoniagenes* are expected to exist.

2.3. Primary Target of Manganese Depletion

Manganese-depleted cells of *B. ammoniagenes* recover from inhibition of DNA formation by subsequent addition of Mn^{2+} ions. Any stimulatory effect of other metal cations (especially iron) occasionally observed in experiments with whole cells could be traced back to their residual contamination with manganese. Remarkably, the manganese-reactivated cells display higher DNA formation activity than manganese-supplied cells (Fig. 2b). Inhibitor studies demonstrated that the recovery of DNA formation ability in such bacteria is insensitive to transcriptional and translational inhibitors, but is completely abolished by hydroxyurea which is a well-known radical scavenger of the iron ribonucleotide reductase (Fig. 3). Thus it was proposed that not DNA replication itself, including initiation at the chromosomal origin, but DNA precursor biosynthesis via ribonucleotide reduction is manganese-dependent in *B. ammoniagenes* and related bacteria [1,31]. Manganese-depleted cells

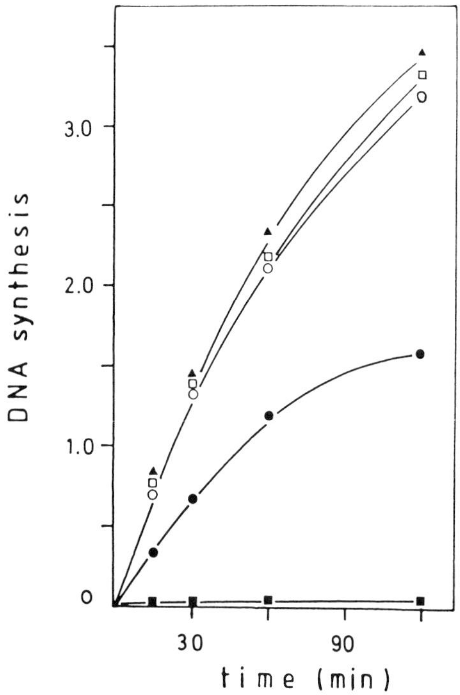

FIG. 3. Influence of inhibitors on the recovery of DNA formation in manganese-depleted-reactivated cells of *Brevibacterium ammoniagenes*. (●) Control culture (regular Mn supply); (▲) depleted-reactivated culture; (■) same, + hydroxyurea; (○) same, + rifampicin; (□) same, + chloramphenicol. Ordinate: radioactive precursor incorporation into DNA (pmol·mL^{-1}).

have an elevated amount of inactive, manganese-free ribonucleotide reductase, which is intracellularly activated upon addition of Mn^{2+} ions. Consequently, this reactivation is noticeable as enhanced DNA synthesis in cells recovering from manganese depletion, and it is insensitive to chloramphenicol or rifampicin. Increased synthesis of ribonucleotide reductase under conditions which lead to DNA synthesis inhibition is a well-known phenomenon in bacteria [32].

Brevibacterium ammoniagenes can by no means circumvent a block in DNA precursor biosynthesis resulting from a nonfunctional metal-free ribonucleotide reductase (Fig. 4). This fact and the low extracellular nucleotidase activity explain why *B. ammoniagenes* has proved particularly useful for industrial production of ribonucleotides (cf. Table 1) under conditions of manganese depletion [26]. It should be noted that perhaps the majority of bacteria cannot escape lethal unbalanced growth upon inhibition of ribonucleotide reduction. *Bacillus*

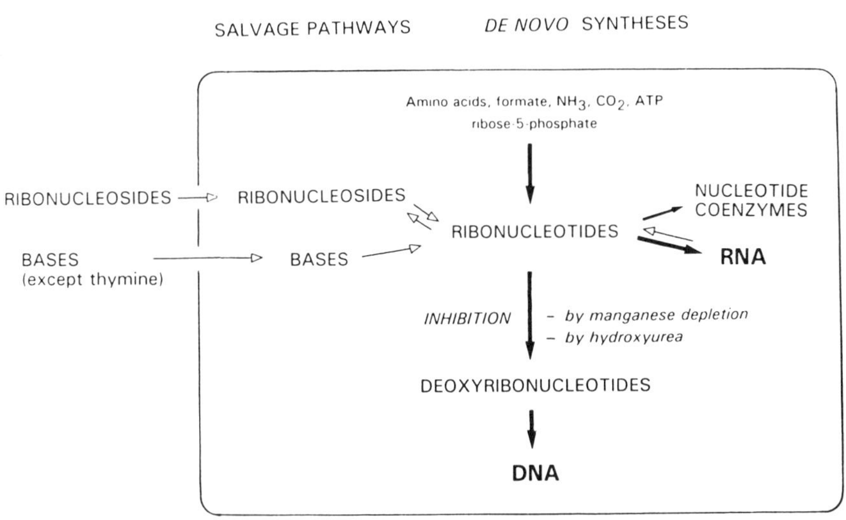

FIG. 4. General scheme of nucleotide metabolism in *Brevibacterium ammoniagenes*. De novo syntheses (bold arrows) include pyrimidine and purine ribonucleotide synthesis (top) and ribonucleotide reduction (center). Salvage pathways (light arrows) indicate utilization of degraded nucleic acids or uptake of extracellular precursors. Note that salvage reactions leading to deoxyribonucleotides (lower left corner) are missing.

subtilis is, to our knowledge, the only bacterium which circumvents blocking of ribonucleotide reduction through uptake of extracellular nucleic acid precursors [33] whereas *B. ammoniagenes* has a limited ability to take up nucleobases and nucleosides (Fig. 4).

The neglect of microbial diversity with regard to salvage pathways has long prevented elucidation of the primary target of manganese depletion in *B. ammoniagenes*. Deoxynucleoside kinases, required for rapid channeling of extracellular precursors into DNA, are also absent in this organism [34,35]. This was confirmed by a survey [36] which demonstrated that absence of thymidine kinase is a general property of microorganisms related to *B. ammoniagenes*.

The primary target of manganese deprivation in *B. ammoniagenes* and similar bacteria was substantiated in experiments with toluene-permeabilized cells, which allow uptake of nucleotides across a weakened membrane to provide the substrates for intracellular enzymes of the DNA replication machinery. Thus, permeabilized cells harvested from manganese-deficient cultures of different bacterial species were still able to synthesize DNA when supplied with all four deoxyribonucleotides [19]. Recently ether-permeabilized cells have served for an enzymatic screening of thermosensitive mutants obtained from *B. ammoniagenes* by chemical mutagenesis [37]. These mutants had been proven to be defective in DNA formation by in vivo labeling with nucleic acid precursors at 40°C, but not at 30°C. Temperature sensitivity of ribonucleotide reductase of three of these mutants was indeed found in assays both with ether-permeabilized cells and with partially purified protein fractions.

3. THE MANGANESE-CONTAINING RIBONUCLEOTIDE REDUCTASE OF *BREVIBACTERIUM AMMONIAGENES*

3.1. Purification, Subunit Composition, and Metal Incorporation

The establishment of ribonucleotide reduction as target of the manganese requirement of DNA synthesis in coryneform bacteria necessitated purification of the Mn-dependent protein system. However, purification of any novel ribonucleotide reductase is seriously hindered by the naturally low and growth-dependent intracellular amount of these enzymes. Several proteins of the diferric iron-tyrosyl radical family present in *E. coli*, animal viruses, and mouse cells have recently been obtained and overexpressed by recombinant DNA techniques [38,39]. Our attempts to identify and clone ribonucleotide reductase genes of *B. ammoniagenes* using in vivo complementation in thermosensitive mutants and in vitro hybridization probes derived from *E. coli* [40] have failed so far, most likely due to lack of homology between the iron and manganese enzymes. The recently isolated thermosensitive mutants of *B. ammoniagenes* [37] offer a more promising approach in the future.

For conventional enzyme preparation we have exploited the physiological overexpression and unbalanced growth phenomena of manganese-deficient bacteria, and their highly active and specific manganese transport systems described above. Cells were first cultured for 10–12 hr in Mn^{2+}-free mineral media, then supplemented with as little as 10 μM Mn^{2+} and cultured for one more hour. Complete removal of manganese ions, e.g., by EDTA treatment of all equipment, and complete DNA synthesis arrest during the initial culture are essential for the successful preparation of active Mn-ribonucleotide reductase from depleted-reactivated bacteria. The same procedure permitted preparation of a ^{54}Mn-labeled enzyme.

A protocol for the purification of ribonucleotide reductase from cell-free extracts of *B. ammoniagenes* has been described [10]. In brief, it has been possible to copurify two nonidentical subunits R1 and R2 (formerly B1, B2), despite their tendency toward dissociation, as fully active “holoenzyme” by carefully controlled chromatography on Sephacryl S-400, DEAE cellulose, and hydroxyapatite columns. The most active fraction, enriched about 600-fold, has a specific activity of 450 nmol cytidine diphosphate (CDP) reduction $\cdot mg^{-1} \cdot hr^{-1}$ and exhibits two protein bands of M_r = 80000 and 50000 on denaturing electrophoresis gels. Alternatively, the enzyme could be separated into two fractions, inactive separately, by affinity chromatography on 2′,5′-ADP Sepharose (Fig. 5). A specifically bound protein R1 was eluted by 1 mM dATP from this column and could then be purified to homogeneity on DEAE cellulose; it possesses a molecular weight of 80000. The second fraction required for activity, R2, does not bind to the immobilized nucleotide ligand and is much more difficult to enrich than R1. Subunit R2 exhibits

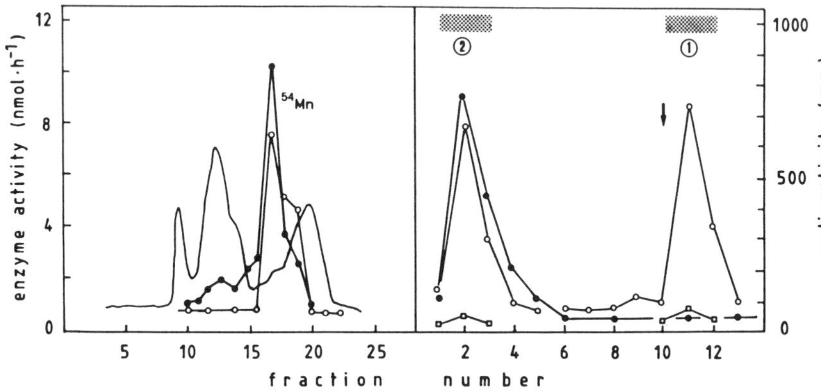

FIG. 5. Purification of ^{54}Mn-labeled *B. ammoniagenes* ribonucleotide reductase on Sephacryl S-400 (left panel; in K-phosphate buffer, pH 6.6, containing 2 mM Mg^{2+} and 2 mM dithiothreitol) and on 2′,5′-ADP Sepharose (right panel). (●) Radioactivity distribution; (○) enzyme activity measured after recombination of subunits R2 and R1; (□) activity of the separated subunits. Fractions containing the subunits are marked on top; the arrow marks elution of R1 with 1 mM dATP.

a molecular weight of 100000 on gel filtration columns but a prominent protein band at M_r = 50000 (besides minor proteins) shows on sodium dodecyl sulfate (SDS) polyacrylamide electrophoresis.

When this purification scheme was adapted to a *B. ammoniagenes* culture reactivated in the presence of 20 μCi $^{54}MnCl_2$ about 2% of the isotope was found in a crude ribonucleotide reductase fraction which contained three major manganoproteins. One of them was clearly identified as reductase subunit R2 by the coelution of enzyme activity and radioactivity from four different chromatography columns, comigration on nondenaturing electrophoresis gels, and spectroscopic evidence (vide infra). In contrast, R1 is not a manganoprotein.

All these properties allow the assignment to *B. ammoniagenes* ribonucleotide reductase of an R1:R2 subunit composition in which R1 is a nucleotide (substrates, effectors) binding protein and R2 a metalloprotein. While this structure bears resemblance to that of *E. coli* ribonucleotide reductase it differs from the $\alpha_2\beta_2$ polypeptide composition common to the iron enzymes in that R1 is monomeric and R2 is dimeric ($\alpha\beta_2$). No evidence could be found under our conditions [10] for larger (dimeric, $M_r \approx 160000$) forms of R1 alone or in the holoenzyme purified from Mn-depleted-reactivated bacteria. An explanation based on the regulatory functions of protein R1 is offered below. The behavior of fully active enzyme preparations on gel filtration media (M_r = 100000 ± 10000) indicates that at least in vitro the two subunits are not associated in a tight complex of $M_r \approx 180000$, in accordance with the assumed role of R2 as dissociable, catalytic cofactor. Magnesium or calcium ions (1–2 mM) stimulate enzyme activity in R1–R2 mixtures up to fivefold and are thought to promote subunit interaction; however, they do not affect the chromatography profiles.

The restoration of DNA synthesis and ribonucleotide reduction in Mn^{2+}-depleted coryneform bacteria is specific for manganese and cannot be achieved by addition of iron or other transition metals. It has been difficult to reproduce manganese incorporation in vitro for a number of reasons. Thus extraction of the metal from partially purified enzyme by treatment with most chelating agents led to irreversible denaturation and loss of activity whereas the Mn^{2+}-complexing macrocyclic antibiotic A 23187 had no effect. Only after limited dialysis vs. 1 mM EDTA and partial inactivation was activity restored by addition of 10 mM Mn^{2+} (100%) or 10 mM Fe^{2+} (75%), but not by Co^{2+} or Zn^{2+} ions [1]; however, these experiments required increased salt concentration (100 mM NaCl) and may not have regenerated the original, native protein. Preparation of metal-free apoenzyme from Mn^{2+}-deficient cells would be another approach. Again, it has not been possible to "blind"-purify a protein fraction which would be activated by Mn^{2+} in vitro. The intracellular process of specific Mn integration into R2 apoprotein must be highly controlled indeed.

We have combined the separated R1 and R2 subunits of the ribonucleotide reductases purified from *B. ammoniagenes* and from *E. coli* in hybrid protein mixtures, with or without Mg^{2+} ions, but found no activity. Likewise, a semisynthetic enzyme of *E. coli* in which the iron centers have been substituted by manganese centers [41] is a stable yet catalytically inactive protein.

3.2. Kinetic Data and Specificities

Manganese ribonucleotide reductase of *B. ammoniagenes* accepts all four common ribonucleotides as substrates [11]. The highly purified, kinase-free enzyme preferentially acts on ribonucleoside 5′-diphosphates but 5′-triphosphates are also reduced at about fivefold lower rates (Table 2). Substrate reduction depends on the presence of 0.1–0.5 mM deoxyribonucleoside 5′-triphosphates or ATP which act as specific allosteric effectors; very little or no activity is observed without effector nucleotides. The pattern of activity modulation is in part similar to that of other ribonucleotide reductases. Thus, the reduction of CDP and UDP is stimulated by dATP and ATP, ADP reduction is stimulated by dGTP, and GDP reduction by dTTP. In contrast to the enzymes from other sources, however, none of the effector nucleotides will *inhibit* substrate reduction even at higher concentration.

The solely positive modulation of deoxyribonucleotide formation correlates very well with the physiological needs of DNA precursor supply in *Brevibacterium*. As described above (Fig. 4) the bacteria have only rudimentary salvage pathways for incorporation of purine and pyrimidine bases and nucleosides into nucleic acids. In particular, they lack deoxyribonucleoside kinases which could interfere with the reductive de novo pathway, rendering negative control of ribonucleotide reduction unnecessary. Considering the specific binding of R1 to immobilized nucleotide ligands, allosteric effector sites must reside on the R1 subunit of *B. ammoniagenes* as in all other ribonucleotide reductases. We conclude that less elaborate regulatory demands, as in the coryneform bacteria and also in B_{12}-dependent lactobacilli, can be met by

TABLE 2

Nucleotide Specificity and Kinetic Data of Mn Ribonucleotide Reductase from *B. ammoniagenes*

Substrate nucleotide	Effector nucleotide and concentration (mM)	K_m (app.) (μM)	Specific activity ($nmol \cdot h^{-1} \cdot mg^{-1}$)
CDP	dATP (0.1)	10	273
	ATP (1.0)		130
UDP	dATP (0.1)	60	195
	ATP (1.0)		108
ADP	dGTP (0.5)	10	254
GDP	dTTP (0.5)	30	200

monomeric R1 proteins whereas the more complex nucleotide metabolism of *E. coli* and eukaryotes requires dimeric R1 subunits for more efficient regulation of ribonucleotide reduction (cf. Fig. 1).

The Mn ribonucleotide reductase is coupled to a thioredoxin–thioredoxin reductase system for generation of reducing equivalents. Thioredoxin from *B. ammoniagenes* (M_r = 12000) has been purified to homogeneity [11] and an NADPH-dependent thioredoxin reductase was also demonstrated. The bacterial protein, chemically reduced, is a much better hydrogen-donating substrate of CDP reduction in vitro [$K_{m(app.)}$ = 4 μM] than dithiothreitol as a low molecular weight dithiol [$K_{m(app.)}$ = 0.5 mM]. The physiological role of bacterial thioredoxins is not always well defined since some fail to serve as reductants in vitro [42], or the glutathione–glutaredoxin system [43] is the more specific reductant in ribonucleotide reduction in vivo. However, the function of thioredoxin in *B. ammoniagenes* as reductase substrate is not in doubt because gram-positive bacteria do not contain glutathione [44].

Ribonucleotide reductases are delicate enzymes, prone to inactivation at their sulfhydryl groups, metal centers, or radical sites. It is true that a "reductase cannot stand a 0.5 M concentration of anything except water" [45]. Nevertheless, the sensitivity to inhibition varies and can in fact provide information about the type of enzyme and its mechanism. The effects of a number of compounds other than nucleotides, metals, and sulfhydryls on Mn ribonucleotide reductase were therefore compared.

Most characteristic is the inhibition produced by hydroxyurea (I_{50} = 2 mM) and by oligomeric aurintricarboxylic acid (I_{50} = 30 μM, based on M_r = 6000) (Fig. 6). Both these compounds also inhibit iron-tyrosyl ribonucleotide reductases, but not the coenzyme B_{12}-dependent enzymes. Hydroxyurea specifically reduces the radical species [46] whereas aurintricarboxylate is thought to act by a combination of metal chelation and radical scavenging [47]. Inhibition by hydroxyurea was instantaneous and not time-dependent as in the yeast enzyme. The HU concentrations required for complete inhibition of ribonucleotide reductase in vitro and of DNA synthesis in growing cultures ($\geq$5 mM) were remarkably similar. Isoquinoline carboxaldehyde thiosemicarbazone, a potent inhibitor of mammalian ribonucleotide reductase, did not affect our bacterial enzyme.

Oxygen is a positive effector of the *B. ammoniagenes* enzyme

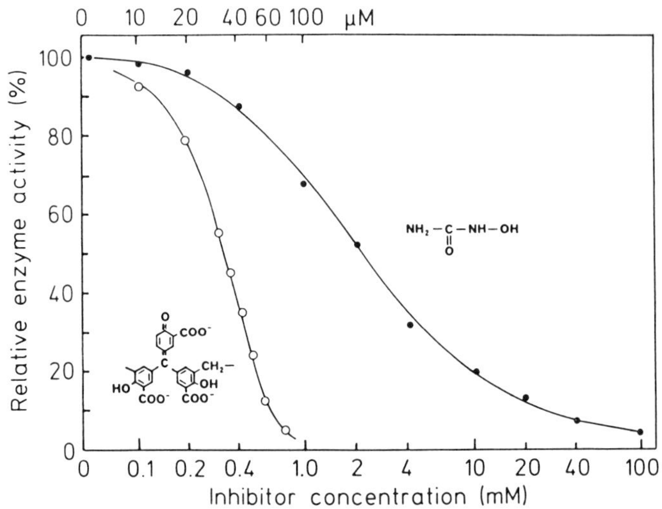

FIG. 6. Inhibition of Mn ribonucleotide reductase by hydroxyurea (●, lower scale) and aurintricarboxylic acid (○, upper scale).

system under certain conditions. Active enzyme can be freshly prepared from the bacteria and assayed under nitrogen. However, when enzyme samples had been inactivated by hydroxyurea and the inhibitor removed by dialysis, partial recovery of activity only occurred under aerobic conditions (Table 3). Magnesium ions have an as yet unexplained augmenting effect on the reactivation reaction. Samples of Mn ribonucleotide reductase "aged" by prolonged storage (>4 weeks, frozen) suffered less inactivation when kept aerobically; in fact, their activity could best be recovered in the presence of a critical, low concentration of bleomycin–O_2 complex, in analogy to the behavior of eukaryotic tyrosyl radical ribonucleotide reductase [48].

3.3. Spectroscopic and Magnetic Characterization

Structure elucidation of the metal center(s) in Mn ribonucleotide reductase is not by far as advanced as in the iron enzymes [8] because of lack of an overexpression system and the limited amount of protein that can be obtained by conventional purification. The following information has been assembled by spectroscopic analysis.

TABLE 3

Effects of Oxygen on Purified Mn Ribonucleotide Reductase from *B. ammoniagenes*

Conditions	Enzyme activity	
	$nmol \cdot h^{-1}$	%
Control sample, standard assay [10]	1.20	100
2-hr 20 mM HU, anaerobic	0.26	22
Dialyzed, anaerobic	0.28	23
Dialyzed, aerobic	0.59	49
Dialyzed, aerobic, 1 mM Mg^{2+}	0.84	70
Aged sample, + 5 μM bleomycin·O_2	1.38	117

The electronic spectrum of a highly purified sample of *B. ammoniagenes* ribonucleotide reductase is shown in Fig. 7 together with that of another bacterial manganese enzyme, pseudocatalase of *Lactobacillus plantarum* [49]. It exhibits absorption maxima at 370, 455, 490, and 616 nm. The two spectra are remarkably similar and bear resemblance to those of binuclear, μ-oxo-, di-μ-carboxylato-bridged complexes of manganese(III) [50,51]. Because a binuclear manganese site has been identified by EPR spectroscopy and extended X-ray absorption fine structure (EXAFS) measurement in the pseudocatalase [52] we conclude that the metal in Mn ribonucleotide reductase is also organized in Mn-O-Mn centers, most likely in the Mn(III) oxidation state.

Packed cells or partially purified extracts of *B. ammoniagenes* exhibit the characteristic six-line EPR spectrum of intracellular, free $Mn(H_2O)_6^{2+}$ ions at $g=2$. This intense spectrum renders the search for paramagnetic species specifically associated with ribonucleotide reductase (Mn centers, or an organic free radical) very difficult. However, it has been shown that highly purified, active enzyme samples do *not* contain any EPR-active species in the temperature range from 4 K to room temperature [10]. In an attempt to produce more concentrated enzyme solutions in larger quantity, albeit at lower purity, a protocol was developed for removal of free Mn^{2+} and other metal ions by extensive dialysis in ultrapure buffers, controlled by inductively coupled

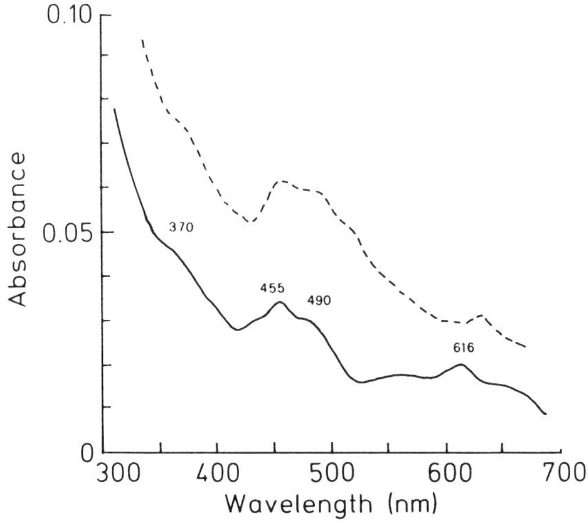

FIG. 7. UV/VIS spectrum of highly purified *B. ammoniagenes* ribonucleotide reductase (R1 + R2; 0.6 mg protein/mL). Dashed line: spectrum of the pseudocatalase from *Lactobacillus plantarum* [49].

plasma (ICP) metal analysis. Samples of 35 mg protein/mL and less than 10 ppb Mn/mg protein were thus obtained which no longer exhibit the six-line EPR spectrum (K. Blasczyk, G. Auling, E. Bill, unpublished). As shown in Fig. 8, the resonance lines of free manganous ions appear after denaturation in trichloroacetic acid (TCA). An analogous experiment with the pure enzyme [10] had been used for a semiquantitative estimate of the released Mn^{2+} ions and had confirmed the presence of (at least) 1 mol Mn per 1 mol protein in subunit R2 of the reductase.

The problems associated with structure predictions of manganoproteins from spectral data in solution have been clearly discussed by Wieghardt [53], namely, loss or exchange of metal ions during purification, uncontrolled redox changes, contaminations, and difficulties of quantification. All these reservations apply to our current knowledge of the Mn ribonucleotide reductase system. Thus, the estimate of metal content by EPR is most likely too low and does not contradict a binuclear manganese site. The failure to detect paramagnetic Mn species in EPR studies of the native protein, even in concentrated solutions, can

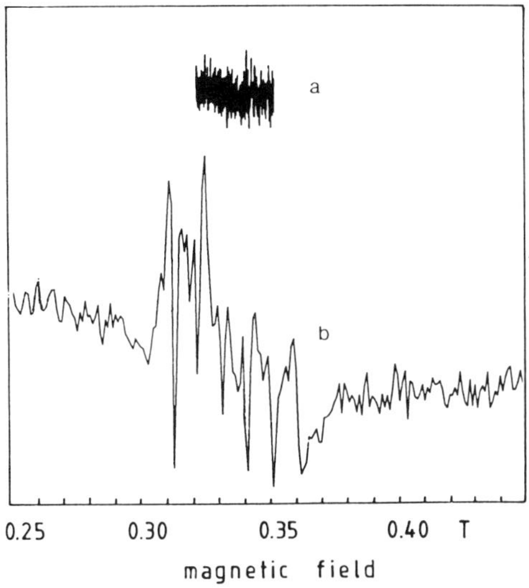

FIG. 8. EPR Spectrum of partially purified ribonucleotide reductase subunit R2 in Mn^{2+}-free solution, measured at 50 K. (a) Native protein; (b) same sample after denaturation in TCA. Microwave frequency, 9.4297 GHz; microwave power, 2.0 mW.

have various causes such as strong protein binding [41] and/or intramolecular magnetic coupling in EPR-silent Mn(II) or Mn(III) dimers. Presence of Mn(III) in the enzyme, suggested by the absorption spectrum (Fig. 7), is chemically compatible with aerobic incorporation of Mn(II) into apoprotein, in analogy to formation of the iron enzyme metal center [54]; the reappearance of Mn(II) upon acid denaturation (Fig. 8), probably formed in a disproportionation reaction, is also not unreasonable. It appears safe to say, however, that mixed valence states, e.g., Mn(II)/Mn(III), with distinct multiline EPR spectra [52], are not present in *B. ammoniagenes* ribonucleotide reductase under native conditions.

3.4. The Elusive Radical

There is no reason to believe that the bacterial manganese enzymes deviate from the common radical mechanism of ribonucleotide reduction depicted in Scheme 1. The inhibition by hydroxyurea and activation by oxygen described above should be specific indicators of a radical involvement in the *B. ammoniagenes* enzyme system too. But contrary to expectation purified, active enzyme samples do not contain any EPR-detectable organic free radical. The tyrosyl radical of *E. coli* ribonucleotide reductase, when admixed to these samples, was clearly measurable under the experimental conditions. Resonance signals present in less purified enzyme preparations decreased during further purification. Addition of one-electron redox agents [hexacyanoferrate(III), ascorbate] also produced no significant EPR effects. It may be noted in this context that Mn-substituted *E. coli* ribonucleotide reductase [41] likewise generated no tyrosyl radical.

The missing organic free radical could be hidden and EPR-silent by magnetic coupling to a manganese center, or be short-lived. The 5′-deoxyadenosyl radical engaged in B_{12}-dependent ribonucleotide reduction is indeed a short-lived species but not involved in the manganese enzyme. The free radicals derived from amino acid residues (tyrosine, tryptophan, glycine) of radical enzymes are usually stable enough to be detected by low-temperature EPR spectroscopy. In the absence of such a species we wish to discuss whether Mn ribonucleotide reductases could perform catalysis *without* an equivalent of the diiron enzymes' tyrosyl radical. Manganese(III) complexes are capable of abstracting hydrogen atoms from organic compounds in polar solvents [55], or in a fungal Mn peroxidase [56], producing secondary or tertiary carbon radicals in distinct analogy to abstraction of H-3 from ribose (Scheme 1). Or a manganic protein site could transiently oxidize an amino acid residue [56] in a mechanism suggested by Stubbe et al. [57] and discussed in Sec. 5 of this chapter.

Whether or not manganese ribonucleotide reductase is a radical enzyme strictu sensu like its iron-tyrosyl counterparts or catalyzes rapid radical reactions like the B_{12} enzymes remains to be established. Preparation of sufficient amounts of proteins R1 and R2 for more detailed structure analysis is still a challenge. However, the chemical basis of our above hypothesis can be put to the test: Examination of the many manganese(III) complexes synthesized to mimic the metal sites of manganese proteins [53] for their ability to abstract hydrogens from sugars, amino acid side chains, or related compounds should permit evaluation of whether the model is viable.

4. DISTRIBUTION OF MANGANESE RIBONUCLEOTIDE REDUCTASES

There are a number of analogies (e.g., activation by oxygen, inhibition by HU) between the manganese ribonucleotide reductase of *B. ammoniagenes* and the well-known diiron enzyme of *E. coli*, but major differences such as subunit structure and absence of a stable tyrosyl radical are also evident. The lack of enzyme activity in hybrid protein mixtures [10] and in vitro metal replacement experiments which result in a nonfunctional protein [41] also indicate that the bacterial Mn and Fe ribonucleotide reductases are two distinctly different enzyme classes. This fact is firmly underlined by a view of their distribution, which gives the Mn, and not the Fe, enzymes a prominent position among prokaryotes.

Elucidation of the target of manganese deprivation in *B. ammoniagenes* and related bacteria was facilitated by the observation that manganese-depleted cells recover from inhibition of DNA formation by subsequent addition of manganese [1,18,19,31]. The degree of stimulation seems to be strain-dependent although one has to bear in mind that both the residual manganese contamination in the growth medium and the culture age will affect DNA formation activity in pulse labeling experiments with whole cells. It is self-evident that a screening for lethal unbalanced growth induced by manganese depletion is limited to those bacteria which can be grown in a defined mineral medium. Nevertheless, the pattern of manganese control of DNA formation activity described for *B. ammoniagenes* was found in nine other coryneform bacteria (Fig. 9). Manganese-type ribonucleotide reduction is present in species of the genera *Arthrobacter*, *Micrococcus*, and *Nocardia*. Interestingly, *Nocardia opaca* was the first bacterium reported to elongate and to have a reduced DNA content under manganese depletion [58]. In addition, manganese requirement for growth and cell elongation upon manganese deprivation has been reported in even more coryneform strains, including *Corynebacterium glutamicum*, *C. callunae*, *Brevibacterium divariaticum*, *B. flavum*, *B. lactofermentum*, and *Arthrobacter citreus*.

If unbalanced growth induced by manganese depletion is a characteristic of organisms which have the manganese type of ribonucleotide

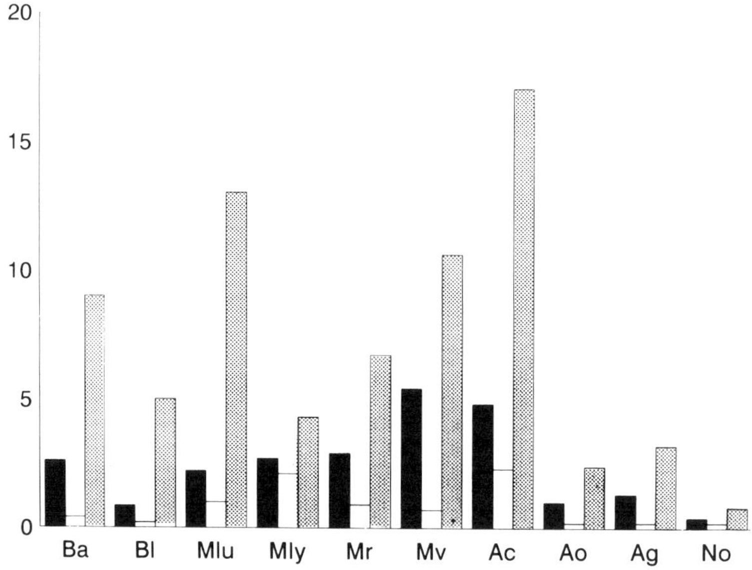

FIG. 9. Distribution of manganese-dependent ribonucleotide reduction in coryneform and related bacteria as established by reversible inhibition of DNA formation in pulse-labeled, manganese-depleted cells. Ordinate: radioactive precursor incorporation into DNA (pmol·mL^{-1}). Full columns: manganese-supplied cells; empty columns: manganese depletion; half-tone columns: recovery from depletion by 1 hr incubation in the presence of 2 μM $MnCl_2$. Ba, *Brevibacterium ammoniagenes*; Bl, *B. lactofermentum*; Ao, *Arthrobacter oxydans*; Ag, *A. globiformis*; Ac, *A. citreus*; Mlu, *Micrococcus luteus*; Mly, *M. lylae*; Mr, *M. roseus*; Mv, *M. varians*; No, *Nocardia opaca*.

reductase detected in *B. ammoniagenes*, more direct confirmation by enzyme activity measurements is required. The data in Table 4 indeed demonstrate a clear-cut correlation between manganese levels in different coryneform bacteria and their ribonucleotide reductase activities. Low or virtually no activity was encountered in deprived cultures with arrested DNA formation. Upon recovery by 1 hr incubation in the presence of manganese the enzyme activity was stimulated three-to tenfold compared with the manganese-sufficient controls.

Thus, the Mn ribonucleotide reductase predominates in different species of the coryneform bacteria and related genera. All of them are located on the high-GC branch of gram-positive bacteria in the new bacterial systematics, which is supposed to reflect natural evolution

TABLE 4

Manganese Control of Ribonucleotide Reductase Activity in Cell-Free Extracts from Coryneform Bacteria

Strain	Culture conditions (nmol GTP reduction·h^{-1} mg $protein^{-1}$)		
	Mn-sufficient	Mn-deficient	Mn-stimulated
B. ammoniagenes	0.9	0.1	2.9
Arthrobacter citreus	0.26	0.18	2.37
A. globiformis	0.09	0.01	0.30
A. oxydans	0.08	0.01	0.95
Micrococcus luteus	0.1	0	0.7

[59]. The impression of a broad, homogeneous distribution of Mn ribonucleotide reductase in this phylogenetic branch (Fig. 10) is not disturbed by previous reports of B_{12}-dependent ribonucleotide reductase in *Micrococcus denitrificans* and in *Corynebacterium nephridii* [60,61]. Both species are now viewed by microbiologists as distant relatives of rhizobia which have a B_{12}-ribonucleotide reductase. For this reason *M. denitrificans* has been renamed *Paracoccus denitrificans* [62] and *C. nephridii* is a member of the α subclass of proteobacteria based on chemotaxonomical data (G. Auling, unpublished results). Within the gram-positive bacteria B_{12}-dependent ribonucleotide reduction prevails, however, in the anaerobic lactobacilli and clostridia which are prominent members of the low-GC branch (Fig. 10).

Considering the new phylogenetic bacterial systematics, we would like to attract attention to the possibility that an even greater biochemical diversity of ribonucleotide reductases than shown in Fig. 1 may be expected in nature. Only 3 out of 11 classes of eubacteria have been investigated with respect to the distribution of these enzymes. Furthermore, the large eukaryotic group of filamentous fungi has not been analyzed for ribonucleotide reductase. Here a number of manganese-controlled effects have been described within the Deuteromycetes [63], which resemble lethal unbalanced growth observed in the coryneform bacteria.

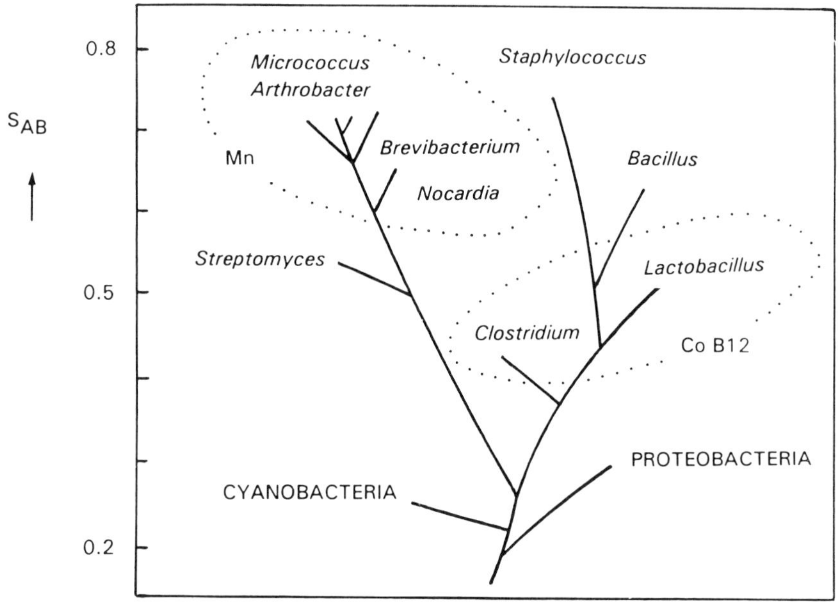

FIG. 10. Correlation of the distribution of manganese-dependent (Mn) and of coenzyme B_{12}-dependent (CoB_{12}) ribonucleotide reductase with the phylogeny of gram-positive bacteria. The dichotomy into a high-GC branch (left) and a low-GC branch (right) is based on ribosomal RNA similarity coefficients (S_{AB} values) [59,67]. (Dendrogram courtesy of E. Stackebrandt, modified.)

5. CONCLUDING REMARKS—RIBONUCLEOTIDE REDUCTION: UNIVERSAL BUT NOT UNIFORM

One might expect that one of the truly universal processes of life, DNA synthesis, proceeds in an almost identical fashion in all types of organisms. This is obviously true for the mechanism but not for the components involved in the formation of DNA precursors. Chemistry dictates that the transition from RNA, likely the most primeval genetic material, to DNA had to occur on the level of nucleotides rather than sugars because free 2-deoxyribose has not been available [64]. The radical mechanism for replacement of the C2′ hydroxyl group in a ribonucleotide by hydrogen (Scheme 1) is probably the only one that can function in an aqueous medium with nothing but protein-associated functional groups for catalysis. To date there exists no nonenzymatic model of the physiological process but it may not be fortuitous that the only reasonable synthetic route from ribo- to 2′-deoxyribonucleosides involves radical chemistry, i.e., use of tributyltin hydride in hot toluene. Modeling ribonucleotide reduction in aqueous solution is overdue.

Reduction of ribonucleoside 5′-diphosphates or triphosphates, regardless of base structure, and allosteric interaction with other nucleotides occurs at the R1 enzyme (subunit) of all ribonucleotide reductases. The universally conserved substrate–effector nucleotide specificities (ADP/dGTP, GDP/dTTP, etc.) suggest that they have a common structural origin. Nucleotide binding domains are, in fact, considered to be among the most ancient protein families.

The variable component of ribonucleotide reduction, then, is the radical protein or cofactor required for ribose deoxygenation, and the mode of generation of the radical species. Unfortunately, the nature of the atom or substituent which specifically abstracts the hydrogen atom from ribose C3′ of a substrate in the first place (the "working radical" [65]; $X^\bullet$ in Scheme 1) is not known in any of the four classes of enzyme. For example, it cannot be identical with the tyrosyl radical in the diiron-R2 subunits because that residue is buried in the protein [8]. An attractive new model has been put forth in which $X^\bullet$ is the thiyl radical of a cysteine residue in R1 [57]. Although this model is derived from studies of the cysteines present in *E. coli* ribonucleotide reductase we hypothesize that it can be applied to all the different enzymes in which similar thiyl working radicals would be generated by electron transfer between thiols and the observable organic free radicals, or directly from an oxidized transition metal center like in Mn peroxidase [56].

Is there an explanation for the multiplicity of radical-generating cofactors and metalloproteins in the enzymatic catalysis of DNA precursor synthesis? The ribonucleotide reductase apparatus must have arisen polyphyletically, combining R1 (likely of common origin) with various small cofactors or transition metal–polypeptide complexes which were available and capable of generating and supporting radical reactions. It cannot be overlooked (cf. Fig. 1) that anaerobic cells utilize the simpler cofactors, probably closer to abiotic chemistry, whereas later, aerobic prokaryotes and eukaryotes have harnessed for ribonucleotide reduction the two most versatile transition metals of oxygen biochemistry: manganese and iron. All these details corroborate the view

[66] that catalytic deoxyribonucleotide formation has been one of the major checkpoints, or even *the* checkpoint, in the early evolution of DNA genomes, and again on the route from anaerobic to aerobic life.

ABBREVIATIONS

ADP	adenosine 5′-diphosphate
ATP	adenosine 5′-triphosphate
CDP	cytidine 5′-diphosphate
CoB_{12}	5′-deoxyadenosylcobalamin
dATP	2′-deoxyadenosine 5′-triphosphate
DEAE	diethylaminoethyl
dGTP	2′-deoxyguanosine 5′-triphosphate
dTTP	thymidine 5′-triphosphate
EDTA	ethylenediamine-N,N,N′,N′-tetraacetic acid
ESR or EPR	electron spin or paramagnetic resonance
EXAFS	extended X-ray absorption fine structure
FAD	flavin adenine dinucleotide
GC	guanine + cytosine content of DNA
GDP	guanosine 5′-diphosphate
GTP	guanosine 5′-triphosphate
HU	hydroxyurea
ICP	inductively coupled plasma
NAD	nicotinamide adenine dinucleotide
NADPH	nicotinamide adenine dinucleotide phosphate, reduced
R1	ribonucleotide reductase subunit 1
R2	ribonucleotide reductase subunit 2
SAM	S-adenosylmethionine
SDS	sodium dodecyl sulfate
TCA	trichloroacetic acid
UDP	uridine 5′-diphosphate

REFERENCES

1. G. Schimpff-Weiland, H. Follmann, and G. Auling, *Biochem. Biophys. Res. Commun., 102*, 1276 (1981).

2. P. Reichard and A. Ehrenberg, *Science*, *221*, 514 (1983).
3. M. Lammers and H. Follmann, *Struc. Bond.*, *54*, 27 (1983).
4. J. A. Stubbe, *J. Biol. Chem.*, *265*, 5329 (1990).
5. H. Follmann and H. P. C. Hogenkamp, *Biochemistry*, *8*, 4372 (1969).
6. J. A. Stubbe and M. Ator, *J. Biol. Chem.*, *258*, 1625 (1983).
7. H. P. C. Hogenkamp, *Pharmacol. Ther.*, *23*, 393, 1984.
8. P. Nordlund, B. M. Sjöberg, and H. Eklund, *Nature*, *345*, 593 (1990).
9. J. R. Cowles, H. J. Evans, and S. A. Russell, *J. Bacteriol.*, *97*, 1460 (1969).
10. A. Willing, H. Follmann, and G. Auling, *Eur. J. Biochem.*, *170*, 603 (1988).
11. A. Willing, H. Follmann, and G. Auling, *Eur. J. Biochem.*, *175*, 167 (1988).
12. H. P. C. Hogenkamp, H. Follmann, and R. Thauer, *FEBS Lett.*, *219*, 197 (1987).
13. R. Eliasson, M. Fontecave, H. Jörnvall, M. Krook, E. Pontis, and P. Reichard, *Proc. Natl. Acad. Sci. USA*, *87*, 3314 (1990).
14. I. Sze, S. C. McFarlan, H. P. C. Hogenkamp, and H. Follmann, *Biochem. Biophys. Res. Commun.*, *184*, 1101 (1992).
15. J. Harder and H. Follmann, *Free Rad. Res. Commun.*, *10*, 281 (1990).
16. L. Sun and J. A. Fuchs, *Mol. Biol. Cell*, *3*, 1095 (1992).
17. G. Auling, M. Thaler, and H. Diekmann, *Arch. Microbiol.*, *127*, 105 (1980).
18. G. Auling, *Eur. J. Appl. Microbiol.*, 6, 379 (1979).
19. J. Plönzing and G. Auling, *Arch. Microbiol.*, *146*, 396 (1987).
20. J. Schmid and G. Auling, *Agric. Biol. Chem.*, *53*, 1783 (1989).
21. A. Furuya, S. Abe, and S. Kinoshita, *Agric. Biol. Chem.*, *34*, 210 (1970).
22. M. Thaler and H. Diekmann, *Eur. J. Appl. Microbiol.*, *6*, 379 (1979).
23. C. Seehaus, H. Siede, and G. Auling, *FEMS Lett.*, *66*, 281 (1990).
24. T. Oka, K. Udagawa, and S. Kinoshita, *J. Bacteriol.*, *96*, 1760 (1968).
25. A. Furuya, in *Microbial Production of Nucleic Acid-Related Substances* (K. Ogata, S. Kinoshita, T. Tsunoda, and K. Aida, eds.), Kodansha, Tokyo, and Wiley, New York, 1976, p. 125.

26. K. Ogata, S. Kinoshita, T. Tsunoda, and K. Aida (eds.), *Microbial Production of Nucleic Acid-Related Substances*, Kodansha, Tokyo, and Wiley, New York, 1976.
27. B. P. Rosen and S. Silver (eds.), *Ion Transport in Prokaryotes*, Academic Press, New York, 1987.
28. G. Auling, in *Metal Ion Homeostasis: Molecular Biology and Chemistry* (D. Winge and D. Hamer, eds.), Alan R. Liss, New York, 1989, p. 459.
29. J. Schmid and G. Auling, *J. Bacteriol.*, *169*, 3385 (1987).
30. J. Schmid, Ph.D. thesis, University of Hannover, 1985.
31. G. Auling, "Habilitationsschrift," Thesis, University of Hannover, 1980.
32. D. Filpula and J. A. Fuchs, *J. Bacteriol.*, *139*, 694 (1979).
33. B. K. Rima and I. Takahashi, *J. Gen. Microbiol.*, *107*, 139 (1978).
34. G. Auling, H. Prelle, and H. Diekmann, *Eur. J. Biochem.*, *121*, 365 (1982).
35. G. Auling and B. Moss, *J. Bacteriol.*, *158*, 733 (1984).
36. H. Saito, H. Tomioka, and S. Ohkido, *J. Gen. Microbiol.*, *131*, 3091 (1985).
37. C. Luo and G. Auling, *BioEngineering*, *9*, 85 (1993).
38. B. M. Sjöberg, S. Hahne, M. Karlsson, H. Jörnvall, M. Göransson, and B. Uhlin, *J. Biol. Chem.*, *261*, 5658 (1986).
39. G. J. Mann, A. Gräslund, E. I. Ochiai, R. Ingemarson, and L. Thelander, *Biochemistry*, *30*, 1939 (1991).
40. A. Platz and B. M. Sjöberg, *J. Bacteriol.*, *143*, 561 (1980).
41. M. Atta, P. Nordlund, A. Aberg, H. Eklund, and M. Fontecave, *J. Biol. Chem.*, *267*, 20682 (1992).
42. M. Meng and H. P. C. Hogenkamp, *J. Biol. Chem.*, *256*, 9174 (1981).
43. J. O. Höög, H. Jörnvall, A. Holmgren, M. Carlquist, and M. Persson, *Eur. J. Biochem.*, *136*, 223 (1983).
44. R. C. Fahey, W. C. Brown, W. B. Adams, and M. M. Worshaw, *J. Bacteriol.*, *133*, 1126 (1968).
45. O. Berglund and B. M. Sjöberg, *J. Biol. Chem.*, *254*, 253 (1979).
46. G. Lassmann, L. Thelander, and A. Gräslund, *Biochem. Biophys. Res. Commun.*, *188*, 879 (1992).
47. H. Baumann, R. Hofmann, M. Lammers, G. Schimpff-Weiland, and H. Follmann, *Z. Naturforsch. 39c*, 276 (1984).

48. R. Hofmann and H. Follmann, *Z. Naturforsch.*, *40c*, 919 (1985).
49. Y. Kono and I. Fridovich, *J. Biol. Chem.*, *258*, 6015 (1982).
50. K. Wieghardt, U. Bossek, D. Ventur, and J. Weiss, *Chem. Commun.* 347 (1985).
51. J. E. Sheats, R. S. Czernuszewicz, G. C. Dismukes, A. L. Rheingold, V. Petrouleas, J. A. Stubbe, W. H. Armstrong, R. H. Beer, and S. J. Lippard, *J. Am. Chem. Soc.*, *109*, 1435 (1987).
52. R. M. Fronko and J. E. Penner-Hahn, *J. Am. Chem. Soc.*, *110*, 7554 (1988).
53. K. Wieghardt, *Angew. Chem. 101*, 1179 (1989); *Angew. Chem. Intl. Ed. Engl.*, *28*, 1153 (1989).
54. J. M. Bollinger Jr., D. E. Edmondson, B. H. Huynh, J. Filley, J. R. Norton, and J. Stubbe, *Science*, *253*, 292 (1991).
55. B. B. Snider and B. O. Buckman, *Tetrahedron*, *45*, 6969 (1989).
56. H. Wariishi, K. Valli, V. Renganathan, and M. H. Gold, *J. Biol. Chem.*, *264*, 14185 (1989).
57. S. S. Mao, T. P. Holler, G. X. Yu, Jr., S. Booker, M. I. Johnston, and J. Stubbe, *Biochemistry*, *31*, 9733 (1992).
58. D. M. Webley, R. B. Duff, and G. Anderson, *J. Gen. Microbiol.*, *29*, 179 (1962).
59. C. R. Woese, *Microbiol. Rev.*, *51*, 221 (1987).
60. F. K. Gleason and H. P. C. Hogenkamp, *Biochim. Biophys. Acta*, *277*, 466 (1972).
61. P. K. Tsai and H. P. C. Hogenkamp, *J. Biol. Chem.*, *255*, 1273 (1980).
62. H. W. van Verseveld and A. H. Stouthamer, in *The Prokaryotes—A Handbook on the Biology of Bacteria: Ecophysiology, Isolation, Identification, Application*, 2nd ed. Vol. 3 (A. Balows, H. G. Trüper, M. Dworkin, W. Harder, and K.-H. Schleifer, eds.), Springer-Verlag, New York, 1992, p. 2321.
63. G. Auling, in *Metal Ions in Fungi* (G. Winkelmann and D. Winge, eds.), Marcel Dekker, New York, 1993.
64. H. Follmann, *Adv. Space Res.*, *6*, 33 (1986).
65. S. P. Ong, L. S. Nelson, and H. P. C. Hogenkamp, *Biochemistry*, *31*, 11210 (1992).
66. H. Follmann, *Naturwissenschaften*, *69*, 75–81 (1982).
67. E. Stackebrandt, B. J. Lewis, and C. R. Woese, *Zbl. Bakt. Hyg., I. Abt. Orig.*, *C1*, 137 (1980).

6
Prostaglandin Endoperoxide Synthases

William L. Smith[1] and Lawrence J. Marnett[2]

[1]Department of Biochemistry, Michigan State University, Biochemistry Building, Room 510, East Lansing, MI 48824-1319 USA

[2]A.B. Hancock Jr. Memorial Laboratory for Cancer Research, Center in Molecular Toxicology, Departments of Biochemistry and Chemistry, Vanderbilt University School of Medicine, Nashville, TN 37232 USA

1. INTRODUCTION

1.1. Eicosanoids—An Overview

Eicosanoids are oxygenated fatty acids which are synthesized from 18-, 20-, and 22-carbon members of the n-6 and n-3 families of essential polyunsaturated fatty acids (Fig. 1) [1–9]. Many of these products exhibit autocrine and/or paracrine activities. The major precursor of the eicosanoids is arachidonic acid [20:4 (n-6)]. There are three pathways for the formation of eicosanoids from arachidonate including the cyclooxygenase, lipoxygenase, and epoxygenase pathways (Fig. 1); collectively, these pathways are known as the *arachidonate cascade*. Several cogent reviews describing the various lipoxygenase [2–6] and epoxygenase [7–9] pathways are available. In this chapter, we will focus on the cyclooxygenase pathway and, in particular, on the prostaglandin H (PGH) synthases which catalyze the committed step in the production of prostanoids. Prostanoids are defined here to include the prostaglandins (PGs) and thromboxanes (TXs).

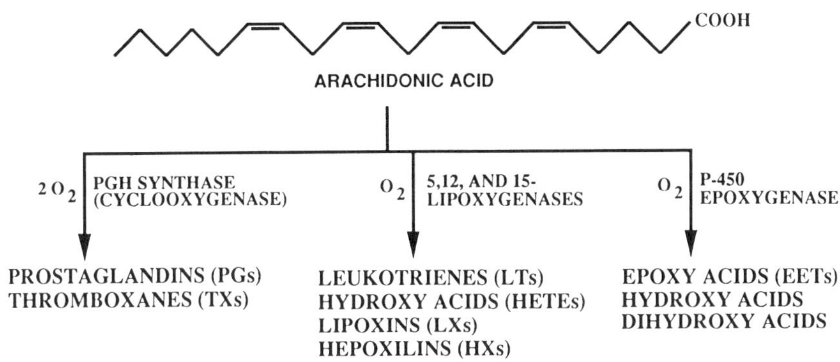

FIG. 1. Cyclooxygenase, lipoxygenase, and epoxygenase pathways for the formation of various eicosanoids from arachidonic acid.

1.2. Prostanoid Biosynthesis and Mechanisms of Action

Figure 2 outlines the biosynthetic pathway for the formation of prostanoids in a model cell [1,6]. Prostanoid biosynthesis is typically initiated by a circulating hormone interacting with its cell surface receptor. For example, both bradykinin and arginine vasopressin act on renal collecting tubule epithelia to stimulate prostaglandin E_2 (PGE_2) formation [10,11]. Hormone-induced prostanoid formation results from a net increase in the concentration of arachidonate within the cell either through the activation of an intracellular lipase or through an increased uptake of arachidonate from the surroundings (see Ref. 6 and 12 for recent reviews). Increases in the concentration of arachidonate to levels which begin to saturate PGH synthase permit this enzyme to catalyze its bis oxygenation reaction and lead to production of prostaglandin endoperoxide H_2 (PGH_2). Newly formed PGH_2 is then acted on by a synthase or by a reductase which catalyzes the production of a major biologically active endproduct [13]. Although all the major metabolic pathways for PGH_2 metabolism are shown in Fig. 2, a prostanoid-forming cell usually forms only one major prostanoid product. For example, platelet cells form almost exclusively thromboxane A_2 (TXA_2), endothelial cells form principally PGI_2 (prostacyclin), and renal collecting tubule epithelia form mainly PGE_2 [1].

Newly formed prostanoids exit the cell probably via carrier-

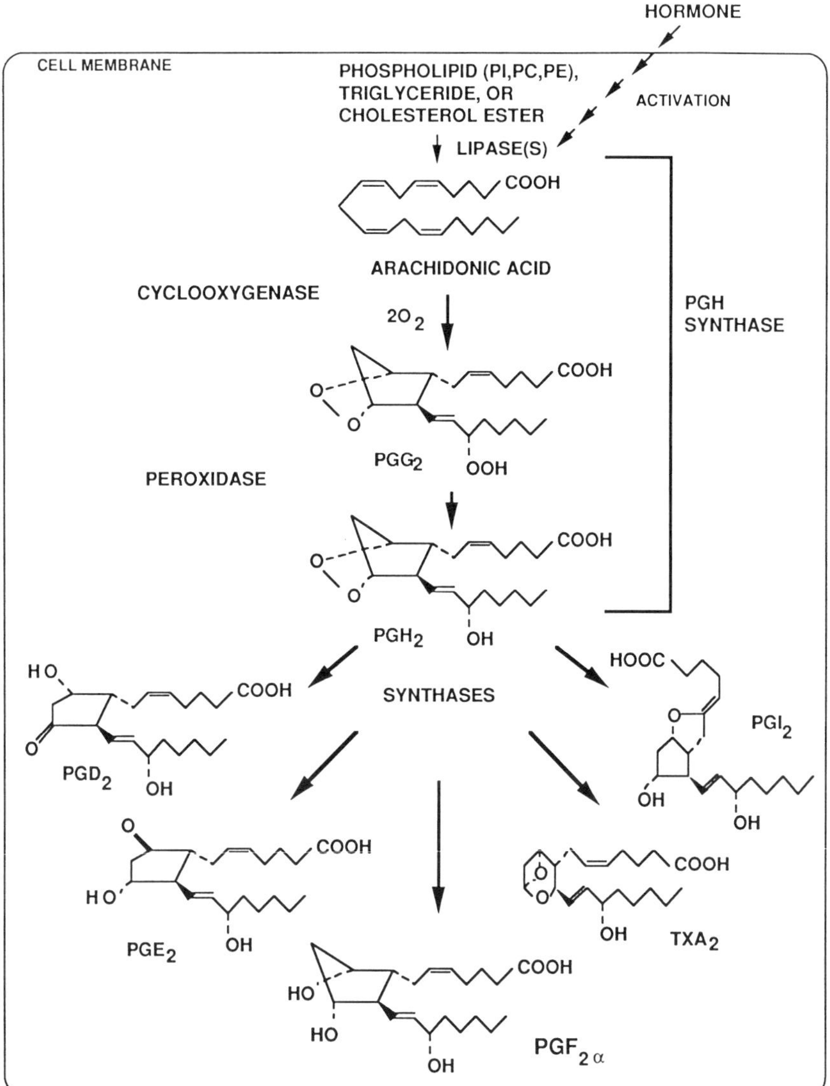

FIG. 2. Prostanoid biosynthetic pathway. (Adapted from Ref. 12.)

mediated diffusion [6]. These prostanoids then act through G-protein–linked receptors on the parent cell (autocrine action) and on adjacent cells having prostanoid receptors (paracrine action) to elicit responses (see Ref. 1, 12, and 13 for recent reviews). Prostanoids are rendered inactive by a combination of dilution in the bloodstream and active catabolism. Prostanoid catabolic enzymes are found mainly in the lung, kidney, and liver (see Ref. 13 for review). Accordingly, prostanoids are viewed as "local hormones" that act at or near their sites of biosynthesis. Prostanoids probably function mainly to coordinate multicellular responses to circulating hormones which stimulate prostanoid synthesis. For example, TXA_2 formed by platelets in response to thrombin acts on platelet cells to cause their adherence to one another and to the subendothelial layer of the arterial vasculature and on vascular smooth muscle cells to stimulate vasoconstriction; PGE_2 formed by renal collecting tubule cells in response to arginine vasopressin acts on the collecting tubule to inhibit water reabsorption and on adjacent thick limb epithelia of Henle's loop to inhibit Na^+ reabsorption [1,11]. More detailed reviews describing the biochemical mechanisms of actions of prostanoids including a discussion of prostanoid receptors are available [1,6,12].

2. PGH SYNTHASE ISOZYMES

2.1. Structure of PGH Synthase Isozymes

There are two PGH synthase isozymes called PGH synthase 1 and 2. Both enzymes exhibit the same two catalytic activities (Fig. 2): (1) a cyclooxygenase which catalyzes a bis oxygenation of arachidonate producing PGG_2 and (2) a peroxidase which catalyzes a net two electron reduction of the 15-hydroperoxyl group of PGG_2 to form PGH_2 [14–19]. Figure 3 compares the deduced amino acid sequences of various PGH synthases 1 and 2. PGH synthase 1 is the enzyme originally purified from ovine and bovine vesicular glands [14,15,26]. cDNAs encoding murine [20], human [23,24], and ovine [27–29] PGH synthase 1 have been cloned. PGH synthase 1 contains a signal peptide which, when cleaved (between amino acid residues 26 and 27 in the case of murine PGH synthase 1), yields a mature protein containing 576 amino acids and having a molecular weight of about 65500 [20]. The deduced amino

```
Chick-2   MLLPCALLAALL------------------AAGHAANPCCSLPCQNRGECMTTGFDRYECD
Human-2   MLARALLLCAVL------------------ALSHTANPCCSHPCQNRGVCMSVGFDQYKCD
Mouse-2   MLFRAVLLCAAL------------------GLSQAANPCCSNPCQNRGECMSTGFDQYKCD
Mouse-1   MSRRSLSLWFPLLLLLLLPPTPSVLLADPGVPSPVNPCCYYPCQNQGVCVRFGLDNYQCD
Human-1   MSR-SLLLRFLLLLLLL-PPLP-VLLADPGAPTPVNPCCYYPCQHQGICVRFGLDRYQCD
Sheep-1   MSRQSISLRFPLLLLLL-SPSP-VFSADPGAPAPVNPCCYYPCQHQGICVRFGLDRYQCD
                    10        20        30        40        50

Chick-2   CTRTGYYGENCTTPEFFTWLKLLIKPTPNTVHYILTHFKGVWNIINNSPFLRDTIMRYVL
Human-2   CTRTGFYGENCSTPEFLTRIKFLLKPTPNTVHYILTHFKGFWNVVNNIPFLRNAIMSYVL
Mouse-2   CTRTGFYGENCTTPEFLTRLKLLLKPTPNTVHYILTHFKGVWNIVNNIPFLRSLTMKYVL
Mouse-1   CTRTGYSGPNCTIPEIWTWLRNSLRPSPSFTHFLLTHGYWLWEFVNAT-FIREVLMRLVL
Human-1   CTRTGYSGPNCTIPGLWTWLRNSLRPSPSFTHFLLTHGRWFWEFVNAT-FIREMLMLLVL
Sheep-1   CTRTGYSGPNCTIPEIWTWLRTTLRPSPSFIHFLLTHGRWLWDFVNAT-FIRDTLMRLVL
          60        70        80        90        100       110

Chick-2   TSRSHLIDSPPTYNSDYSYKSWEAYSNLSYYTRSLPPVGHDCPTPMGVKGKKELPDSKLI
Human-2   TSRSHLIDSPPTYNADYGYKSWEAFSNLSYYTRALPPVPDDCPTPLGVKGKKQLPDSNEI
Mouse-2   TSRSYLIDSPPTYNVHYGYKSWEAFSNLSYYTRALPPVADDCPTPMGVKGNKELPDSKEV
Mouse-1   TVRSNLIPSPPTYNSAHDYISWESFSNVSYYTRILPSVPKDCPTPMGTKGKKQLPDVQLL
Human-1   TVRSNLIPSPPTYNSAHDYISWESFSNVSYYTRILPSVPKDCPTPMGTKGKKQLPDAQLL
Sheep-1   TVRSNLIPSPPTYNIAHDYISWESFSNVSYYTRILPSVPRDCPTPMGTKGKKQLPDAEFL
           120       130       140       150       160       170

Chick-2   VEKFLLRRKFIPDPQGTNVMFTFFAQHFTHQFFKTDHKRGPGFTKAYGHGVDLNHIYGET
Human-2   VGKLLLRRKFIPDPQGSNMMFAFFAQHFTHQFFKTDHKRGPAFTNGLGHGVDLNHIYGET
Mouse-2   LEKVLLRREFIPDPQGSNMMFAFFAQHFTHQFFKTDHKRGPGFTRGLGHGVDLNHIYGET
Mouse-1   AQQLLLRREFIPAPQGTNILFAFFAQHFTHQFFKTSGKMGPGFTKALGHGVDLGHIYGDN
Human-1   ARRFLLRRKFIPDPQGTNLMFAFFAQHFTHQFFKTSGKMGPGFTKALGHGVDLGHIYGDN
Sheep-1   SRRELLRRKFIPDPQGTNLMFAFFAQHFTHQFFKTSGKMGPGFTKALGHGVDLGHIYGDN
           180       190       200       210       220       230

Chick-2   LERQLKLRLRKDGKLKYQMIDGEMYPPTVKDTQAEMIYPPHVPEHLQFSVGQEVFGLVPG
Human-2   LARQRKIRLFKDGKMKYQIIDGEMYPPTVKDTQAEMIYPPQVPEHLRFAVGQEVFGLVPG
Mouse-2   LDRQHKLRLFKDGKLKYQVIGGEVYPPTVKDTQVEMIYPPHIPENLQFAVGQEVFGLVPG
Mouse-1   LERQYHLRLFKDGKLKYQVLDGEVYPPSVEQASVLMRYPPGVPPERQMAVGQEVFGLLPG
Human-1   LERQYQLRLFKDGKLKYQVLDGEMYPPSVEEAPVLMHYPRGIPPQSQMAVGQEVFGLLPG
Sheep-1   LERQYQLRLFKDGKLKYQMLNGEVYPPSVEEAPVLMHYPRGIPPQSQMAVGQEVFGLLPG
           240       250       260       270       280       290

Chick-2   LMMYATIWLREHNRVCDVLKQEHPEWDDEQLFQTTRLILIGETIKIVIDDYVQHLSGYHF
Human-2   LMMYATIWLREHNRVCDVLKQEHPEWGDEQLFQTSRLILIGETIKIVIDDYVQHLSGYHF
Mouse-2   LMMYATIWLREHNRVCDILKQEHPEWGDEQLFQTSRLILIGETIKIVIDDYVQHLSGYHF
Mouse-1   LMLFSTIWLREHNRVCDLLKEEHPTWDDEQLFQTTRLILIGETIKIVIEEYVQHLSGYFL
Human-1   LMLYATLWLREHNRVCDLLKAEHPTWGDEQLFQTTRLILIGETIKIVIEEYVQQLSGYFL
Sheep-1   LMLYATIWLREHNRVCDLLKAEHPTWGDEQLFQTARLILIGETIKIVIEEYVQQLSGYFL
           300       310       320       330       340       350

Chick-2   KLKFDPELLFNQRFQYQNRIAAEFNTLYHWHPLLPDTFQIHNQEYTFQQFLYNNSIMLEH
Human-2   KLKFDPELLFNKQFQYQNRIAAEFNTLYHWHPLLPDTFQINDQKYNYQQFIYNNSILLEH
Mouse-2   KLKFDPELLFNQQFQYQNRIASEFNTLYHWHPLLPDTFNIEDQEYSFKQFLYNNSILLEH
Mouse-1   QLKFDPELLFRAQFQYRNRIAMEFNHLYHWHPLMPNSFQVGSQEYSYEQFLFNTSMLVDY
Human-1   QLKFDPELLFGVQFQYRNRIATEFNHLYHWHPLMPDSFKVGSQEYSYEQFLFNTSMLVDY
Sheep-1   QLKFDPELLFGAQFQYRNRIAMEFNQLYHWHPLMPDSFRVGPQDYSYEQFLFNTSMLVDY
           360       370       380       390       400       410

Chick-2   GLSHMVKSSKRQIAGRVAGGKNVPAAVQKVAKASIDQSRQMRYQSLNEYRKRFMLKPFKS
Human-2   GITQFVESFTRQIAGRVAGGRNVPPAVQKVSQASIDQSRQMKYQSFNEYRKRFMLKPYES
Mouse-2   GLTQFVESFTRQIAGRVAGGRNVPIAVQAVAKASIDQSREMKYQSLNEYRKRFSLKPYTS
Mouse-1   GVEALVDAFSRQRAGRIGGGRNFDYHVLHVAVDVIKESREMRLQPFNEYRKRFGLKPYTS
Human-1   GVEALVDAFSRQIAGRIGGGRNMDHHILHVAVDVIRESREMRLQPFNEYRKRFGMKPYTS
Sheep-1   GVEALVDAFSRQPAGRIGGGRNIDHHILHVAVDVIKESRVLRLQPFNEYRKRFGMKPYTS
           420       430       440       450       460       470

Chick-2   FEELTGEKEMAAELEELYGDIDAMELYPGLLVEKPRPGAIFGETMVEIGAPFSLKGLMGN
Human-2   FEELTGEKEMSAELEALYGDIDAVELYPALLVEKPRPDAIFGETMVEVGAPFSLKGLMGN
Mouse-2   FEELTGEKEMAAELKALYSDIDVMELYPALLVEKPRPDAIFGETMVELGAPFSLKGLMGN
Mouse-1   FQELTGEKEMAAELEELYGDIDALEFYPGLLLEKCQPNSIFGESMIEMGAPFSLKGLLGN
Human-1   FQELVGEKEMAAELEELYGDIDALEFYPGLLLEKCHPNSIFGESMIEIGAPFSLKGLLGN
Sheep-1   FQELTGEKEMAAELEELYGDIDALEFYPGLLLEKCHPNSIFGESMIEMGAPFSLKGLLGN
           480       490       500       510       520       530

Chick-2   TICSPEYWKPSTFGGKVGFEIINTASLQSLICNNVKGSPFTAFHVLNPEPTETATINVST
Human-2   VICSPAYWKPSTFGGEVGFQIINTASIQSLICNNVKGCPFTSFSVPDPELIKTVTINASS
Mouse-2   PICSPQYWKPSTFGGEVGFKIINTASIQSLICNNVKGCPFTSFNVQDPQPTKTATINASA
Mouse-1   PICSPEYWKPSTFGGDVGFNLVNTASLKKLVCLNTKTCPYVSFRVPDYPGDDGSVIV---
Human-1   PICSPEYWKPSTFGGEVGFNIVKTATLKKLVCLNTKTCPYVSFRVPDASQDDGPAVE---
Sheep-1   PICSPEYWKASTFGGEVGFNLVKTATLKKLVCLNTKTCPYVSFHVPDPRQEDRPGVE---
           540       550       560       570       580       590

Chick-2   SNFAMEDINPTLLLKEQSAEL
Human-2   SRSGLDDINPTVLLKERSTEL
Mouse-2   SHSRLDDINPTVLIKRRSTEL
Mouse-1   ---------------RRSTEL
Human-1   ---------------RPSTEL
Sheep-1   ---------------RPPTEL
                            600
```

acid sequences of PGH synthases 1 from sheep, human, and mouse are about 90% identical at the amino acid level, with the major differences being found in the signal peptides and the 12 amino acids immediately preceding a putative endoplasmic reticulum (ER) retention signal [(Pro/Ser)ThrGluLeu, or (P/S)TEL] at the C terminus.

PGH synthase 2 was originally isolated as a v-*src*-inducible gene product from chicken fibroblasts [22,30] and as a phorbol ester–inducible immediate early gene product called TIS10 from murine 3T3 cells [21]. PGH synthase 2 has not been purified and what is currently known about the enzyme comes from characterization and expression of its cDNA in vitro. There is about 60% amino acid identity between the deduced amino acid sequences of PGH synthases 1 and 2 from the same species (Fig. 3). PGH synthase 2 differs significantly from PGH synthase 1 at positions prior to amino acid residue 30; mature PGH synthase 2, which has undergone processing in the ER, has the N-terminal sequence ANPCC corresponding to cleavage of a 17-amino-acid signal peptide from the N terminus of the deduced sequence (Fig. 3; [18]). In addition, PGH synthase 2 contains an 18-amino-acid insert near the C terminus of the enzyme; this insert is absent from PGH synthase 1; antibodies prepared against this peptide insert are specific for PGH synthase 2 [31,32]. Excluding sequences near the N and C termini, the sequences of PGH synthases 1 and 2 are about 75% identical. As will be discussed in more detail below, those residues which have been identified as important in the catalytic functions of PGH synthase 1 are all conserved in PGH synthase 2 (Fig. 3). The C-terminal (P/S)TEL sequence is also conserved in both PGH synthase isozymes. This latter sequence may be at least partly responsible for retaining the two proteins in the ER [31–34].

PGH synthase 1 is present in numerous cell lines and in extracts of almost all mammalian tissues [35]. In contrast, PGH synthase 2 is

FIG. 3. Sequence comparison of various PGH synthases 1 and 2. Included are chicken PGH synthase 2 [22], murine PGH synthase 2 [21,22], human PGH synthase 2 [25], murine PGH synthase 1 [20], human PGH synthase 1 [23,24], and ovine PGH synthase 1 [27–29]. Numbering is for ovine PGH synthase 1 and begins with the methionine at the translation start site. Shown in bold letters are the signal peptides, asparagine residues which are N-glycosylated, catalytically essential histidine and tyrosine residues, the serine residue which is acetylated by aspirin, and the characteristic 18-amino-acid insert close to the C termini of PGH synthase 2.

present in prostate, brain, testis, and lung, but is absent from most other major organs [22]. PGH synthase 2 is found in RAT1 cells [21,22], mitogen-stimulated Swiss mouse 3T3 fibroblasts [21,22,36,37] and C127 mouse fibroblasts [38], rat granulosa cells [18], human endothelial cells [25], and activated macrophages [39–41], but absent from many common cell lines [21,22].

2.2. Expression of PGH Synthase 1 and 2 Genes

The mRNA for PGH synthase 1 is about 2.8 kb and that for PGH synthase 2 about 4.5 kb with minor variations between species [21–43]. In Swiss mouse 3T3 cells, mRNA levels for both PGH synthase 1 and 2 genes are increased by serum, growth factors, phorbol esters, and cAMP [21,22,36,42]. However, PGH synthase 2 mRNA levels are upregulated more dramatically than those of PGH synthase 1. Furthermore, the time course for upregulation is different for the two mRNAs. In serum-stimulated quiescent 3T3 cells, PGH synthase 2 mRNA rises from undetectable levels to peak at about 2 hr poststimulation and then decreases to baseline levels within 8 hr [17,22,36,42]. PGH synthase 1 mRNA levels increase only two- to threefold from basal levels peaking about 4 hr postserum stimulation, and remain elevated for 8 hr [36,42]. Nuclear run-on assays indicate that serum causes increases in the rates of transcription of both genes [42]; however, as determined by Western blotting only the level of PGH synthase 2 protein increases.

Preincubation of 3T3 cells with antiinflammatory glucocorticoids such as dexamethasone attenuates the serum-induced increases in the levels of mRNAs for both PGH synthase 1 and 2 [17,36,42]; this inhibitory effect is also observed in nuclear run-off assays suggesting that glucocorticoids can inhibit the transcription of the genes for both PGH synthase 1 and 2 [42]. Another effect of dexamethasone seems to be to destabilize the mRNA for PGH synthase 2 [22]; this latter effect is observed when serum-starved 3T3 cells are treated simultaneously with dexamethasone and serum.

Although induction of PGH synthase isozymes has been studied most extensively in 3T3 cells, there is evidence for regulation of PGH synthase 1 and 2 expression in other cell types. Interleukin-1α (IL-1α) stimulates but acidic fibroblast growth factor 1 inhibits the expression of PGH synthase 1 in cultured endothelial cells derived from human umbilical veins [43]. Human chorionic gonadotropin stimulates the expression of PGH synthase 2 in rat ovarian follicles; stimulation of PGH synthase 2 in follicles occurs in association with ovulation [18].

2.3. Genes for PGH Synthases 1 and 2

The intron/exon structures of the murine [44] and human [23] PGH synthase 1 genes and the murine PGH synthase 2 gene [17] are shown in Fig. 4. The PGH synthase 1 gene is 22 kb and has 11 exons. The intron/exon structures of the human and murine PGH synthase 1 genes are almost identical [23,44]. The murine PGH synthase 2 gene is considerably smaller (~8 kb), characteristic of immediate early genes, but, except for the first exon, the locations of the intron/exon boundaries are the same [17].

The nucleotide sequences of the 5′ end of the genes for PGH synthases 1 and 2 have been determined [17,44]; the transcriptional start sites for both genes have also been defined [17,44]. There are numerous consensus response elements present in the 5′-flanking regions of each gene. For example, there are three AP-1-like elements (for the binding of *fos/jun* dimers) in the 5′-flanking region of the PGH synthase 1 gene; there is also a putative negative glucocorticoid response element (nGRE) neighboring an AP-1 site. The functional relevance of these putative response elements has not been determined. A region containing approximately 1 kb upstream of the transcriptional start site of murine PGH synthase 2 has been sequenced and shown to contain AP-1 and SP-1 sites and a putative TATA box. When this upstream regulatory region of the PGH synthase 2 gene is coupled to a luciferase reporter gene and transfected into NIH 3T3 cells, luciferase activity is induced in response to both serum and phorbol ester treatments [17]. Curiously, although it has been established that serum will induce the expression of PGH synthase 2, there are no obvious serum response elements (SREs) near the 5′ end of the gene.

3. CYCLOOXYGENASE CATALYSIS

3.1. Cyclooxygenase Substrate Specificities

As noted earlier, PGH synthases exhibit both cyclooxygenase and peroxidase activities converting arachidonate to PGG_2 and PGG_2 to PGH_2,

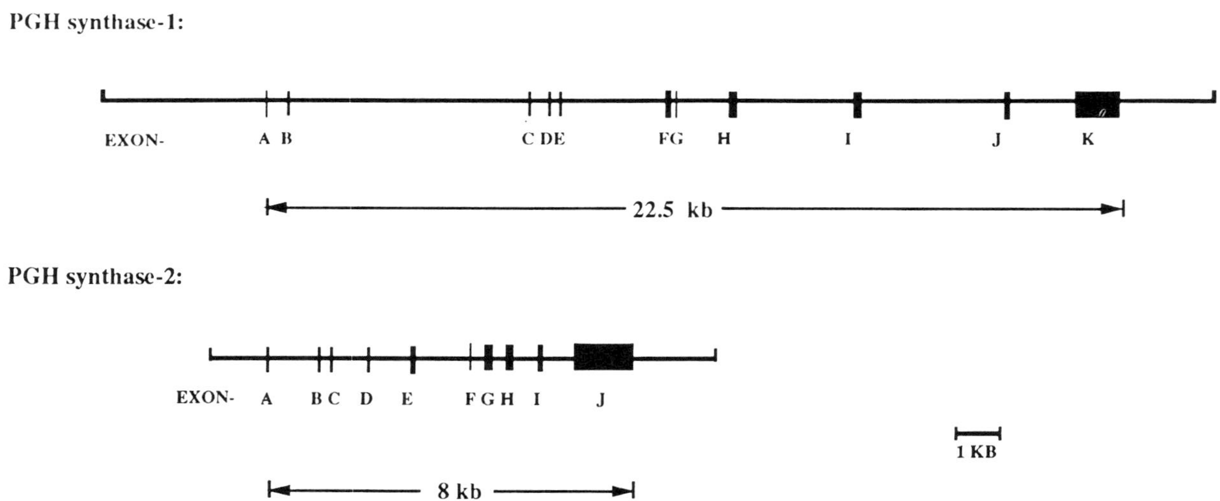

FIG. 4. Comparison of the structures of the genes encoding murine PGH synthases 1 and 2. Exons are denoted with letters. PGH synthase 1 and 2 gene structures are from Kraemer et al. [44] and Fletcher et al. [17], respectively.

respectively (Fig. 2). The cyclooxygenase activities of PGH synthases 1 and 2 utilize arachidonate [20:4 (n-6)] and dihomo-γ-linolenic acid [20:3 (n-6)] most efficiently as substrates ($K_m \sim 5$ μM; $V_{max} \sim 150$ μmol O_2/min/mg; [45–47]). PGH synthase 2 catalyzes the oxygenation of eicosapentaenoic acid, a member of the n-3 family of essential fatty acids, at about one fourth the rate of arachidonate; however, eicosapentaenoic acid is oxygenated by PGH synthase 1 at less than 2% of the rate of arachidonate [46]. A variety of other common fatty acids having different combinations of double-bond positions and orientations have been tested as substrates and inhibitors for the cyclooxygenase activity of PGH synthase 1. In general, n-3 and n-9 polyunsaturated fatty acids containing 18–22 carbons are poor substrates but efficient cyclooxygenase inhibitors [45].

Three fatty acids are noteworthy with regard to enzyme inhibition. A polyacetylenic fatty acid, eicosatetraynoic acid, known as ETYA, is an excellent inhibitor of PGH synthase 1. In addition to its ability to bind the cyclooxygenase active site and act as an instantaneous competitive inhibitor, ETYA can also function as a suicide substrate but only in the presence of O_2 [48]. Docosahexaenoic acid (22:6 (n-3); [45,49,50]) and 5c,8c,12t,15c-eicosatetraenoic acid [51] are effective competitive inhibitors of PGH synthase 1. Docosahexaenoic acid is equally effective with PGH synthase 1 and 2 [46,50]. It is important to note that docosahexaenoic acid, like other competitive cyclooxygenase inhibitors, inhibits the cyclooxygenase activity without affecting the peroxidase activity of PGH synthases [49].

3.2. Mechanism of the Cyclooxygenase Reaction

The formation of PGG_2 from arachidonate and two molecules of O_2 is depicted in Fig. 5 [52]. PGH synthases are believed to produce or interact with arachidonate having a kink in the carbon chain due to rotation about C-9/C-10 [52]. This permits the enzyme to abstract the pro-S hydrogen from C-13 in the initial step in the cyclooxygenase reaction [52–54]. This is followed by the insertion of an O_2 molecule coming from the solvent side at C-11. Serial cyclization of the 11-peroxyl radical produces an endoperoxide with trans-aliphatic chains. Addition of a second O_2 on the carbon radical at C-15 then also occurs from the solvent

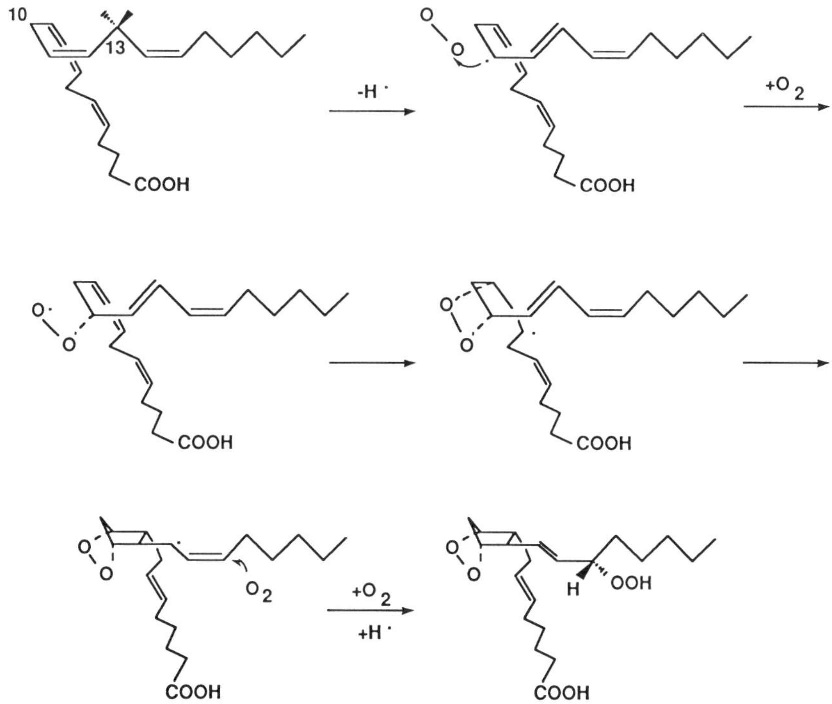

FIG. 5. Proposed mechanism for the formation of PGG_2 from arachidonic acid catalyzed by PGH synthases (adapted from Ref. 52.)

side to yield PGG_2 having the appropriate stereochemical configuration.

An arachidonyl radical is formed when arachidonate and PGH synthase 1 are incubated anaerobically suggesting that the 13-pro-S hydrogen is removed as a hydrogen atom [55,56]. Also supporting this concept is the finding that PGH synthase 1 converts 10,10-difluoroarachidonate to (11S)-10,10-difluoro-11-hydroxy-5(Z),8(Z),12(E),14(Z)-eicosatetraenoic acid (11-HETE) and epimeric 8,15-diHETEs; cyclization to prostaglandins does not occur with this substrate [57]. This product profile can only be explained if the 13-pro-S hydrogen is abstracted from 10,10-difluoroarachidonate as an atom and not as a hydride ion or a proton.

3.3. Kinetics of Cyclooxygenase Catalysis—Hydroperoxide Activation and Suicide Inactivation

The cyclooxygenase activity of PGH synthase is typically measured by monitoring the uptake of O_2 with an O_2 electrode. Following the addition of enzyme to an assay mixture containing O_2, arachidonate, exogenous heme, and an electron donor (e.g., phenol) for the peroxidase, there is a rapid increase in the rate of O_2 uptake which decreases to near zero within about 5 min (Fig. 6). This decrease in O_2 uptake is not due to exhaustion of substrates or to product inhibition but rather to "suicide inactivation" of the enzyme [58–60]. Each mole of purified enzyme can catalyze the synthesis of about 1300 mol of PGG_2 before the cyclooxygenase activity disappears [61]. The chemical details of the suicide inactivation are not known, but it appears that an unstable cyclooxygenase protein intermediate decays irreversibly to a catalytically incompetent form. As discussed in further detail below, suicide inactivation may be associated with formation of singlet protein tyrosyl radical signals.

Although the initial rate of the cyclooxygenase reaction is the

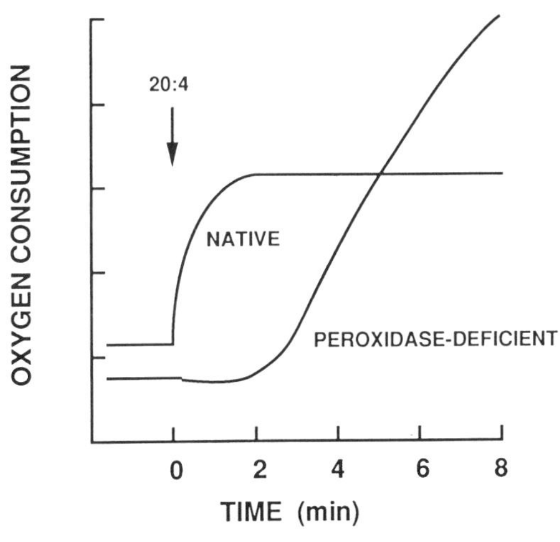

FIG. 6. Schematic comparison of O_2 consumption profiles for cyclooxygenase activities of native and peroxidase-deficient forms of PGH synthase 1. (From Ref. 130 with permission.)

maximal rate, there are maneuvers which induce a kinetic lag to the reaction (Fig. 6). Holoenzyme incubated with NaCN [62], Mn^{3+}-heme–containing PGH synthase 1 [63], and an H386A mutant of PGH synthase 1 [64] are all peroxidase-deficient, and all exhibit pronounced lag phases (Fig. 6). Furthermore, addition of reduced glutathione (GSH) plus GSH peroxidase to the reaction mixture will slow or prevent initiation of the cyclooxygenase reaction [58,62]; inhibition by GSH peroxidase can be overcome by the addition of N-ethylmaleimide. The lag phase induced by GSH peroxidase and NaCN can be overcome by adding low concentrations of fatty acid hydroperoxides [59,62]. These observations have led to the conclusion that hydroperoxide is necessary for oxidation of the cyclooxygenase to a form which is capable of abstracting the 13-pro-S hydrogen from arachidonate. A potential mechanism for hydroperoxide-dependent activation is discussed in further detail below in describing the interrelationships between the cyclooxygenase and peroxidase activities of PGH synthases.

3.4. Nonsteroidal Antiinflammatory Drugs as Inhibitors of Cyclooxygenase Activity

Aspirin and most other nonsteroidal antiinflammatory drugs (NSAIDs) are potent inhibitors of the cyclooxygenase activities of PGH synthases 1 and 2 (Fig. 7) [50,65,66]. It is still not clear whether it is inhibition of PGH synthase 1 and/or 2 that is responsible for the antiinflammatory and/or analgesic activities of these compounds observed in vivo. The antithrombogenic activity of aspirin [67,68], however, appears to be due solely to inhibition of the cyclooxygenase activity of platelet PGH synthase 1.

Aspirin and most other NSAIDs interact with PGH synthases at their cyclooxygenase but not their peroxidase active sites. Some of these compounds are rapidly reversible inhibitors which compete with arachidonate for binding to the enzyme; included in this group are piroxicam [69], flufenamate [70], sulindac sulfide [50], and ibuprofen [50,70]. Other compounds including indomethacin, flurbiprofen, and meclofenamate dissociate from the cyclooxygenase active site only very slowly although they do not appear to cause covalent modification of the enzyme [70,71].

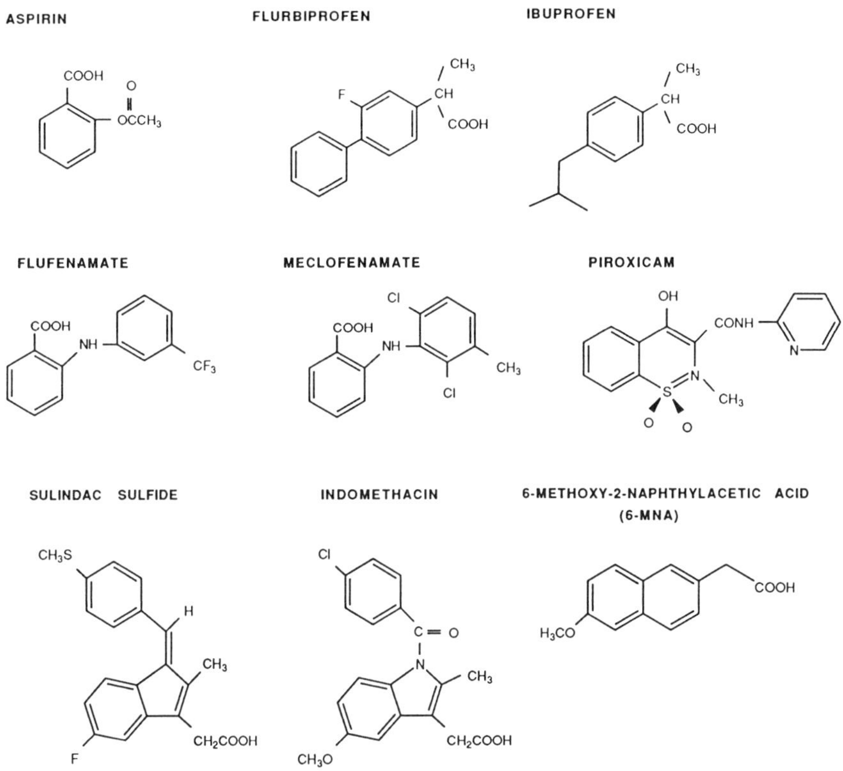

FIG. 7. Structures of some common nonsteroidal antiinflammatory drugs (NSAIDs) which inhibit PGH synthases 1 and 2.

Aspirin is unique among NSAIDs because upon binding to the cyclooxygenase active site, it causes acetylation of an "active site" serine residue (Fig. 8) [72]. In the ovine PGH synthase 1, this is Ser530 [27–29]; homologous serine residues in PGH synthases 1 and 2 from other species (Fig. 3) are also acetylated (M. Lecomte and W. L. Smith, unpublished results). In the case of PGH synthase 1, acetylation by aspirin causes complete inhibition of cyclooxygenase activity with no apparent effect on peroxidase activity (Fig. 8) [73,74] (M. Lecomte and W. L. Smith, unpublished results). In the case of PGH synthase 2, aspirin acetylation produces an enzyme which catalyzes formation of 15-hydroperoxyeicosatetraenoic acid (15-HPETE) instead of PGG_2 (Fig. 8)

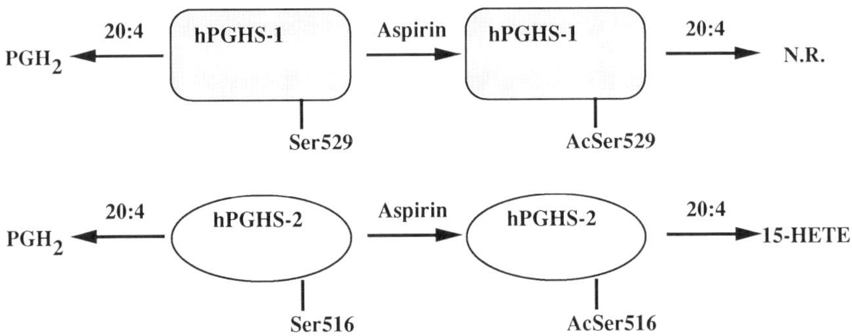

FIG. 8. Schematic representation of the acetylation of human (h) PGH synthases 1 and 2 by aspirin and the effects of acetylation on product formation by each of the enzymes.

[50,75] (M. Lecomte and W. L. Smith, unpublished results). Studies in which Ser530 of ovine PGH synthase 1 was replaced with a variety of other amino acids indicate (1) that the hydroxyl group of Ser530 is not required for either cyclooxygenase activity or arachidonate binding and (2) that the effect of aspirin acetylation is simply to prevent arachidonate binding [20,76]. Thus, replacement of Ser530 with an alanine yields a mutant enzyme which retains both cyclooxygenase and peroxidase activity, but replacement of Ser530 with an asparagine residue (which is roughly isosteric with an acetylated serine) produces an enzyme which lacks cyclooxygenase but retains peroxidase activity [20,76].

As shown in Table 1 some striking differences have also been observed in comparing the effects of NSAIDs other than aspirin on murine PGH synthases 1 and 2 [50]. Most of the agents identified to date are equipotent or more potent on PGH synthase 1 than PGH synthase 2. A possible exception is 6-methylnaphthylacetic acid, which is a slightly more potent inhibitor of PGH synthase 2 [50]. These observations are of potential therapeutic interest because (1) PGH synthase 2 may be the key isoenzyme which needs to be inhibited in inflammation [39–41] and (2) PGH synthase 1 but not PGH synthase 2 is the predominant enzyme in stomach (W. L. Smith, unpublished results). The major side effect of the use of NSAIDs is gastrointestinal bleeding and ulcer formation. Recognition of pharmacological differences between the two PGH syn-

TABLE 1

Inhibition of Murine PGHS-1 and -2 by Various Nonsteroidal Antiinflammatory Drugs

Inhibitor	ID_{50} for PGHS-1 (μM)	ID_{50} for PGHS-2 (μM)
Flurbiprofen	0.46–0.50	2.1–3.4
Ibuprofen	190–690	100–610
Indomethacin	4.9–8.1	130–160
Sulindac sulfide	0.33–0.48	11–14
Meclofenamic acid	2.0–2.5	13–18
Piroxicam	9–24	70–240
Docosahexaenoic acid	11	13–17

Source: From Ref. 50, with permission.

thase isozymes may be the first step in identifying new NSAIDs specific for PGH synthase 2 which lack ulcerogenic activity.

4. PEROXIDASE CATALYSIS

4.1. Peroxide Substrate Specificity

PGH synthases 1 and 2 exhibit peroxidase activities which catalyze reductions of a variety of hydroperoxides to their corresponding alcohols with concomitant oxidations of reducing cosubstrates. The peroxidase activities of ovine and bovine PGH synthase 1 copurify with cyclooxygenase activity [14,15]; moreover, ovine PGH synthase 1 immunoprecipitates with antibodies raised against the cyclooxygenase [16]. In addition, transfection of *cos*-1 cells with cDNAs for either PGH synthase 1 [20] or PGH synthase 2 [17,19] leads to enhanced expression of both cyclooxygenase and peroxidase activities. A heme prosthetic group is required for peroxidase activity (as well as cyclooxygenase activity). Apo-PGH synthase 1 reconstituted with Fe^{3+}-protoporphyrin IX (PPIX) exhibits peroxidase activity, whereas apoprotein reconstituted with

Mn^{3+}-PPIX exhibits about 1% of the peroxidase activity of the native enzyme [63,77,78]. In contrast, both Fe^{3+}-PPIX and Mn^{3+}-PPIX support comparable levels of cyclooxygenase activity [63,77–79].

The peroxidase activity of PGH synthase 1, which preferentially reduces fatty acid hydroperoxides including PGG_2 [79], is much less active toward H_2O_2 and virtually inactive with tertiary hydroperoxides [79]. Importantly, the peroxide specificity of the peroxidase activity is similar to that observed for peroxide activation of the cyclooxygenase activity of PGH synthase 1 [80]. The products of the reduction of 5-phenyl-4-pentenylhydroperoxide (PPHP) or 10-hydroperoxy-8,12-octadecadienoic acid by PGH synthase 1 indicate that the hydroperoxide is reduced simultaneously by two electrons [81,82]. This is also typical of reduction of hydroperoxides by other heme peroxidases including horseradish and cytochrome c peroxidases [53].

4.2. Spectral Intermediates

Classic heme peroxidases undergo well-known spectral changes when incubated with peroxide substrates. These spectral intermediates represent higher oxidation states of the heme group [83–86]. Addition of a variety of hydroperoxides to ovine PGH synthase 1 in the absence of reducing agents causes a rapid decrease in the Soret absorbance of the native enzyme and formation of a featureless visible spectrum (Fig. 9) [87]. This spectral intermediate I is analogous to compound I of horseradish peroxidase [88] and corresponds to a two-electron oxidized heme in which the iron is in the 4+ state and the porphyrin group is oxidized to a radical cation. Upon standing ($\geqslant$100 msec), the Soret absorbance increases and the peak shifts from 412 nm to 420 nm and α and β absorption maxima at 527 and 557 nm become apparent (Fig. 10) [87]. The spectrum of this second intermediate, called intermediate II (Fig. 10), is analogous to the spectra of compound II of horseradish peroxidase and compound ES of cytochrome c peroxidase. These latter intermediates represent one-electron oxidized heme groups in which the porphyrin is neutral and the iron is in its 4+ state [85,89,90].

Formation of an EPR signal at $g = 2.005$ occurs concomitantly with the formation of the intermediate II absorption spectrum when ovine PGH synthase 1 is incubated with hydroperoxides [91]. The

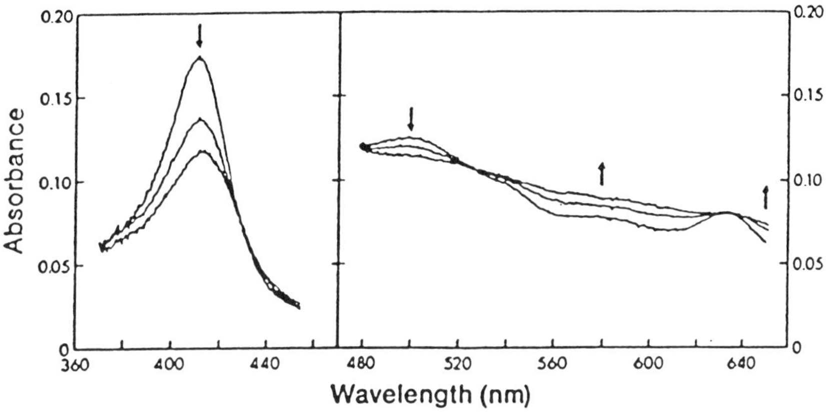

FIG. 9. Spectrum of PGH synthase 1 compound I. PGH synthase 1 was mixed with 5-phenyl-4-pentenylhydroperoxide (PPHP). Concentrations after mixing were 1.5 μM PGH synthase 1 for the Soret spectra, 12.1 μM PGH synthase 1 for the visible spectra, and 17 μM PPHP for each region. Spectra in the Soret region are shown for native enzyme, 1 and 3 msec after mixing. Spectra in the visible region are shown for native enzyme, 2 and 16 msec after mixing. Arrows indicate the direction of the absorbance change with increasing time. The spectra were collected over 1 msec in 96-nm-wavelength intervals. (From Ref. 87 with permission.)

hyperfine splitting pattern of the EPR signal suggests that it is due to formation of a tyrosyl radical. Subsequent experiments with PGH synthase 1 having deuterated tyrosine residues have confirmed this assignment (A.-L. Tsai, L. C. Hsi, R. J. Kulmacz, and W. L. Smith, unpublished results). Since no reducing agents are required for production of the EPR signal at $g = 2.005$, the spectral changes must result from an intramolecular electron transfer from a tyrosyl residue on the protein to the porphyrin π cation radical of intermediate I [91,92]. Accordingly, intermediate II of PGH synthase 1 is two oxidizing equivalents above the resting enzyme with one of the oxidizing equivalents present as a protein tyrosyl radical. This type of intramolecular electron transfer also occurs with cytochrome c peroxidase. As described in further detail below, a tyrosyl radical present in PGH synthase may be involved in hydrogen abstraction from arachidonate.

The catalytic turnover of the peroxidase activity of PGH synthase 1

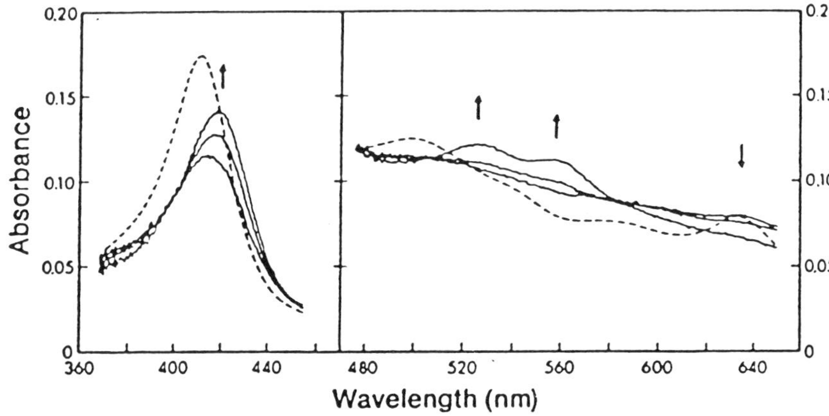

FIG. 10. Spectrum of PGH synthase 1 compound II. Experimental conditions as in Fig. 9. Spectra in the Soret region are shown at 11, 22, and 132 msec after mixing. Spectra in the visible region are shown at 16, 34, and 166 msec after mixing. Arrows indicate the direction of the absorbance change with increasing time. The broken line is the native enzyme spectrum. (From Ref. 87 with permission.)

has been estimated to be as high as 4000 [93]. This value is similar to the turnover numbers for other heme peroxidases. Thus, the peroxidase of PGH synthase 1, like other heme peroxidases, catalyzes two-electron hydroperoxide reduction, forms a higher oxidation state containing a ferryl–oxo porphyrin cation radical, and exhibits high catalytic efficiency.

4.3. Peroxidase-Reducing Cosubstrates

As with most peroxidases, a variety of compounds can serve as reducing cosubstrates for PGH synthase 1 peroxidase. Two general mechanisms for reduction of the enzyme are electron transfer and oxygen transfer (Fig. 11). Electron transfer is the most common mechanism; however, oxygen transfer has been observed with alkylaryl sulfides [94]. Electron transfer occurs by two sequential one-electron transfers (Fig. 11). One-electron oxidized derivatives of several reducing substrates have been detected by EPR spectroscopy [95]. Electron transfer occurs on the

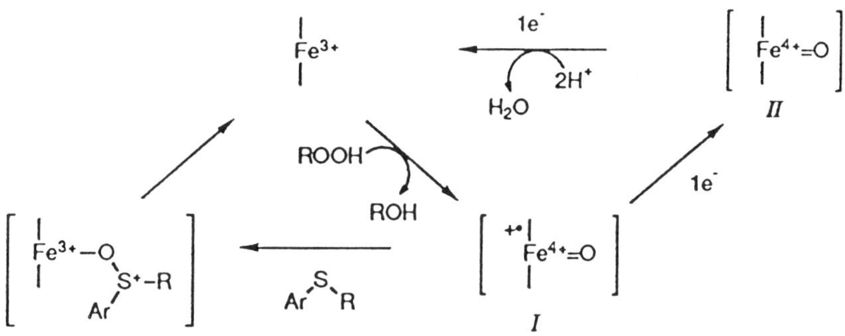

FIG. 11. Catalytic cycle of heme-containing peroxidase of PGH synthases 1 and 2. Intermediates I and II represent the spectroscopically detectable higher oxidation states described in the text. (From Ref. 54 with permission.)

protein, but the electron-deficient forms of the reducing substrates diffuse into solution, where they react to form stable products. The mechanism of oxygen transfer to sulfides is uncertain. ^{18}O-labeling studies indicate that sulfoxide oxygen is derived from the hydroperoxide oxidant [94]. This is consistent with reaction of the sulfide with the oxo ligand of intermediate I. It is possible that electron transfer is the first step followed by coupling of the sulfur radical cation to the ferryl–oxo complex (Fig. 12); however, it is also possible that intermediate I inserts oxygen in a concerted fashion. Spectroscopic studies of horseradish peroxidase have shown that sulfides reduce compound I by one electron [96]. If PGH synthase 1 behaves in a similar fashion, oxygen transfer may result from the electron transfer process.

There are substantial differences in the abilities of various compounds to serve as reducing cosubstrates for the peroxidase activity of PGH synthase 1. Electron-rich molecules are the best reductants, although polar substituents in these molecules can affect their activities [82,93]. Oxygenation of para-substituted alkylaryl sulfides exhibits a linear free energy relationship to σ^+ with a ρ value comparable to that of other peroxidases. Steric effects are also important in the oxygenation of alkylaryl sulfides. Studies on the effect of the size of the alkyl group on the oxygenation of alkylaryl sulfides has suggested that the heme pocket of PGH synthase is intermediate in size between that of horseradish peroxidase and that of cytochrome P-450 [93].

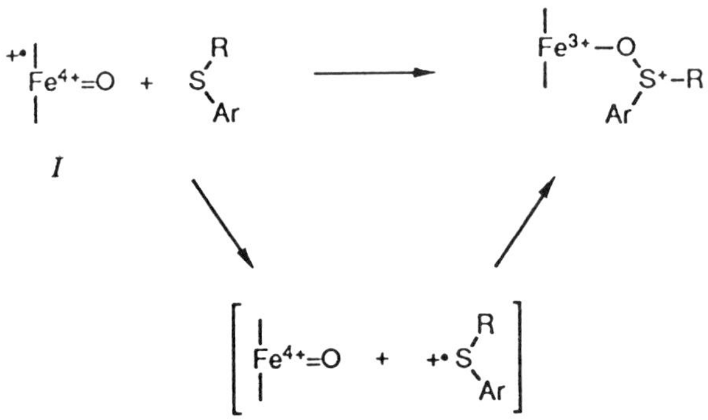

FIG. 12. Two possible mechanisms of sulfide oxidation by PGH synthase 1 peroxidase intermediate I. (From Ref. 54 with permission.)

The most likely physiological reducing cosubstrates of the peroxidase activity of PGH synthases are epinephrine and uric acid [82,97]. Although uric acid is a relatively inefficient reducing cosubstrate [97], it is present in high concentrations (~300 μM) in human plasma [98]. Similarly, reduced glutathione is a relatively poor reducing cosubstrate but is present in high concentrations in many cells (~6 mM) [61,99].

Recent reviews are available on the peroxidatic oxidation of xenobiotics by PGH synthase [53,95]. PGH synthase–dependent cooxidations may prove to be important in extrahepatic tissues having low levels of cytochrome P-450. Compounds cooxidized through the peroxidase activity of PGH synthase include aromatic amines, polycyclic hydrocarbons, nitrofurans, mycotoxins, synthetic estrogens, bisulfite, phenols, heterocyclic amines, hydantoins, and indoles [95].

5. CYCLOOXYGENASE–PEROXIDASE INTERRELATIONSHIPS

5.1. Hydroperoxide-Dependent Activation of Cyclooxygenase Activity

As discussed earlier, the dependence of cyclooxygenase activity on hydroperoxide, the appearance of oxidized heme intermediates and a tyrosyl radical signal, and the removal of the 13-pro-S hydrogen atom from arachidonate appear to be related. Ruf and coworkers developed the model depicted in Fig. 13 to explain these interrelationships [92]. According to their model, two-electron oxidation of the heme group at the peroxidase active site of PGH synthase by a hydroperoxide causes formation of an intermediate I. This intermediate I can then abstract a hydrogen from the phenolic side chain of a protein tyrosine residue to produce an intermediate II, containing a protein tyrosyl radical. The tyrosyl radical was envisioned to be the species that then abstracts the 13-pro-S hydrogen from arachidonate thereby initiating cyclooxygenase catalysis.

The Ruf tyrosyl radical model (Fig. 13) is predicated on the following findings. Alkyl hydroperoxides are required for cyclooxygenase activity [58,62,100,101]. Incubation of alkyl hydroperoxides with ovine PGH synthase 1 leads to sequential formation of two different visible spectral intermediates (intermediates I and II) with characteristics closely resembling those of classic peroxidase compound I and compound ES intermediates, respectively [87,91,102,103]. Protein tyrosyl radicals are formed rapidly in up to 80% yield based on the heme content of ovine PGH synthase 1 when the native enzyme is incubated with a hydroperoxide [91,104]. Tyrosyl radical formation and formation

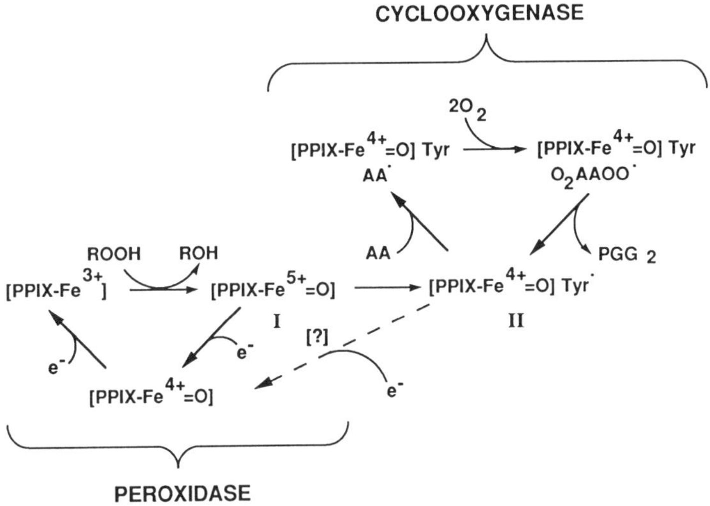

FIG. 13. Model for peroxide-dependent activation of the cyclooxygenase activity of PGH synthase 1 by formation of an intermediate tyrosyl radical. Intermediates I and II are designated. PPIX-Fe^{3+}, heme; AA, arachidonic acid. (Adapted from Dietz et al. [92].)

of the compound ES–like peroxidase intermediate (i.e., intermediate II) occur synchronously [91,92]. Modification of the enzyme with tetranitromethane, which involves nitration of tyrosine groups, inhibits cyclooxygenase activity, and alters the EPR spectrum of the tyrosyl radical [104–106]. Tyr385 appears to be at the cyclooxygenase active site of PGH synthase (Fig. 3) [106].

The Ruf model serves to integrate the cyclooxygenase and peroxidase functions of PGH synthases. However, the question of the role of a tyrosyl radical in cyclooxygenase catalysis remains controversial. A broad (~35 gauss) doublet tyrosyl radical near $g = 2$ in the EPR spectrum is formed rapidly when PGH synthase 1 is incubated with a hydroperoxide or with arachidonate [91,102,107–109]; the peak of doublet formation occurs at roughly the same time or slightly after the peak of cyclooxygenase activity [107]. A second radical signal, a broad singlet, appears subsequent to the doublet and after the rate of cyclooxygenase activity drops to zero. Both the broad doublet and the broad singlet have similar power saturation behaviors indicating that the radical signals are approximately the same distance from the heme and could well be due to a radical on the same tyrosyl group. A third radical signal, a narrow singlet, occurs when PGH synthase 1 is inactivated by indomethacin or incubated with a large excess of substrate [104,107–109]. DeGray et al. [108] provided evidence that the broad singlet signal is a composite of the broad doublet and narrow singlet signals. The rate of formation of the broad doublet is not inconsistent with its being involved in cyclooxygenase catalysis. In contrast, the broad singlet and narrow singlet appear later when much of the activity has been lost and may be markers of suicide inactivation. A potential precedent for protein modification involving tyrosyl radicals exists in the case of met-myoglobin. Peroxide-dependent oxidation of the heme group of met-myoglobin leads to generation of tyrosyl radicals, which can in turn react with the heme [110].

6. PGH SYNTHASE STRUCTURE

6.1. General Chemical Properties of PGH Synthases

As noted earlier, PGH synthases 1 and 2 have subunit molecular weights of approximately 65500 including the signal peptides. Native PGH synthase 1 has an $M_r = 72000$ on sodium dodecyl sulfate–

polyacrylamide gel electrophoresis (SDS-PAGE). The difference between the calculated and experimentally determined molecular weight is due, at least in part, to the presence of asparagine-linked oligosaccharides. There are three high-mannose oligosaccharides, one $Man_7(NAcGln)_2$ and two $Man_9(NAcGln)_2$ groups, coupled to PGH synthase 1 [111]. In contrast to PGH synthase 1, PGH synthase 2 appears as a doublet on SDS-PAGE with M_r = 72000 and 74000 [18,19,32,112]. The 72-kD species appears to contain three N-linked oligosaccharides while the 74-kD species probably contains four N-linked oligosaccharides [112]. The sites of N-glycosylation in ovine PGH synthase 1 are Asn68, Asn144, and Asn410 [112]. Curiously, if glycosylation at Asn410 is prevented using various maneuvers, the enzyme appears not to fold into its active conformation. Asparagine residues homologous to those which are glycosylated in PGH synthase 1 are also present and apparently glycosylated in PGH synthase 2. In addition, there is another site of N-glycosylation in PGH synthase 2 located near the C terminus (i.e., at Asn580 in murine PGH synthase 2; Fig. 3); this latter site is glycosylated in about 50% of PGH synthase 2 molecules but glycosylation at this position is not required for enzyme activity [112].

PGH synthase 1 solubilized from membranes with nonionic detergents appears as a dimer as judged by chemical crosslinking studies [113] and sedimentation analysis [15]. Because various attempts to generate a monomeric form of the enzyme which retains activity have been unsuccessful, it is assumed that the enzyme functions as a dimer. The interactions responsible for dimer formation have not been determined.

6.2. Subcellular Location of PGH Synthases

PGH synthases 1 and 2 have both been localized to the endoplasmic reticulum (ER) and the nuclear membrane using immunocytochemistry [31,32,34]. PGH synthase 1 can only be solubilized by treatment of membranes with detergents [14,15,26,114] indicating that the enzyme is an integral membrane protein. Various computer algorithms suggest that there is a single transmembrane involving amino acids 285–306 of ovine PGH synthase 1 [28]. The current model for the orientation of PGH synthase 1 in the endoplasmic reticulum is shown in Fig. 14 [112]. This model, which also applies to PGH synthase 2, is based on the following reasoning:

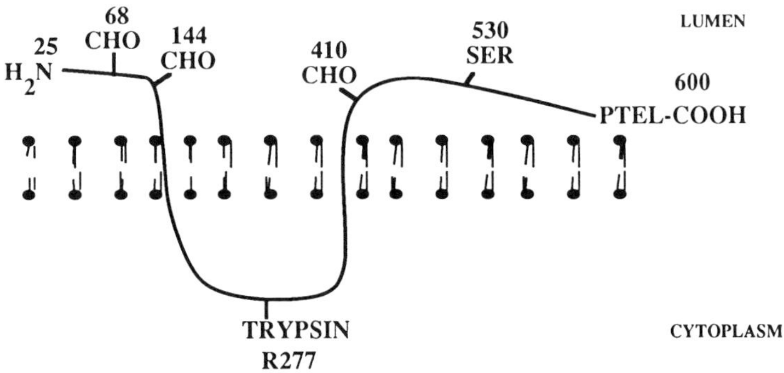

FIG. 14. Model for the orientation of ovine PGH synthase 1 in the endoplasmic reticulum. Sites of N-glycosylation (CHO) and trypsin cleavage are designated. Also shown are the N-terminal Ala25 of the processed protein, the position of aspirin acetylation at Ser530, and the putative ER retention signal PTEL. (Adapted from Ref. 112.)

1. The N termini of processed PGH synthases have an epidermal growth factor (EGF)-like domain (e.g., residues 34–72 of ovine PGH synthase 1 [115]), and there is no obvious transmembrane sequence between the N terminus of the mature protein (Ala25) and the first site of N-glycosylation at Asn68.
2. The C termini of all mammalian PGH synthases have an (S/P)TEL sequence which is probably an ER retention signal and found on the luminal side of the ER [33,116].
3. N-glycosylation sites are located at Asn68, Asn144, and Asn410, and are usually found on the luminal side of the ER [117].
4. A trypsin cleavage site is present at Arg277 of ovine PGH synthase 1 [81] and this site appears to be on the cytoplasmic side of the ER [81,118,119].
5. Several unidentified antigenic determinants of PGH synthase 1 are known to be on the cytoplasmic surface of the ER [119].

6.3. Heme Binding to PGH Synthase

Heme is required for both the cyclooxygenase and peroxidase activities of PGH synthase 1. Titration of apoenzyme with heme coordinately restores both activities. Quantitation of heme binding to purified PGH synthase 1 by titration of the apoenzyme has yielded values of 0.5–2.0 hemes per subunit [14,26,120–123]. The difficulty in obtaining a consistent value has been attributed to both the presence of inactive enzyme protein in apoenzyme preparations and the presence of a relatively high-affinity site for nonspecific heme binding [124]. There is a general consensus that there is likely to be one heme per enzyme subunit, and recent X-ray crystallography data [125] have also suggested that one heme per subunit (i.e., two per dimer) is the correct stoichiometry.

There are VIS and EPR spectral data suggesting that the heme group is liganded by imidazole groups at the fifth and sixth coordination positions of the iron [87,102]. Magnetic circular dichroism of native PGH synthase 1 and its cyano derivative indicates that 80% of the holoenzyme is high spin at 24°C but only 50% at low temperature [102]. This suggests that the heme iron equilibrates between five- and six-coordinate with the five-coordinate form predominating at room temperature.

There are 10 conserved histidine residues among various PGH synthases (Fig. 3). Substitution of His207, His309, or His388 with either glutamine or alanine residues yields mutant ovine PGH synthases 1 lacking both cyclooxygenase and peroxidase activities [64]. Substitutions of the other conserved histidines yield proteins which retain cyclooxygenase and peroxidase activities. His309 was originally proposed to be the proximal heme ligand based on information coming from sequence comparisons and on "heme protection" measurements [64]. For example, His309 is present in a decapeptide (TIWLREHNRV; residues 303–312 of ovine PGH synthase 1) which is conserved in all PGH synthases (Fig. 3) and is present in several other mammalian peroxidases including myeloperoxidase and thyroid peroxidase [126–128]. Moreover, the H309A mutant of ovine PGH synthase 1 cannot be protected from cleavage by trypsin in the presence of exogenous heme although neither the H207A and H388A mutants are protected from trypsin cleavage by heme; thus, His309 appears to be particularly critical for heme binding [64]. Recently, the crystal structure of canine myeloperoxidase has been solved [129]. The His residue in myeloperoxidase (His257), which is analogous to His309 of ovine PGH synthase, is not a heme ligand. Instead, His336 is the proximal heme ligand of myeloperoxidase. This latter histidine residue corresponds most closely to His388 of ovine PGH synthase 1. Therefore, His388 now becomes the best candidate as the proximal heme ligand of PGH synthase 1 [129] and by default His207 is the best candidate to be the distal heme ligand.

6.4. Structure of PGH Synthase Active Sites

Figure 15 is a model of the active site of ovine PGH synthase 1 [130]. An alkyl hydroperoxide is shown associated with the heme group bound to the peroxidase active site. The heme is shown liganded at the proximal position by His388 and at the distal position by His207. Oxidation of the hydroperoxide to an alcohol causes formation of a compound I–like oxidized heme (cf. intermediate I; Fig. 13). An intramolecular electron transfer from a neighboring Tyr385 to the heme yields intermediate II having a tyrosyl radical centered on Tyr385. Although this point is still controversial, the tyrosyl radical may abstract the hydrogen atom from the arachidonate bound to the cyclooxygenase active site to initiate the cyclooxygenase reaction. Ser530, the site of aspirin acetylation, is shown in close proximity to the cyclooxygenase active site.

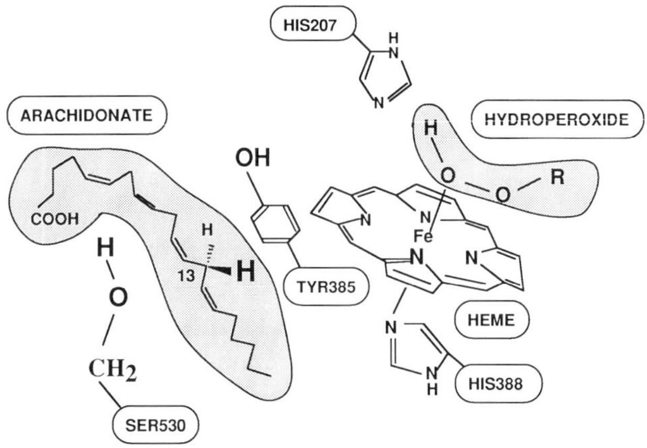

FIG. 15. Model for the cyclooxygenase and peroxidase active sites of ovine PGH synthase 1. Residues involved in heme binding and the relative locations of the groups neighboring the arachidonate (cyclooxygenase) and hydroperoxide (peroxidase) binding sites are illustrated. (Adapted from Ref. 54.)

Five general features are known about the active site of PGH synthase 1 (Fig. 15). First, the cyclooxygenase active site is distinct from the peroxidase active site. This conclusion is based on the findings that aspirin [73,74], related NSAIDs [73], and docosahexaenoic acid [49] are competitive inhibitors of the cyclooxygenase activity of PGH synthase 1 but that none of these agents appreciably affects the peroxidase reaction. Second, the cyclooxygenase active site is large enough to accommodate a structurally diverse group of NSAIDs and fatty acid substrates [66,118,130]. Third, the cyclooxygenase active site becomes more accessible to NSAIDs when heme is bound to PGH synthase 1. For example, only holo-PGH synthase 1 is covalently labeled at appreciable rates with [^{3}H]aspirin [131]. Fourth, binding of NSAIDs induces global structural changes in PGH synthase. When indomethacin is bound to PGH synthase 1, the enzyme becomes even less susceptible to trypsin cleavage [118] and to covalent labeling of nonactive site tyrosines (Tyr355 and Tyr417 of ovine PGH synthase 1) by tetranitromethane [104–106]. And fifth, two residues, Ser530 and Tyr385, lie within or closely neighbor the cyclooxygenase active site. Ser530 is acetylated by aspirin causing irreversible inactivation of cyclooxygenase activity [72]. Tyr385 is required for cyclooxygenase activity and is protected from nitration by tetranitromethane by the presence of indomethacin or ibuprofen [106].

Only two of the residues making up the cyclooxygenase active site have been identified—Tyr385 and Ser530. Thus, residues separated by over 100 amino acids come together to form at least part of the cyclooxygenase active site. The most direct approach to identifying other residues and other regions of the protein which make up the cyclooxygenase active site would be to prepare and use reactive NSAID or fatty acid substrate analogs to label the protein.

ACKNOWLEDGMENTS

Studies in the authors' laboratories described in this chapter were supported in part by NIH grants DK42509 (W.L.S.), DK22042 (W.L.S.), and CA47479 (L.J.M.). We are grateful to Drs. Thomas Eling, Ah-lim Tsai, Richard Kulmacz, David DeWitt, and Harvey Herschman for preprints of manuscripts to be published.

ABBREVIATIONS

cAMP	3′,5′-cAMP
AP-1	AP-1 transcription factor
EGF	epidermal growth factor
EPR	electron paramagnetic resonance
ER	endoplasmic reticulum
ETYA	eicosatetraynoic acid
nGRE	negative glucocorticoid response element
GSH	reduced glutathione
11-HETE	11-hydroxy-5(Z),8(Z),12(E),14(Z)-eicosatetraenoic acid
15-HPETE	15-hydroperoxyeicosatetraenoic acid
IL	interleukin
NIH	National Institutes of Health
NSAID	nonsteroidal antiinflammatory drug
PG	prostaglandin
PGE_2	prostaglandin E_2
PGH	prostaglandin H
PGH_2	prostaglandin H_2
PGI_2	prostaglandin I_2
PPHP	5-phenyl-4-pentenylhydroperoxide
PPIX	protoporphyrin IX
SDS-PAGE	sodium dodecyl sulfate–polyacrylamide gel electrophoresis
SRE	serum response element
TATA	TATA box
(P/S)TEL	(Pro/Ser)ThrGluLeu
TX	thromboxane
VIS	visible spectrum

REFERENCES

1. W. L. Smith, *Biochem. J.*, *259*, 315 (1989).
2. B. Samuelsson, S.-E. Dahlen, J. A. Lindgren, C. A. Rouzer, and C. H. Serhan, *Science*, *237*, 1171 (1987).
3. T. Shimizu, *Int. J. Biochem.*, *20*, 661 (1988).

4. R. A. Lewis, K. F. Austen, and R. J. Soberman, *N. Engl. J. Med.*, *323*, 645 (1990).
5. S. Yamamoto, *Biochim. Biophys. Acta*, *1128*, 117 (1992).
6. W. L. Smith and O. Laneuville, in *Eicosanoids and Their Inhibitors in Cancer Immunology and Immunotherapy* (K. M. Anderson and J. Harris, eds.), CRC Press, Boca Raton, (in press).
7. F. A. Fitzpatrick and R. C. Murphy, *Pharmacol. Rev.*, *40*, 229 (1989).
8. J. H. Capdevila, A. Karara, D. J. Waxman, M. V. Martin, J. R. Falck, and F. P. Guenguerich, *J. Biol. Chem.*, *265*, 10865 (1990).
9. J. C. McGiff, *Ann. Rev. Pharmacol. Toxicol.*, *31*, 339 (1991).
10. M. A. Kirschenbaum, A. G. Lowe, W. Trizna, and L. G. Fine, *J. Clin. Invest.*, *70*, 1193 (1982).
11. W. L. Smith, W. K. Sonnenburg, M. L. Allen, T. Watanabe, J. Zhu, and E. A. El-Harith, in *Renal Eicosanoids* (M. J. Dunn, C. Patrono, and G. A. Cinotti, eds.), Plenum Press, New York, 1989, pp. 131–147.
12. W. L. Smith, *Am. J. Physiol.*, *263*, F181 (1992).
13. W. L. Smith, L. J. Marnett, and D. L. DeWitt, *Pharmacol. Ther.*, *49*, 153 (1991).
14. T. Miyamoto, N. Ogino, S. Yamamoto, and O. Hayaishi, *J. Biol. Chem.*, *251*, 2629 (1976).
15. F. J. van der Ouderaa, M. Buytenhek, D. H. Nugteren, and D. A. van Dorp, *Biochim. Biophys. Acta*, *487*, 315 (1977).
16. W. R. Pagels, R. J. Sachs, L. J. Marnett, D. L. DeWitt, J. S. Day, and W. L. Smith, *J. Biol. Chem.*, *258*, 6517 (1983).
17. B. S. Fletcher, D. A. Kujubu, D. M. Perrin, and H. R. Herschman, *J. Biol. Chem.*, *267*, 4338 (1992).
18. J. Sirois and J. S. Richards, *J. Biol. Chem.*, *267*, 6382 (1992).
19. E. A. Meade, W. L. Smith, and D. L. DeWitt. *J. Lipid Mediators*, *6*, 119 (1993).
20. D. L. DeWitt, E. A. El-Harith, S. A. Kraemer, M. J. Andrews, E. F. Yao, R. L. Armstrong, and W. L. Smith, *J. Biol. Chem.*, *265*, 5192 (1990).
21. D. A. Kujubu, B. S. Fletcher, B. C. Varnum, R. W. Lim, and H. R. Herschman, *J. Biol. Chem.*, *266*, 12866 (1991).

22. D. L. Simmons, W. Xie, J. G. Chipman, and G. E. Evett, in *Prostaglandins, Leukotrienes, Lipoxins, and PAF* (J. M. Bailey, ed.), Plenum Press, New York, 1991, pp. 67 ff.

23. C. Yokoyama and T. Tanabe, *Biochem. Biophys. Res. Commun.*, *165*, 888 (1989).

24. C. D. Funk, L. B. Funk, M. E. Kennedy, A. S. Pong, and G. A. FitzGerald, *FASEB J.*, *5*, 2304 (1991).

25. T. Hla and K. Neilson, *Proc. Natl. Acad. Sci. USA*, *89*, 7384 (1992).

26. M. Hemler, W. E. M. Lands, and W. L. Smith, *J. Biol. Chem.*, *251*, 5575 (1976).

27. D. L. DeWitt and W. L. Smith, *Proc. Natl. Acad. Sci. USA*, *85*, 1412 (1988).

28. J. P. Merlie, D. Fagan, J. Mudd, and P. Needleman, *J. Biol. Chem.*, *263*, 3550 (1988).

29. C. Yokoyama, T. Takai, and T. Tanabe, *FEBS Lett.*, *231*, 347 (1988).

30. W. Xie, J. G. Chipman, D. L. Robertson, R. L. Erikson, and D. L. Simmons, *Proc. Natl. Acad. Sci. USA*, *88*, 2692 (1991).

31. M. K. Regier, D. L. DeWitt, M. S. Schindler, and W. L. Smith, *Arch. Biochem. Biophys.*, *301*, 439 (1993).

32. D. A. Kujubu, S. T. Reddy, B. S. Fletcher, and H. R. Herschman, *J. Biol. Chem.*, (in press).

33. R. Noiva and W. J. Lennarz, *J. Biol. Chem.*, *267*, 3553 (1992).

34. T. E. Rollins and W. L. Smith, *J. Biol. Chem.*, *255*, 4872 (1980).

35. W. L. Smith, in *Handbook of Eicosanoids: Prostaglandins and Related Lipids*, Vol. 1, Part A (A. L. Willis, ed.), CRC Press, Boca Raton, 1987, pp. 175–184.

36. D. L. DeWitt, S. A. Kraemer, and E. A. Meade, in *Prostaglandins and Related Compounds II*, Vol. 21 (B. Samuelsson, ed.), Raven Press, New York, 1991, pp. 75–90.

37. A. L. Lin, M. J. Bienkowski, and R. R. Gorman, *J. Biol. Chem.*, *264*, 17379 (1989).

38. M. K. O'Banion, H. B. Sadowski, V. Winn, and D. A. Young, *J. Biol. Chem.*, *266*, 23261 (1991).

39. S. H. Lee, E. Soyoola, P. Chanmugam, S. Hart, W. Sun, H. Zhong, S. Liou, D. L. Simmons, and D. Hwang, *J. Biol. Chem.*, *267*, 25934, 1992.

40. M. G. O'Sullivan, F. H. Chilton, E. M. Huggins, Jr., and C. E. McCall, *J. Biol. Chem.*, *267*, 14547 (1992).
41. M. G. O'Sullivan, E. M. Huggins, Jr., E. A. Meade, D. L. DeWitt, and C. E. McCall, *Biochem. Biophys. Res. Commun.*, *187*, 1123 (1992).
42. E. A. Meade and D. L. DeWitt, *Arch. Biochem. Biophys.* (in press).
43. T. Hla and T. Maciag, *J. Biol. Chem.*, *266*, 24059 (1991).
44. S. A. Kraemer, E. A. Meade, and D. L. DeWitt, *Arch. Biochem. Biophys.*, *293*, 391 (1992).
45. W. E. M. Lands, P. R. LeTellier, L. H. Rome, and J. Y. Vanderhoek, in *Advances in the Biosciences*, Vol. 9, Pergamon Press, New York, 1973, pp. 15–28.
46. E. A. Meade, Ph.D. thesis, "Kinetic and Pharmacological Comparison of Murine PGH Synthase-1 and PGH Synthase-2," Michigan State University, East Lansing, 1992.
47. R. J. Flower, *Pharmacol. Rev.*, *26*, 33 (1974).
48. J. Y. Vanderhoek and W. E. M. Lands, *Biochim. Biophys. Acta*, *296*, 374 (1973).
49. P. J. Marshall and R. J. Kulmacz, *Arch. Biochem. Biophys.*, *266*, 162 (1988).
50. E. A. Meade, W. L. Smith, and D. L. DeWitt, *J. Biol. Chem.*, *268*, 6610 (1993).
51. D. H. Nugteren, *Biochim. Biophys. Acta*, *210*, 171 (1970).
52. M. Hamberg and B. Samuelsson, *J. Biol. Chem.*, *242*, 5334 (1967).
53. L. J. Marnett and K. R. Maddipati, in *Peroxidases: Chemistry and Biology* (J. Everse, K. Everse, and M. Grisham, eds.), CRC Press, Boca Raton, 1991 pp. 293–334.
54. W. L. Smith and L. J. Marnett, *Biochim. Biophys. Acta*, *1083*, 1 (1991).
55. R. P. Mason, B. Kalyanaraman, B. E. Tainer, and T. E. Eling, *J. Biol. Chem.*, *255*, 5019 (1980).
56. J. Schreiber, T. E. Eling, and R. P. Mason, *Arch. Biochem. Biophys.*, *249*, 126 (1986).
57. P-Y. Kwok, F. W. Muellner, and J. Fried, *J. Am. Chem. Soc.*, *109*, 3692 (1987).
58. W. L. Smith and W. E. M. Lands, *Biochemistry*, *11*, 3276 (1972).

59. M. E. Hemler, H. W. Cook, and W. E. M. Lands, *Arch. Biochem. Biophys.*, *193*, 340 (1979).
60. R. J. Kulmacz, *Prostaglandins*, *34*, 225 (1987).
61. P. J. Marshall, R. J. Kulmacz, and W. E. M. Lands, *J. Biol. Chem.*, *262*, 3510 (1987).
62. M. E. Hemler and W. E. M. Lands, *J. Biol. Chem.*, *255*, 6253 (1980).
63. N. Ogino, S. Ohki, S. Yamamoto, and O. Hayaishi, *J. Biol. Chem.*, *252*, 5061 (1978).
64. T. Shimokawa and W. L. Smith, *J. Biol. Chem.*, *266*, 6168 (1991).
65. J. R. Vane and R. Botting, *FASEB J.*, *1*, 89 (1987).
66. W. L. Smith, D. L. DeWitt, T. Shimokawa, S. A. Kraemer, and E. A. Meade, *Stroke*, *21*, IV24 (1990).
67. C. Patrono, C. Ciabattoni, and F. Davi, *Stroke*, *21*, IV130 (1990).
68. C. H. Hennekens, J. E. Buring, P. Sandercock, R. Collins, and R. Peto, *Circulation*, *80*, 749 (1989).
69. T. J. Carty, J. S. Stevens, J. G. Lombardino, M. J. Parry, and M. J. Randall, *Prostaglandins*, *19*, 671 (1980).
70. L. H. Rome and W. E. M. Lands, *Proc. Natl. Acad. Sci. USA*, *72*, 4863 (1975).
71. R. W. Walenga, S. F. Wahl, B. N. Y. Setty, and M. J. Stewart, *Prostaglandins*, *31*, 625 (1986).
72. G. J. Roth, E. T. Machuga, and J. Ozols, *Biochemistry 22*, 4672 (1983).
73. K. Mizuno, S. Yamamoto, and W. E. M. Lands, *Prostaglandins*, *23*, 743 (1982).
74. F. J. van der Ouderaa, M. Buytenhek, D. H. Nugteren, and D. A. van Dorp, *Eur. J. Biochem.*, *109*, 1 (1980).
75. M. J. Holtzman, J. Turk, and L. P. Shornick, *J. Biol. Chem.*, *267*, 21438 (1992).
76. T. Shimokawa and W. L. Smith, *J. Biol. Chem.*, *267*, 12387 (1992).
77. R. Odenwaller, K. R. Maddipati, and L. J. Marnett, *J. Biol. Chem.*, *267*, 13863 (1992).
78. S. Strieder, K. Schaible, H.-J. Scherer, R. Dietz, and H. H. Ruf, *J. Biol. Chem.*, *267*, 13870 (1992).
79. S. Ohki, N. Ogino, S. Yamamoto, and O. Hayaishi, *J. Biol. Chem.*, *254*, 829 (1979).

80. R. J. Kulmacz and W. E. M. Lands, *Prostaglandins*, *25*, 531 (1983).

81. L. J. Marnett, Y-N. P. Chen, K. R. Maddipati, P. Ple, and R. Labeque, *J. Biol. Chem.*, *263*, 16532 (1988).

82. C. M. Markey, A. Alward, P. E. Weller, and L. J. Marnett, *J. Biol. Chem.*, *262*, 6266 (1987).

83. B. Chance, *J. Biol. Chem.*, *151*, 553 (1943).

84. P. George, *J. Biol. Chem.*, *201*, 413 (1953).

85. H. B. Dunford and J. S. Stillman, *Coord. Chem. Rev.*, *19*, 187 (1976).

86. B. Chance, L. Powers, Y. Ching, T. Poulos, G. R. Schonbaum, I. Yamazaki, and K. G. Paul, *Arch. Biochem. Biophys.*, *235*, 596 (1984).

87. A. M. Lambeir, C. M. Markey, H. B. Dunford, and L. J. Marnett, *J. Biol. Chem.*, *260*, 14894 (1985).

88. D. Dolphin and R. H. Felton, *Acc. Chem. Res.*, *7*, 26 (1974).

89. W. Nastainczyk, *Eur. J. Biochem.*, *144*, 381 (1984).

90. T. Yonetani, in *The Enzymes*, Vol. 13 (P. Boyer, ed.), Academic Press, New York, 1976, pp. 345–362.

91. R. Karthein, R. Dietz, W. Nastainczyk, and H. H. Ruf, *Eur. J. Biochem.*, *171*, 313 (1988).

92. R. Dietz, W. Nastainczyk, and H. H. Ruf. *Eur. J. Biochem.*, *171*, 321 (1988).

93. P. Ple and L. J. Marnett, *J. Biol. Chem.*, *264*, 13983 (1989).

94. R. W. Egan, P. H. Gale, W. J. Vandenheuvel, E. M. Baptista, and F. A. Kuehl, *J. Biol. Chem.*, *255*, 323 (1980).

95. T. E. Eling, D. C. Thompson, G. L. Foureman, J. F. Curtis, and M. R. Hughes, *Ann. Rev. Pharmacol. Toxicol.*, *30*, 1 (1990).

96. U. Perez and H. B. Dunford, *Biochim. Biophys. Acta*, *1038*, 98 (1990).

97. N. Ogino, S. Yamamoto, O. Hayaishi, and T. Tokuyama, *Biochem. Biophys. Res. Commun.*, *87*, 184 (1979).

98. B. N. Ames, R. Cathcart, E. Schwiers, and P. Hochstein, *Proc. Natl. Acad. Sci. USA*, *78*, 6858 (1981).

99. D. J. Reed, in *Bioactivation of Foreign Compounds* (M. W. Anders, ed.), Academic Press, New York, 1985, pp. 71–108.

100. M. E. Hemler, C. G. Crawford, and W. E. M. Lands, *Biochemistry*, *17*, 1772 (1978).

101. M. E. Hemler, G. Graff, and W. E. M. Lands, *Biochem. Biophys. Res. Commun.*, *85*, 1325 (1978).

102. R. J. Kulmacz, A.-L. Tsai, and G. Palmer, *J. Biol. Chem.*, *262*, 10524 (1987).

103. Y. Hsuanyu and H. B. Dunford, *Biochem. Cell Biol.*, *68*, 965 (1990).

104. R. J. Kulmacz, Y. Ren, A.-L. Tsai, and G. Palmer, *Biochemistry*, *29*, 8760 (1990).

105. W. L. Smith, D. L. DeWitt, S. A. Kraemer, M. J. Andrews, T. Hla, T. Maciag, and T. Shimokawa, *Adv. Prostaglandin Thromboxane Leukotriene Res.*, *20*, 14 (1990).

106. T. Shimokawa, R. J. Kulmacz, D. L. DeWitt, and W. L. Smith, *J. Biol. Chem.*, *265*, 20073 (1990).

107. A.-L. Tsai, G. Palmer, and R. J. Kulmacz, *J. Biol. Chem.*, *267*, 17753 (1992).

108. J. A. DeGray, G. Lassmann, J. F. Curtis, T. A. Kennedy, L. J. Marnett, T. E. Eling, and R. P. Mason, *J. Biol. Chem.*, *267*, 23583 (1992).

109. G. Lassmann, R. Odenwaller, J. F. Curtis, J. A. DeGray, R. P. Mason, L. J. Marnett, and T. E. Eling, *J. Biol. Chem.*, *266*, 20045 (1991).

110. C. E. Catalano, Y. S. Choe, and P. R. Ortiz de Montellano, *J. Biol. Chem.*, *264*, 10534 (1989).

111. J. H. G. M. Mutsaers, H. van Halbeek, J. P. Kamerling, and J. F. G. Vliegenthart, *Eur. J. Biochem.*, *147*, 569 (1985).

112. J. C. Otto and W. L. Smith, *J. Biol. Chem.* (in press).

113. G. J. Roth, C. J. Siok, and J. Ozols, *J. Biol. Chem.*, *255*, 1301 (1980).

114. F. J. G. van der Ouderaa and M. Buytenhek, *Meth. Enzymol.*, *86*, 60 (1982).

115. H. Toh, *FEBS Lett.*, *258*, 317 (1989).

116. M. Robbi and H. Beaufay *J. Biol. Chem.*, *266*, 20498 (1991).

117. G. W. Hart, R. S. Haltiwanger, G. D. Holt, and W. G. Kelly, *Ann. Rev. Biochem.*, *58*, 841 (1989).

118. R. J. Kulmacz, *J. Biol. Chem., 264*, 14136 (1989).
119. D. L. DeWitt, T. E. Rollins, J. S. Day, J. A. Gauger, and W. L. Smith, *J. Biol. Chem., 256*, 10375 (1981).
120. G. J. Roth, E. T. Machuga, and P. Strittmatter, *J. Biol. Chem., 256*, 10018 (1981).
121. F. J. van der Ouderaa, M. Buytenhek, F. J. Slikkerveer, and D. A. van Dorp, *Biochim. Biophys. Acta, 572*, 29 (1979).
122. H. H. Ruf, D. Schuhn, and W. Nastainczyk, *FEBS Lett., 165*, 293 (1984).
123. R. J. Kulmacz and W. E. M. Lands, *J. Biol. Chem., 259*, 6358 (1984).
124. Y.-N. P. Chen, M. J. Bienkowski, and L. J. Marnett, *J. Biol. Chem., 252*, 16892 (1987).
125. D. Picot and R. M. Garavito, in *Cytochrome P450: Biochemistry and Biophysics* (I. Schuster, ed.), Taylor and Francis, Washington, D.C., 1989, pp. 29–36.
126. K. R. Johnson, W. M. Nauseef, A. Care, J. J. C. Weelock, and G. Rovera, *Nucl. Acids Res., 15*, 2013 (1987).
127. K. Morishita, N. Kubota, S. Asano, Y. Kaziro, and S. Nagata, *J. Biol. Chem., 262*, 3844 (1987).
128. S. Kimura, T. Kotani, O. W. McBride, K. Umeki, K. Hirai, T. Kakayama, and S. Ohtaki, *Proc. Natl. Acad. Sci. USA, 84*, 5555 (1987).
129. J. Zeng and R. E. Fenna, *J. Mol. Biol., 226*, 185 (1992).
130. W. L. Smith, T. E. Eling, R. J. Kulmacz, L. J. Marnett, and A.-L. Tsai, *Biochemistry, 31*, 3 (1992).
131. Y-N.P. Chen and L. J. Marnett, *FASEB J., 3*, 2294 (1989).

7

Diol Dehydrase from *Clostridium glycolicum*: The Non-B_{12}-Dependent Enzyme

Maris G. N. Hartmanis

Kabi Pharmacia BioScience Center,
Strandbergsgatan 49, S–11287 Stockholm, Sweden

1. INTRODUCTION

1.1. Bacterial Growth on 1,2-Ethanediol and 1,2-Propanediol

Various aerobic and facultatively anaerobic bacteria, such as a *Flavobacterium* [1,2], *Klebsiella pneumoniae* [3], and other *Klebsiella* species [4], *Propionibacterium freudenreichii* [5,6], *Arthrobacter oxydans* [7], and other unidentified bacteria [8], can utilize 1,2-ethanediol or 1,2-propanediol as the sole carbon or energy source. Under anaerobic conditions these diols are fermented to their corresponding acids and alcohols [3,6]. Products formed during growth under aerobic conditions include lactate and pyruvate [7,8].

During anaerobic conditions in the bacteria above, the initial reaction in the diol fermentation is conversion of 1,2-ethanediol or 1,2-propanediol to their corresponding aldehydes [9] by the enzyme diol dehydrase (EC 4.2.1.28). This enzyme has been isolated from *Klebsiella pneumoniae*, purified to homogeneity, and studied extensively [9–12]. Lee and Abeles have shown that this enzyme is coenzyme B_{12}–dependent [9].

In this chapter I will summarize the current knowledge on the non–B_{12}-dependent diol dehydrase, isolated from the strictly anaerobic bacterium *Clostridium glycolicum*, and make comparisons with the B_{12}–dependent enzymes isolated from the bacteria mentioned above.

1.2. *Clostridium glycolicum*

The strictly anaerobic organism *C. glycolicum* was isolated more than 30 years ago by Gaston and Stadtman [13] from a specimen of mud obtained from a stagnant pond near the Chesapeake and Ohio Canal outside Washington, D.C. At that time they were studying vitamin B_{12} coenzyme function and hoped that a bacterium which can utilize 1,2-ethanediol as its carbon and energy source would be a richer source of vitamin B_{12}–dependent enzymes than the organisms available to them at that time.

The anaerobic bacterium they isolated is a long, slender, motile, gram-positive, spore-forming rod, which grows well from 22 to 37°C at pH 7.4–7.6 [13]. In addition to utilizing 1,2-ethanediol and 1,2-propanediol it can grow anaerobically in a minimal medium on glucose, fructose, sorbitol, dulcitol, and xylose. Other diols, however, such as butanediol and 1,3-propanediol, do not support growth of this organism [13]. *C. glycolicum* also differs biochemically from aerobic bacteria in that, although it can utilize 1,2-ethanediol and 1,2-propanediol, it cannot grow on glycerol as the sole carbon and energy source [13].

2. DIOL METABOLISM IN *C. GLYCOLICUM*

2.1. Metabolic Pathway of Diol Fermentation in *C. glycolicum*

Gaston and Stadtman [13] found that *C. glycolicum* can utilize 1,2-ethanediol and 1,2-propanediol as sole carbon and energy source. During fermentation the diols are converted to equal amounts of their corresponding alcohols and acids. Specifically, they showed that 2 mol of 1,2-ethylene glycol is converted to 1 mol each of acetate and ethanol in a balanced redox reaction. To elucidate the metabolic pathway of 1,2-ethanediol and 1,2-propanediol fermentations in *C. glycolicum*, cell-free extracts and membrane fractions were assayed for the enzyme activities outlined in Fig. 1. All enzyme activities with the exception of diol dehydrase were found to be soluble and cytosolic [14]. The diol dehydrase activity is located in the membrane fraction and cannot be released by sonication or other common cell disintegration techniques. The enzyme was found to be inducible, completely absent in bacteria grown on glucose as carbon and energy source [14].

The next two enzymes in the pathway, alcohol and aldehyde dehydrogenase, are both nicotinamide adenine dinucleotide (NAD)-dependent, constitutive enzymes. The last two enzymes in the fermen-

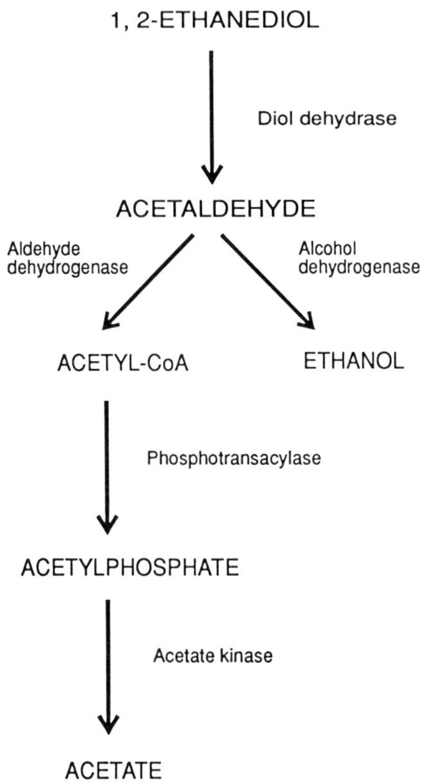

FIG. 1. Metabolic pathway of 1,2-ethanediol fermentation in *C. glycolicum*. (Reproduced from Ref. 14 by permission of Academic Press.)

tative pathway, phosphotransacylase and kinase, were also found to be constitutive. Also, the acetate/propionate kinase is most likely catalyzing the only ATP generation reaction in *C. glycolicum* during growth on 1,2-ethanediol and 1,2-propanediol. Here 1 mol of ATP is generated per 2 mol of diol fermented [14].

2.2. Enzyme Levels

The levels of all enzymes involved in the fermentation of 1,2-ethanediol and 1,2-propanediol have been determined using cells grown on both

TABLE 1

Specific Activities of Enzymes Involved in Anaerobic Diol Metabolism in *C. glycolicum*

		Activity (U/mg)				
Cells grown on	Substrate in assay	Diol dehydrase	Alcohol dehydrogenase	Aldehyde dehydrogenase	Phosphotrans-acylase	Acetate kinase
1,2-Ethanediol	2-Carbon	0.13 (1.5)[a]	0.030	0.19	1.70	9.0
	3-Carbon	0.05	0.028	0.37	1.30	6.6
1,2-Propanediol	2-Carbon	0.11	0.012	0.13	0.95	8.9
	3-Carbon	0.05	0.018	0.67	0.80	6.5
Glucose	2-Carbon	0	0.009	0.08	0.50	8.6
	3-Carbon	0	0.009	0.06	0.40	5.9

[a]Specific activity of enzyme prepared in the anaerobic laboratory with an oxygen level of less than 10 ppm.
Source: Reproduced from Ref. 14 by permission of Academic Press.

diols and glucose (Table 1) [14]. Diol dehydrase activity, found only in the membrane fractions, was assayed under anaerobic conditions by following aldehyde formation [15]. Diol dehydrase exhibited approximately a 2.5-fold higher activity with the two-carbon substrate than with the three-carbon substrate in both 1,2-ethanediol- and 1,2-propanediol-grown cells, while it was completely absent in cells cultured on glucose [14].

Due to its extreme oxygen sensitivity [14], diol dehydrase from 1,2-ethanediol-grown cells was also assayed inside the anaerobic nitrogen laboratory at the National Institutes of Health [16] at oxygen levels of less than 10 ppm. Here, a 12-fold higher specific activity was obtained [14].

3. PROPERTIES OF MEMBRANE-BOUND DIOL DEHYDRASE FROM *C. GLYCOLICUM*

3.1. Assays and Substrate Specificity

The diol dehydrase from *C. glycolicum* differs from other bacterial diol dehydrases [12] in that it is strongly associated with the cell membrane, exhibits an extreme oxygen sensitivity, and does not seem to utilize a cobamide coenzyme as cofactor [14,17].

Due to its extreme oxygen sensitivity and its strong association with the membrane, the diol dehydrase has never been studied in a purified state. Instead all characterization experiments have been carried out with solubilized membrane preparations from *C. glycolicum* [17]. Enzyme activity was followed anaerobically in the physiological direction by measuring aldehyde formation with the 2,4-dinitrophenylhydrazine method under reducing conditions at pH 8.0 [15]. Crude membrane preparations obtained after sonication and centrifugation under anaerobic conditions were used for determination of substrate specificity (Table 2) [14]. The enzyme exhibits a strict specificity for 1,2-ethanediol and 1,3-propanediol. Essentially no activity could be detected with other substrates. In contrast, the coenzyme B_{12}–dependent enzyme from *Klebsiella pneumoniae* has a broader substrate specificity and in addition utilizes glycerol as substrate [12]. A K_m value of 3 mM for 1,2-ethanediol was obtained using membranes from cells cultured on

TABLE 2

Substrate Specificity of Membrane-Bound Diol Dehydrase from 1,2-Ethanediol-Grown Cells

Substrate (20 mM)	Relative activity (%)
1,2-Ethanediol	100
1,2-Propanediol	40
1,2-Butanediol	<1
2,3-Butanediol	<1
1,2-Pentanediol	0
Glycerol	0
Ethanolamine	0
2-Fluoroethanol	<1

Source: Reproduced from Ref. 14 by permission of Academic Press.

this substrate [14], which is in the same range as previously reported for the B_{12}–dependent enzyme from *K. pneumoniae* [18,19].

3.2. Diol Dehydrase Inhibitors

The enzyme exhibited a marked inhibition with several different compounds (Table 3) [14]. Inhibition with *p*-chloromercuriphenylsulfonic acid and iodoacetamide indicate that essential sulfhydryl groups are present at the catalytic site. Metal-chelating compounds such as ethylene glycol bis(β-aminoethyl ether)-N,N,N′,N′-tetraacetate (EGTA), ethylenediamine-N,N,N′,N′-tetraacetate (EDTA), and o-phenanthroline gave a strong inhibition at low levels, suggesting that a metal ion is needed for catalysis. Addition of Fe^{2+} could partially reverse the inhibition, while other divalent metal ions had no effect. Hydroxyurea and hydroxylamine, known radical scavengers, also inhibited diol dehydrase, indicating the possible involvement of a radical in the catalytic mechanism of the enzyme [14].

TABLE 3

Effect of Inhibitors on Membrane-Bound Diol Dehydrase Activity

Inhibitor	Concentration	Inhibition (%)
Hydroxylamine	0.1 μM	15
	1 μM	51
	10 μM	97
Hydroxyurea	1 mM	97
EGTA	50 μM	62
	100 μM	85
EDTA	500 μM	95
p-Chloromercuriphenylsulfonic acid	100 μM	100
Iodoacetamide	5 mM	100
o-Phenanthroline	1 mM	85

Source: Reproduced from Ref. 14 by permission of Academic Press.

3.3. General Properties of Diol Dehydrase

As mentioned earlier, the enzyme is extremely sensitive to exposure to oxygen and the activity is rapidly destroyed in air, even in the presence of a reducing agent [14]. Such a sensitivity has not been seen with the other diol dehydrases studied [12].

In order to detect any enzyme activity, all steps, including breaking the cells, centrifugation, and assaying, had to be carried out under strictly anaerobic conditions either in the anaerobic room at National Institutes of Health [16] or using sparging with nitrogen or argon [14]. In addition, a reducing agent was needed for maximal activity. The activity without a reducing agent present was only 11% compared to using 1 mM dithiothreitol (DTT) (Table 4) [14].

The *C. glycolicum* enzyme preparations cannot be frozen since this procedure causes a complete loss of activity [14]. Storage at 4°C under nitrogen and a reducing agent, the best conditions found for stability, led to a loss of 70% of the activity after 4 days [14]. Therefore all experiments had to be carried out using fresh membrane preparations.

As is the case with several B_{12}–dependent enzymes, the diol dehydrases from both *K. pneumoniae* and *Propionibacterium freuden-*

TABLE 4

Effect of DTT on Membrane-Bound Diol Dehydrase Activity

Concentration (mM)	Relative activity (%)
0	11
0.5	82
1	100
5	70
10	57

Source: Reproduced from Ref. 14 by permission of Academic Press.

reichii require the presence of a monovalent cation, such as potassium or sodium, for maximal activity [12]. These monovalent cations had no detectable effect on the dehydrase from *C. glycolicum* [14].

The activity of the enzyme from *C. glycolicum* shows a fairly sharp pH optimum at pH 8.0 [14], whereas the *K. pneumoniae* enzyme exhibits a constant activity between pH 6.0 and 10.0 [9].

C. glycolicum cells grown on 1,2-ethanediol have a low cobamide content [20]. This suggested that the diol dehydrase might not be B_{12}–dependent. Addition of the intrinsic factor, a B_{12}–binding protein, to the reaction mixture and irradiation with visible light, procedures known to inhibit coenzyme B_{12}–dependent enzymes [21,22], had no effect on enzyme activity in washed membrane preparations [14]. Also, addition of adenosylcobalamin had no effect on the reaction rate [14]. In contrast, the diol dehydrases from both *K. pneumoniae* and *P. freudenreichii* are completely inactive in the absence of added coenzyme [3,6,9].

4. CHARACTERIZATION OF DIOL DEHYDRASE

4.1. Solubilization of Diol Dehydrase

The enzyme is strongly associated with the cell membrane and therefore is most likely an integral membrane protein [14,17]. It could not be liberated using various procedures including treatment with high-ionic-strength organic solvents, osmotic shock, and different ionic and nonionic detergents. Most detergents, with the exception of Tween 80, caused a complete inhibition of enzymatic activity and in no case could any activity be detected in the supernatant after treatment with detergent [14].

It was found, however, that sonication of the membrane pellet anaerobically in 100 mM 2-(N-cyclohexylamino)ethanesulfonic acid (CHES) buffer, pH 8.6–9.0, containing 2 mM DTT, solubilized diol dehydrase with retention of enzyme activity [17]. This treatment essentially dissolved the whole membrane pellet fraction. Essentially all activity was released after a 3-min sonication (Fig. 2) [17]. CHES is an amphiphilic buffer substance with a hydrophobic cyclohexyl group and hydrophilic amino and sulfonic acid groups which make the molecule zwitterionic. The buffer has a pK_a value of 9.27 in water at 25°C [23]. Here, the detergent-like properties of CHES disrupted the membrane structure and released diol dehydrase with retention of enzymatic activity. In these experiments, the recovery of active enzyme in the supernatant was approximately 10% of the activity originally present in

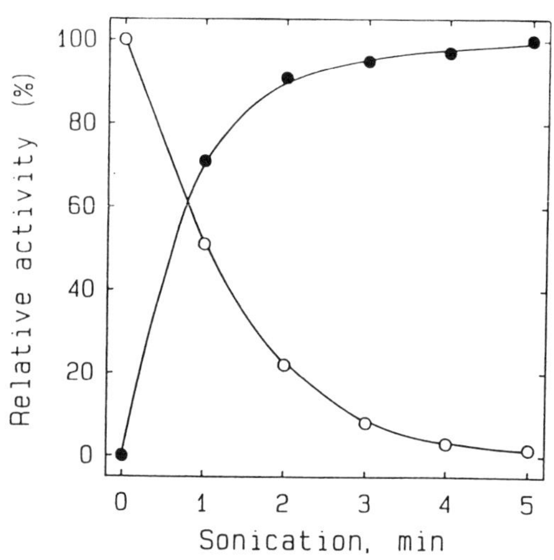

FIG. 2. Solubilization of diol dehydrase by sonication of crude membrane preparations in 0.1 M CHES, pH 8.6/2 mM DTT. (○) Enzyme activity in the resuspended pellet; (●) enzyme activity released into the supernatant. Maximum activity released equals 5% of the activity present in the original membrane preparation.

the membrane preparation prior to CHES treatment. The solubilized enzyme exhibited the same poor stability as the membrane bound [17].

4.2. Effect of Additives

To improve the solubilization procedure, to recover more active enzyme in the supernatant following solubilization, and to improve the stability of the enzyme, different compounds were added to the CHES buffer before sonication of the membrane fraction. Addition of varying concentrations of cell extracts or lipids from *C. glycolicum* did not improve the recovery. Inclusion of glycerol or the substrate 1,2-ethanediol almost completely inhibited the activity. Catalase and superoxide dismutase for decomposition of active oxygen species, phosphoglycerides and organic solvents such as dimethylformamide and tetrahydrofuran all decreased the recovery of activity [17].

A significant increase in the recovery of active diol dehydrase was seen during sonication in the presence of varying concentrations of dimethyl sulfoxide. Addition of 30% dimethyl sulfoxide improved the recovery of activity in the supernatant more than 3.5-fold, with very little activity remaining in the pellet fraction after centrifugation [17]. Inclusion of 0.15 mg/mL of phosphatidylcholine in a solubilization mixture containing 30% dimethyl sulfoxide, 0.1 M CHES, pH 9.0, and 2 mM DTT increased recovery even further. Here 50% of the total diol dehydrase activity present in the membrane pellet could be recovered in a soluble form. This procedure did not, however, improve the stability properties of the enzyme [17].

4.3. EPR Experiments

It has been shown that both the membrane-bound and the solubilized diol dehydrase preparations exhibit a characteristic electron paramagnetic resonance (EPR) signal with a strong correlation with enzyme activity (Fig. 3) [17]. Membrane preparations, prepared under anaerobic conditions, from cells grown on 1,2-ethanediol exhibited a characteristic EPR signal at $g = 2.02$ (Fig. 3A), whereas membrane preparations from cells grown solely on glucose virtually lack the characteristic EPR signal (Fig. 3C). The latter cells also lack a measurable diol

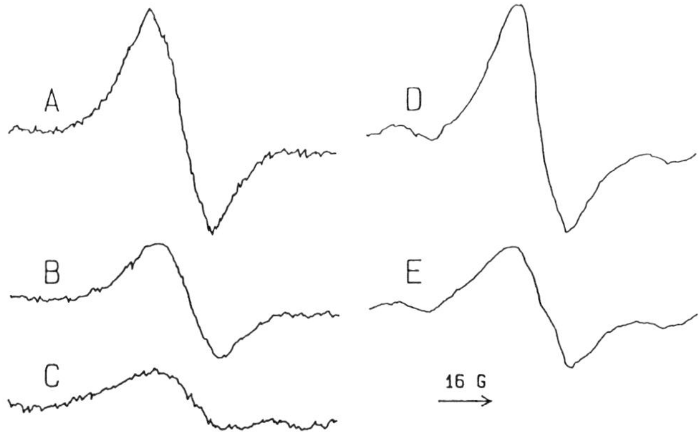

FIG. 3. EPR spectra of crude membranes and solubilized membrane preparations of *C. glycolicum*. All spectra are centered around a *g* value of 2.02. (A) Crude membrane preparations from 1,2-ethanediol-grown cells; (B) as for spectrum A, but after a 15 min incubation with 10 mM hydroxyurea; (C) crude membrane preparations from glucose grown cells; (D) membrane preparation after solubilization with 0.1 M CHES, pH 9.0/2 mM DTT and centrifugation; (E) as for spectrum D, but after incubation with 10 mM hydroxyurea.

dehydrase activity, since the diol dehydrase is induced by growth on diols [14].

A strong correlation can also be seen between both the membrane-bound and the solubilized diol dehydrase activity and the amplitude of the EPR signal in inhibition experiments with a radical scavenger, hydroxyurea (Fig. 3). A 50% decrease in the EPR signal amplitude at $g = 2.02$ could be seen following a 15 min incubation with 10 mM hydroxyurea (Fig. 3B,E). The decrease in EPR signal amplitude was accompanied by a virtually complete loss of enzyme activity. Addition of hydroxylamine gave a similar effect [17]. Other agents which previously were shown to completely inhibit diol dehydrase activity, such as p-chloromercuriphenylsulfonate and oxygen, also abolished the EPR signal [17].

4.4. Is a Radical Involved in the Reaction Mechanism?

The lack of sensitivity of the diol dehydrase to irradiation with visible light, addition of intrinsic factor, or addition of adenosylcobalamin suggests that coenzyme B_{12} is not involved in the reaction mechanism of diol dehydrase from *C. glycolicum* [14,17].

As described above, a marked sensitivity to two radical scavengers, hydroxyurea and hydroxylamine, indicated the possibility of a radical mechanism. Further, the correlation of enzyme activity from both membrane-bound and solubilized diol dehydrase with a characteristic EPR signal suggested the presence of an enzyme-bound radical species involved in the catalytic mechanism [17]. The radical is possibly an integral part of this clostridial diol dehydrase, thereby acting as a substitute for the coenzyme B_{12}, present in other organisms, such as *K. pneumoniae* [9].

For comparison, in the coenzyme B_{12}–independent ribonucleotide reductase from *E. coli*, a tyrosyl radical is involved in the reaction mechanism [24]. This enzyme exhibits a similar sensitivity to hydroxyurea [24,25].

5. CONCLUSIONS

Relatively little is known about the membrane-bound non–B_{12}-dependent diol dehydrase from the strictly anaerobic bacterium *C. glycolicum*, compared to the B_{12}–dependent enzyme, first isolated from *K. pneumoniae* [9]. The former enzyme has never been purified beyond solubilization due to its inherent instability under both anaerobic and aerobic conditions. Various attempts to stabilize the enzyme with a variety of additives have so far been unsuccessful [17]. Therefore all experiments have been carried out with either membrane preparations or solubilized enzyme fractions [14,17].

The most interesting property of the dehydrase from *C. glycolicum* is that, despite its lack of coenzyme B_{12} dependence, it still seems to involve some type of radical mechanism [17]. Initial experiments with membrane preparations showed a pronounced sensitivity to low concentrations of radical scavengers, such as hydroxyurea and hydroxylamine [14,17]. Subsequently, EPR experiments with both membrane preparations and solubilized enzyme showed a strong correlation between EPR signal and enzyme activity [17]. Also, membranes prepared from glucose-grown cells, lacking dehydrase activity, exhibited virtually no EPR signal [17].

Hopefully future research in this area will find ways to stabilize the dehydrase after solubilization so that further studies can shed more light on the currently elusive mechanism of the non–B_{12} radical reaction.

ABBREVIATIONS

ATP	adenosine 5′-triphosphate
CHES	2-(N-cyclohexylamino)ethanesulfonic acid
DTT	dithiothreitol
EDTA	ethylenediamine-N,N,N′,N′-tetraacetate
EGTA	ethylene glycol bis(β-aminoethyl ether)-N,N,N′,N′-tetraacetate
EPR	electron paramagnetic resonance
NAD	nicotinamide adenine dinucleotide
U	unit

REFERENCES

1. A. Willetts, *Biochim. Biophys. Acta*, *588*, 302 (1979).
2. A. Willetts, *Biochim. Biophys. Acta*, *677*, 194 (1981).
3. T. Toraya, S. Honda, and S. Fukui, *J. Bacteriol.*, *139*, 39 (1979).
4. N. Nishio, T. Kawagishi, R. Matsuno, and T. Kamikubo, *Agric. Biol. Chem.*, *42*, 1095 (1978).
5. Y. Ichikawa, N. Hosoi, and Y. Kitamoto, *Hakkokogaku Kaishi*, *55*, 15 (1977).
6. N. Hosoi, K. Morimoto, C. Ozaki, Y. Kitamoto, and Y. Ichikawa, *J. Ferment. Technol.*, *56*, 566 (1978).
7. O. Yagi and K. Yamada, *Agr. Biol. Chem.*, *33*, 1587 (1969).
8. Y. Tanaka, K. Fujii, A. Tanaka, and S. Fukui, *J. Ferment. Technol.*, *53*, 354 (1975).
9. H. A. Lee and R. H. Abeles, *J. Biol. Chem.*, *238*, 2367 (1963).
10. A. A. Poznanskaya, K. Tanizawa, K. Soda, T. Toraya, and S. Fukui, *Arch. Biochem. Biophys.*, *194*, 379 (1979).

11. D. E. McGee and J. H. Richards, *Biochemistry*, *20*, 4293 (1981).
12. T. Toraya and S. Fukui, in B_{12}, Vol. 2 (D. Dolphin, ed.), John Wiley and Sons, New York, 1982, p. 233 ff.
13. L. W. Gaston and E.R. Stadtman, *J. Bacteriol.*, *85*, 356 (1963).
14. M. G. N. Hartmanis and T. C. Stadtman, *Arch. Biochem. Biophys.*, *245*, 144 (1986).
15. H. Böhme and O. Winkler, *Z. Anal. Chem.*, *142*, 1 (1954).
16. J. M. Poston, T.C. Stadtman, and E. R. Stadtman, in *Methods in Enzymology*, Vol. 22 (W. B. Jakoby, ed.), Academic Press, New York, 1971, p. 49 ff.
17. M. G. N. Hartmanis and T. C. Stadtman, *Proc. Natl. Acad. Sci. USA*, *84*, 76 (1987).
18. T. Toraya, T. Shirakashi, T. Kosuga, and S. Fukui, *Biochem. Biophys. Res. Commun.*, *69*, 475 (1976).
19. W. W. Bachovchin, R. G. Eagar, K. W. Moore, and J. H. Richards, *Biochemistry*, *16*, 1082 (1977).
20. E. R. Stadtman, in *Current Aspects of Biochemical Energetics* (N. O. Kaplan and E. P. Kennedy, eds.), Academic Press, New York, 1966, p. 39 ff.
21. H. Weissbach, J. Toohey, and H.A. Barker, *Proc. Natl. Acad. Sci. USA*, *45*, 521 (1959).
22. T. C. Stadtman, *J. Biol. Chem.*, *237*, 2409 (1962).
23. M. J. Taylor and Y. Pignat, *Cryobiology*, *19*, 99 (1982).
24. B.-M. Sjöberg, P. Reichard, A. Gräslund, and A. Ehrenberg, *J. Biol. Chem.*, *252*, 536 (1977).
25. N. C. Brown, R. Eliasson, P. Reichard, and L. Thelander, *Eur. J. Biochem.*, *9*, 512 (1969).

8

Diol Dehydrase and Glycerol Dehydrase, Coenzyme B_{12}-Dependent Isozymes

Tetsuo Toraya

Department of Biotechnology, Faculty of Engineering,
Okayama University, 3-1-1 Tsushima-Naka,
Okayama 700, Japan

1. INTRODUCTION

1,3-Propanediol is known to be produced by a number of microorganisms including bacteria of Enterobacteriaceae during glycerol fermentation. Voisenet postulated that glycerol is first dehydrated to β-hydroxypropionaldehyde [1]. This was confirmed with *Klebsiella oxytoca* (formerly *Aerobacter aerogenes* and *Klebsiella pneumoniae*) ATCC 8724 [2] by Abeles and coworkers, who found that the cells grown anaerobically on glycerol convert glycerol to β-hydroxypropionaldehyde, 1,2-propanediol to propionaldehyde, and 1,2-ethanediol to acetaldehyde. They isolated adenosylcobalamin (AdoCbl) (vitamin B_{12} coenzyme)–dependent diol dehydrase (diol dehydratase or DL-1,2-propanediol hydro-lyase, EC 4.2.1.28), which is responsible for these dehydrations [3] (1). Although in intact cells glycerol and 1,2-propanediol reacted at about the same rate,

$$\underset{\displaystyle OH}{\overset{}{R{-}\underset{|}{C}HCH_2OH}} \longrightarrow R{-}CH_2CHO + H_2O \qquad (R = CH_3,\ H,\ \text{or}\ HOCH_2) \quad (1)$$

glycerol was dehydrated at less than 5–10% the rate of 1,2-propanediol with cell-free extracts [4] or the purified diol dehydrase [3] in the presence of added AdoCbl. The present author and coworkers solved this apparent contradiction by demonstrating that glycerol not only serves as a good substrate but also causes rapid and irreversible inactivation of the enzyme [5].

Smiley and Sobolov demonstrated the in vitro conversion of glycerol to β-hydroxypropionaldehyde by extracts of an acrolein-forming *Lactobacillus* sp. [6]. The enzyme, glycerol dehydrase (glycerol dehydratase or glycerol hydro-lyase, EC 4.2.1.30), of *K. pneumoniae* ATCC 25955 (formerly *Aerobacter aerogenes* PZH 572, Warsaw) [7] was purified by Pawelkiewicz and coworkers [8,9]. Although the two dehydrases were isofunctional and showed similar features, Toraya and Fukui demonstrated immunochemically that they are indeed different enzymes, i.e., these are isozymes which are induced specifically [10].

In this chapter, the current status of AdoCbl-dependent diol dehydrase and glycerol dehydrase are reviewed (for previous reviews on diol dehydrase, see Refs. 11–14).

2. PURIFICATION, PROPERTIES, AND STRUCTURES

2.1. Assays

Diol dehydrase and glycerol dehydrase are assayed by either colorimetric or coupled methods. The aldehydic products are converted either to the 2,4-dinitrophenylhydrazones or to the azine derivatives in the 2,4-dinitrophenylhydrazine (DNPH) method [3,15] and in the 3-methyl-2-benzothiazolinonehydrazone (MBTH) method [9,16], respectively. The MBTH method is superior to the DNPH method in sensitivity by a factor of about 3 [16]. The tryptophan method is also employed for the colorimetric assay of glycerol dehydrase [6]. This method involves conversion of β-hydroxypropionaldehyde to acrolein, which forms a colored product with tryptophan.

In the coupled assay, the aldehydes formed are reduced to the corresponding alcohols by the reduced form of nicotinamide adenine dinucleotide (NADH) in the presence of excess alcohol dehydrogenase [17,18]. The decrease in the absorbance of NADH at 340 nm per unit time is a measure of the dehydrase activities. This method is advantageous in collecting the kinetic data.

2.2. Purification

The apoenzyme of diol dehydrase was purified to near homogeneity by Abeles and coworkers from cell-free extracts of *K. oxytoca* ATCC 8724 [3] grown without aeration in a glycerol-1,2-propanediol medium [19]. Poznanskaja et al. purified the enzyme to homogeneity by the conventional method [20]. Although these procedures were developed on the supposition that the enzyme exists in the cytosol, McGee and Richards pointed out that this enzyme is membrane-associated [21]. They purified the enzyme from detergent-extract of membrane to homogeneity in a yield of 76%.

Affinity chromatography was useful for purification of diol dehydrase [22]. Several types of immobilized corrinoids were prepared and used as affinity adsorbents. Diol dehydrase adsorbed on a corrinoid gel was eluted by removal of K^+ from the eluting solution.

Glycerol dehydrase of *K. pneumoniae* ATCC 25955 was first purified to homogeneity as a stable complex with hydroxocobalamin (OH-Cbl) [8]. This inactive complex can be partially converted to the active holoenzyme by replacement of OH-Cbl with AdoCbl in the presence of Mg^{2+} and SO_3^{2-}. The almost homogeneous preparation of apoglycerol dehydrase was obtained from cell-free extracts by Stroinski et al. [9].

2.3. Relationship Between Diol Dehydrase and Glycerol Dehydrase

Since diol dehydrase and glycerol dehydrase were isofunctional enzymes and showed similar protein chemical and catalytic properties, the relationship between these enzymes was immunochemically studied [10]. Glycerol dehydrase of *K. pneumoniae* ATCC 25955 did not cross-react with rabbit antiserum raised against the purified diol dehydrase of *K. oxytoca* ATCC 8724, although the *K. pneumoniae* ATCC 25955 diol enzyme was immunoprecipitated with the antiserum. Furthermore, these two enzymes were distinguishable in monovalent cation–selectivity pattern [10], in substrate specificity [10,23], and in affinity for AdoCbl [24]. This indicates that diol dehydrase and glycerol dehydrase are different enzymes. Since many strains of bacteria of Enterobacteriaceae form both diol and glycerol dehydrases depending on the growth substrate (1,2-propanediol or glycerol, respectively) [10,23,24], these enzymes can be considered as isozymes that are specifically inducible.

2.4. Subunits and Molecular Weights

The molecular weight of diol dehydrase purified from cell-free extracts was determined to be approximately 230K ± 10K by the sedimentation equilibrium method [20] and 220K ± 5K [20], or 220–240K [25] by gel filtration. One mole of AdoCbl is bound per mol of enzyme [25]. In contrast, the enzyme purified from detergent-extract of the membrane gave higher molecular weights upon gel filtration, i.e., 250K [21,26] and 500K [21] in the presence and absence of detergents, respectively.

Diol dehydrase purified from cell-free extracts was separated into two dissimilar protein components, designated F and S, upon DEAE-cellulose column chromatography in the absence of substrate [27]. None of them alone possessed any appreciable activity, and the enzyme activity was restored when both components were recombined. Both components were required for the tight binding of cobalamins. Reassociation of F and S was strongly promoted by a substrate, i.e., 1,2-propanediol [20], glycerol, 1,2-butanediol, or styrene glycol (T. Toraya, unpublished). The molecular weights of F and S were approximately 26K and 200K, respectively [20]. It was suggested from sodium dodecyl sulfate (SDS) gel electrophoresis that F is composed of a single polypeptide (M_r 25.5K), whereas S consists of at least four different kinds of polypeptide chains (M_r 60K, 23K, 15.5K, and 14K).

In contrast, McGee and Richards demonstrated that the enzyme purified from the detergent-extract of the membrane consists of four subunits (M_r 60K, 51K, 29K, and 15K) in the ratio of 2:1:2:2, respectively [21,28]. They claimed that the apparent discrepancy is due to a limited proteolysis during purification of the enzyme from cell-free extracts. However, Tanizawa et al. reported that the enzyme purified from the detergent-extract of the membrane contains an additional subunit (M_r 26K) which is stained poorly with Coomassie Brilliant Blue [26].

The molecular weight of the glycerol dehydrase–OH-Cbl complex was determined to be 188K by the sedimentation equilibrium method [8]. One mole of OH-Cbl is bound by 1 mol of enzyme. Glycerol dehydrase was also separated into two dissimilar components, designated A (M_r 189K) and B (M_r 22K) [29,30]. Both components were necessary for enzymic activity and for binding of AdoCbl. B dissociated further into subunits (M_r 90K + 25K) in 0.1 M KCl [9], and A dissociates into subunits (M_r 11K) at lower ionic strength and at more acidic pHs, e.g., 6.7 [30].

```
                28,071 Polypeptide (270 amino acid residues)
  1 MSSNELVEQI MAQVIARVAT PEQQAIPTEN HPIRETAMAE KSCSLTEFVG TAIGDTLGLV  60
 61 IANVDTALLD AMKLEKRYRS IGILGARTGA GPHIMAADEA VKATNTEVVS IELPRDTKGG 120
121 AGHGSLIILG GNDVSDVKRG IEVALKELDR TFGDVYGNEA GHIELQYTAR ASYALEKAFG 180
181 APIGRACGII VGAPASVGVL MADTALKSAN VEVVAYSSPA HGTSFSNEAI LVISGDSGAV 240
241 RQAVTSAREI GKTVLATLGA EPKNDRPSYI

                60,348 Polypeptide (554 amino acid residues)
  1 MRSKRFEALA KRPVNQDGFV KEWIEEGFIA MESPNDPKPS IKIVNGAVTE LDGKPVSDFD  60
 61 LIDHFIARYG INLNRAEEVM AMDSVKLANM LCDPNVKRSE IVPLTTAMTP AKIVEVVSHM 120
121 NVVEMMMAMQ KMRARRTPSQ QAHVTNVKDN PVQIAADAAE GAWRGFDEQE TTVAVARYAP 180
181 FNAIALLVGS QVGRPGVLTQ CSLEEATELK LGMLGHTCYA ETISVYGTEP VFTDGDDTPW 240
241 SKGFLASSYA SRGLKMRFTS GSGSEVQMGY AEGKSMLYLE ARCIYITKAA GVQGLQNGSV 300
301 SCIGVPSAVP SGIRAVLAEN LICSSLDLEC ASSNDQTFTH SDMRRTARLL MQFLPGTDFI 360
361 SSGYSAVPNY DNMFAGSNED AEDFDDYNVI QRDLKVDGGL RPVREEDVIA IRNKAARALQ 420
421 AVFAGMGLPP ITDEEVEAAT YAHGSKDMPE RNIVEDIKFA QEIINKNRNG LEVVKALAQG 480
481 GFTDVAQDML NIQKAKLTGD YLHTSAIIVG DGQVLSAVND VNDYAGPATG YRLQGERWEE 540
541 IKNIPGALDP NEID

                24,113 Polypeptide (224 amino acid residues)
  1 MEINEKLLRQ IIEDVLSEMK GSDKPVSFNA PAASAAPQAT PPAGDGFLTE VGEARQGTQQ  60
 61 DEVIIAVGPA FGLAQTVNIV GIPHKSILRE VIAGIEEEGI KARVIRCFKS SDVAFVAVEG 120
121 NRLSGSGISI GIQSKGTTVI HQQGLPPLSN LELFPQAPLL TLETYRQIGK NAARYAKRES 180
181 PQPVPTLNDQ MARPKYQAKS AILHIKETKY VVTGKNPQEL RVAL

                19,173 Polypeptide (173 amino acid residues)
  1 MNTDAIESMV RDVLSRMNSL QGEAPAAAPA AGGASRSARV SDYPLANKHP EWVKTATNKT  60
 61 LDDFTLENVL SNKVTAQDMR ITPETLRLQA SIAKDAGRDR LAMNFERAAE LTAVPDDRIL 120
121 EIYNALRPYR STKEELLAIA DDLESRYQAK ICAAFVREAA TLYVERKKLK GDD
```

FIG. 1. Deduced amino acid sequences of the polypeptides [31]. The 60348, 24113, and 19173 polypeptides are the functional subunits of diol dehydrase.

2.5. Cloning and Sequence Analysis of the Genes

The genes encoding diol dehydrase were cloned by screening a plasmid library of *K. oxytoca* ATCC 8724 by colony hybridization with a synthetic oligonucleotide probe followed by measurement of diol dehydrase activity [31]. From the nucleotide sequence analyzed, protein sequences of the subunits were deduced. As shown in Fig. 1, four open reading frames found in the plasmid encoded polypeptides of 270-, 554-, 224-, and 173-amino-acid residues with predicted molecular weights of 28071, 60348, 24113, and 19173, respectively. The deduced amino acid sequences of the 60348, 24113, and 19173 polypeptides contained N-terminal sequences essentially identical to those reported by McGee and Richards [28] for 60K, 29K, and 15K subunits, respectively. These three polypeptides were present in the extracts of a recombinant *E. coli* strain overexpressing the enzyme genes, indicating that they are the functional subunits of diol dehydrase. The N-terminal sequence for the 51K subunit reported [28] was not found in the deduced sequences.

The structural gene encoding glycerol dehydrase has not yet been characterized, although the *dha* regulon of *K. pneumoniae* and *Citrobacter freundii* was cloned in *E. coli* (see Sec. 6.3).

3. CATALYTIC PROPERTIES AND MECHANISM OF ACTION

3.1. Substrate Specificity

Diol dehydrase catalyzes the dehydration reactions of 1,2-propanediol, 1,2-ethanediol, glycerol, and other glycols [5,18,32,33] (Table 1). The initial velocity with glycerol was the highest. However, the enzyme preferred 1,2-propanediol to glycerol, as judged from a 1-min assay [10], since glycerol brings about rapid inactivation of the enzyme [5,18] (see Sec. 5.1). (S)-1,2-propanediol reacted at a faster rate than the (R) isomer in competitive reactions, but when the diols were run independently, the rate with the (R) isomer was 1.7–1.8 times higher than that with the (S) isomer [18,34–36]. Jensen and Neese [36] suggested that this reversal is due to the ratio of K_m values ($K_{m(R)}/K_{m(S)}$ = 3.1–3.2). Although

TABLE 1

Substrate Specificity of Diol Dehydrase

Substrate	K_m or K_i (mM)	k_{cat} (s^{-1})[a]	k_{inact} (min^{-1})	Ref.
(RS)-1,2-Propanediol	0.030, 0.18	250, 337[b]	~0	5,18
(R)-1,2-Propanediol	0.060	340	~0	18
(S)-1,2-Propanediol	0.0186	191	~0	18
(RS)-3-Fluoro-1,2-propanediol	0.307	128		32
(R)-3-Fluoro-1,2-propanediol	0.147	104		32
(S)-3-Fluoro-1,2-propanediol	1.32	340		32
3-Chloro-1,2-propanediol	0.41	>0	0.36	18
Trifluoro-1,2-propanediol	0.124	>0	0.016	18
Thioglycerol	0.01	~0	1.74	18
Glycerol	1.60, 1.54	412, 371[b]	1.30	5,18
1,2-Ethanediol	0.66, 1.08	150, 286[b]	0.098	5,18
2-Methyl-1,2-propanediol	0.88	18	0.47	18
1,2-Butanediol	0.59, 2.5	>0, 7.2[b]	0.054	5,18
2,3-Butanediol	0.20, 0.64	4.0, 15.1[b]	0.068	5,18
meso-2,3-Butanediol	0.20	4	0.068	33
d-2,3-Butanediol	0.55		0	33
dl-2,3-Butanediol	0.107		0	33

[a]Based on a molecular weight of 250K and specific activity of 60 U/mg with 1,2-propanediol as substrate [18,32,33].
[b]Based on a molecular weight of 230K and specific activity of 88 U/mg [5].

substitution for hydrogen at C-3 by chlorine, a mercapto group, or three fluorines diminished substrate activity, 3-fluoro-1,2-propanediol was as effective a substrate as unsubstituted 1,2-propanediol. 1,2-Butanediol and 2-methyl-1,2-propanediol also served as substrates but to a much lesser degree. 2,3-Butanediol was converted into methyl ethyl ketone. McGee and Richards reported that the meso isomer acts as a substrate, while the *d* and *dl* compounds are purely competitive inhibitors [33], in consistency with the stereochemistry of the hydrogen transfer (see Sec. 3.4).

Glycerol dehydrase also converts glycerol, 1,2-propanediol, and

1,2-ethanediol to the corresponding deoxyaldehydes [7]. As judged from a 1-min assay, this enzyme preferred glycerol to 1,2-propanediol as substrate, in contrast to the diol enzyme [10].

3.2. Requirement for Monovalent Cations

Diol dehydrase requires certain monovalent cations for catalytic activity [3,37]. K^+, NH_4^+, Tl^+, and Rb^+ were the most effective cofactors. Cs^+ and Na^+ also activated the enzyme but to a lesser extent. Li^+ was essentially without activating effect. There is a good relationship between effectiveness as cofactor and the crystal ionic radii of the monovalent cations [37]. Of the substituted ammonium ions tested, $CH_3NH_3^+$ was 92% as active as K^+ while $(CH_3)_2NH_2^+$, $(CH_3)_3NH^+$, $C_2H_5NH_3^+$, and $(C_2H_5)_2NH_2^+$ were not effective [37].

The complexes of diol dehydrase with AdoCbl, cyanocobalamin (CN-Cbl), and methylcobalamin (MeCbl) were resolved into the apoenzyme and cobalamin when subjected to gel filtration on a Sephadex G-25 in the absence of K^+ [37], but not at all in the presence of both K^+ and substrate. Upon gel filtration in the presence of K^+ only, the holoenzyme dissociated gradually according to the first-order kinetics. Therefore, K^+ seems essential for the binding of cobalamins to diol dehydrase. Other monovalent cations of similar size (NH_4^+ and Rb^+) replaced K^+ in participation in the binding of AdoCbl to the enzyme [37].

The requirement of glycerol dehydrase for monovalent cations was also reported, but the pattern of monovalent cation selectivity was substantially different from that of diol dehydrase [10]. K^+, NH_4^+, and Rb^+ were most effective, whereas Cs^+, Na^+, and Li^+ were unable to activate this enzyme at all. The presence of high concentrations of K^+ brought about dissociation of the apoenzyme into subunits [30].

3.3. Amino Acid-Residues Essential for Catalysis

Chemical modification studies demonstrated that cysteine [3,38], arginine [39], lysine [40], and histidine [41] are essential amino acid-residues for catalytic activity. Of the two reactive mercapto groups of cysteine, one was essential for catalytic activity and located at or in close proximity to the coenzyme binding site, while the other was nonessential [38]. An essential guanido group of arginine may be involved in the coenzyme binding site [39]. Imidazole group(s) of histidine were required for the coenzyme binding [41]. One lysine residue, which was located in F, was essential for association of subunits [40].

Glycerol dehydrase also contains essential mercapto group(s) in B, which is sensitive to N-ethylmaleimide and pCMB [7].

3.4. Mechanism of Action

3.4.1. General Aspects and Hydroxyl Group Transfer

The irreversible conversion of vic-diols to aldehydes by diol dehydrase and glycerol dehydrase involves the elimination of water. Since the reaction catalyzed by diol dehydrase is one of the simplest, this enzyme has been widely used to study the mechanism of action of AdoCbl [11–14]. The mechanism of this dehydration is, however, not simple but rather complex.

AdoCbl participates as coenzyme in 10 enzymatic reactions involving isomerization, rearrangement, elimination, or reduction. Although these reactions are seemingly quite different, all the reactions except one catalyzed by ribonucleotide reductase have in common that in general the AdoCbl-dependent rearrangement reactions involve the migration of a hydrogen from one carbon atom of the substrate to an adjacent carbon atom in exchange for a group X that moves in the opposite direction [13,42] (2). In the case of the diol dehydrase reaction,

(2)

X is a hydroxyl group on C-2, and an aldehyde product is formed by elimination of H_2O from 1,1-gem-diol [43]. The evidence for the migration of a hydroxyl group forming 1,1-gem-diol was obtained by Arigoni and Retey [43]. They observed that [^{18}O]propionaldehyde is obtained from the (S) isomer of [1-^{18}O]1,2-propanediol, while unlabeled propionaldehyde is formed from the (R) isomer. This indicated that the initial migration of a hydroxyl group is stereospecific and that the stereospecific dehydration of the gem-diol is sterically controlled by the

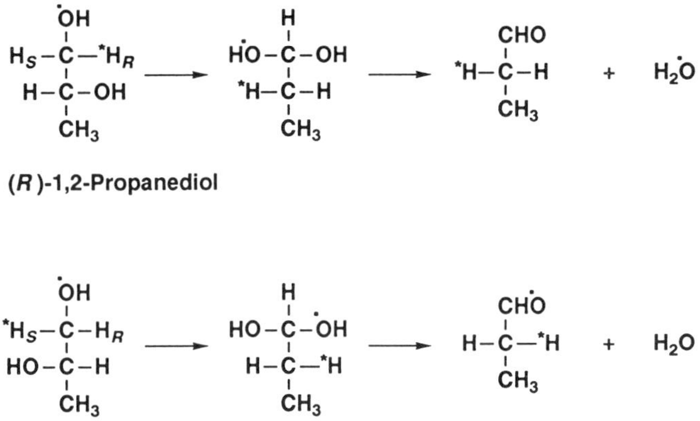

(*R*)-1,2-Propanediol

(*S*)-1,2-Propanediol

FIG. 2. Stereochemistry of the hydrogen and the hydroxyl group transfer in the diol dehydrase reaction [34,43,45]. The configuration of C-1 of 1,1-gem-diol has not yet been determined.

enzyme with only one of the two prochiral hydroxyl groups being eliminated (Fig. 2).

3.4.2. *Stereochemistry of the Hydrogen Transfer*

In the reaction catalyzed by diol dehydrase, the hydrogen was transferred to the adjacent carbon without exchange with solvent protons [44]. Both (R)- and (S)-1,2-propanediol serve as substrate [18,32,34,35]. Abeles and coworkers [34] demonstrated that which hydrogen atom migrates depends on the chirality of the carbon to which it migrates. The pro-S hydrogen migrated from (S)-1,2-propanediol and the pro-R hydrogen from the (R) substrate. Furthermore, it was established that the hydrogen atom replaces the migrating hydroxyl group with inversion of C-2 configuration [34,45] (Fig. 2).

3.4.3. *Role of Adenosylcobalamin in the Hydrogen Transfer*

Abeles and coworkers observed the formation of tritiated acetaldehyde when the enzyme was allowed to act on a mixture of unlabeled 1,2-ethanediol and [1,1-$^3H_2$]1,2-propanediol [46,47]. This indicated that the hydrogen transfer is not necessarily intramolecular but can occur by an intermolecular process as well. Furthermore, they demonstrated that both hydrogens at C-5′ of the coenzyme are replaced by tritium from C-1 of [1,1-$^3H_2$]1,2-propanediol and that the coenzyme could transfer tritium from either of the C-5′ positions to C-2 of the product [47–50]. Although the overall reaction is irreversible, a partial reaction was shown to be reversible by tritium exchange between the product and the enzyme-bound coenzyme [47,49]. It was thus established that the enzyme-bound AdoCbl serves as an intermediate hydrogen carrier, first accepting a hydrogen from C-1 of the substrate and then, in a subsequent step, giving a hydrogen back to C-2 of the product.

The deuterium isotope effect (k_H/k_D) for the overall diol dehydrase reaction was determined by Abeles and coworkers to be 10 [34]. This value indicated that breaking of the C-H bond of the substrate is rate determining. They also reported that the tritium isotope effects (k_H/k_T) for the transfer of hydrogen from substrate to coenzyme and from coenzyme to product are 20 and 125, respectively [25]. The k_H/k_T value of 125 corresponds to k_H/k_D of 28, which shows an unusually large isotope effect.

Kinetic experiments performed by Abeles and coworkers indicated that in the hydrogen transfer process there exists the enzyme-bound intermediate in which the hydrogen abstracted from the substrate and the two hydrogens at C-5′ of the coenzyme all become equivalent [25,47]. These results led to the idea that 5′-deoxyadenosine may be formed as a postulated intermediate from the adenosyl moiety of the coenzyme. Actually, its formation was first shown by Wagner et al. [51] in the reaction of diol dehydrase with glycolaldehyde, and then in the reactions with chloroacetaldehyde [52] and glycerol [53; T. Toraya, T. Tobimatsu, H. Kawamura, and S. Fukui, unpublished]. 5′-Deoxyadenosine was also formed upon suicide inactivation with coenzyme analogs [54–56]. It was demonstrated that the formation of 5′-deoxyadenosine is reversible in the ethanolamine ammonia-lyase–catalyzed conversion of 2-amino-1-propanol to propionaldehyde [57].

3.4.4. *Mechanism of the Overall Reaction*

Since earlier observations indicated formation of 5′-deoxyadenosine and cob(II)alamin (Cbl^{II}) (or OH-Cbl) in the reaction of diol dehydrase with substrate analogs or O_2 [51,52], it has been generally accepted that the reversible cleavage of the Co-C bond of the coenzyme is involved in the catalysis. Evidence was accumulated by optical and electron paramagnetic resonance (EPR) spectroscopy which shows that breaking the Co-C bond during catalysis is homolytic forming Cbl^{II} and the adenosyl radical. As illustrated in Fig. 3A, the optical spectrum of the holoenzyme in the presence of 1,2-propanediol (reacting holoenzyme) was similar to that of Cbl^{II} [16,51,58,59], although the spectrum of the holoenzyme itself was essentially identical to that of free AdoCbl. The holoenzyme of both diol dehydrase [58,60–62] and glycerol dehydrase [63] reacting with 1,2-propanediol gave the X-band EPR spectra shown in Fig. 3B, which were theoretically interpreted [64,65]. An asymmetric doublet signal in the high field with g values of 1.95 and 2.04 was assigned to an organic free radical, and a broad resonance in the low field with a g value of 2.2–2.4 (a prominent resonance at $g = 2.21$) was attributed to the low-spin Co(II) of Cbl^{II}. A model proposed for the system suggested an interaction of the free radical with the Co(II) center $\geqslant 6$ Å away [64]. Abeles and coworkers demonstrated by EPR spectroscopy that the radical species are formed at a kinetically competent rate in the conversion of 1,2-propanediol to propionaldehyde [62] and that the distance of closest approach of water to Co(II) is 10.2 Å [60]. Thus, it was established that the radical species formed by homolysis of the Co-C bond of the coenzyme is involved in the diol dehydrase reaction. It was therefore suggested that a hydrogen migrates as a hydrogen atom ($H^{\bullet}$).

Figure 4 shows a reaction mechanism proposed by Abeles and coworkers for diol dehydrase [12]. Salient features of the mechanism are as follows: The interaction between enzyme and coenzyme leads to the activation and cleavage of the Co-C bond of the coenzyme to form Cbl^{II} and the adenosyl radical. In the absence of substrate, only a small fraction of the coenzyme is present in the dissociated form. The addition of substrate to the complex shifts the equilibrium so that a major fraction of the coenzyme is converted to the dissociated form. The adenosyl radical that results from the homolysis of the Co-C bond then abstracts a hydrogen atom from the substrate to produce a substrate-derived radical and 5′-deoxyadenosine. The substrate radical rearranges to the product radical. The product radical then abstracts a

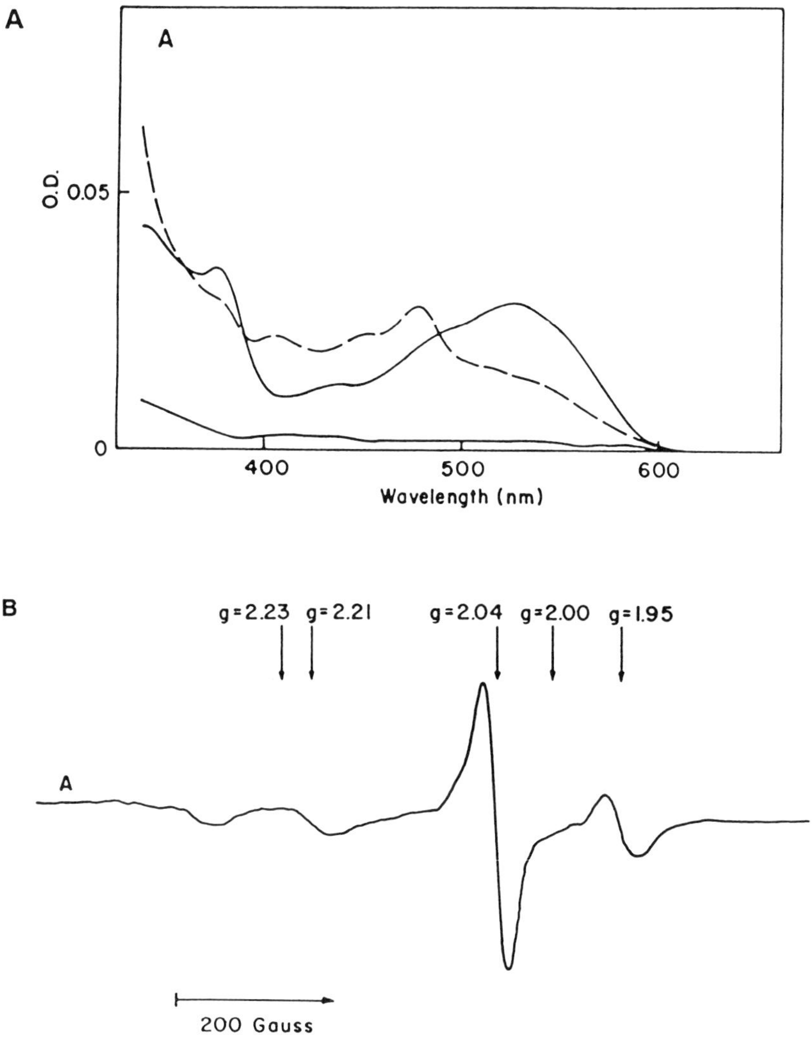

FIG. 3. Optical (A) and electron paramagnetic resonance (B) spectra of holodiol dehydrase in the presence of 1,2-propanediol (reacting holoenzyme). A, Measured at 30°C; (– – –), reacting holoenzyme; (——) free AdoCbl. B; Measured at −155° ± 5°C. (Reprinted with permission from Ref. 58. Copyright 1979 American Chemical Society.)

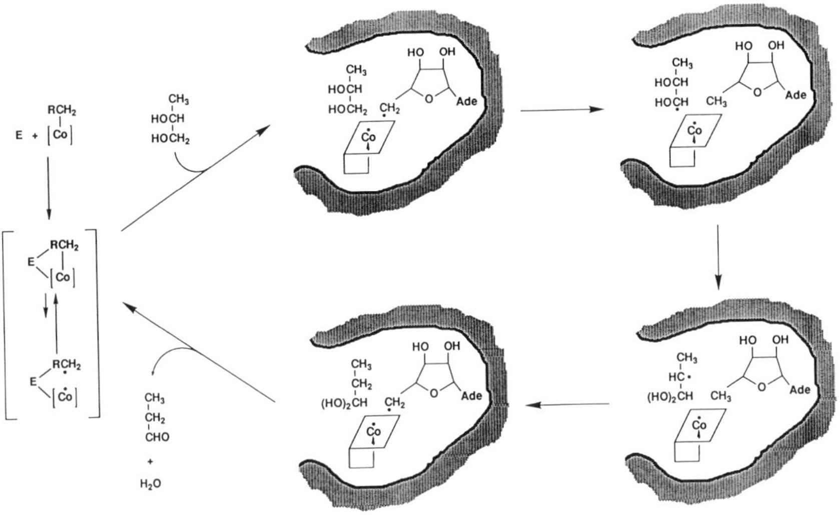

FIG. 4. Mechanism of the diol dehydrase reaction [12]. RCH_2−, adenosyl; [Co], cobalamin.

hydrogen atom from 5′-deoxyadenosine. This leads to the formation of the final product and regeneration of the coenzyme.

The mechanism of the rearrangement of the substrate radical to the product radical has not yet been clarified. Formation of an intermediate with covalent Co-C bond between substrate and cobalamin seems unlikely because of the ability of 2,3-butanediol to serve as a substrate [5] and the relatively long distance between the free radical and Co(II) as analyzed by EPR [64]. Although the mechanism of the hydroxyl group migration is difficult to study at an enzymic level, radical rearrangement was proposed by Golding and coworkers (for reviews, see Refs. 66 and 67). Model studies and calculations demonstrated that the migration of the hydroxyl group is facilitated by its protonation through radical cation intermediates [66]. Cbl^{II} is considered to be a "spectator" in this hypothesis.

4. FORMATION OF FREE RADICALS

4.1. Cleavage of the Co-C Bond of AdoCbl by Binding to the Apoenzyme

Diol dehydrase and other AdoCbl-requiring enzymatic reactions proceed by a free radical mechanism (see Sec. 3.4). The catalysis is triggered by homolytic cleavage of the Co-C bond of enzyme-bound AdoCbl. Since the Co-C bond of AdoCbl is sufficiently weak (homolytic bond dissociation energy reported as ~26 [68] and ~30 [69] kcal/mol), only a modest labilization by interaction with the apoprotein would be required for homolysis. The apoenzyme elicits the function of the coenzyme by activating (labilizing) its organometallic bond.

The Co-C bond gets activated upon binding of AdoCbl to the apoenzyme, but in the absence of substrate only a small fraction of the coenzyme was in the dissociated form. When the substrate was added to the system, most of the Co-C bond underwent homolysis with a half-life ($t_{1/2}$) <3 msec at 5°C [62]. This rate was faster than that calculated from a turnover number of 337 for 1,2-propanediol at 37°C ($t_{1/2}$ = 2–3 msec). The substrate shifted the equilibrium to the dissociated form, most probably by inducing conformational changes of the enzyme. Activation parameters determined by Hay and Finke for base-on AdoCbl yield a 25° base-on rate constant of $10^{-9\pm1}$ s^{-1} ($t_{1/2}$ = 22 years) [70]. Therefore, enzymic acceleration of $\geqslant 10^{11}$ for the Co-C bond cleavage is attained in the presence of substrate, when the coenzyme is bound to the active site of the enzyme. This magnitude of acceleration corresponds to lowering $\Delta\Delta G^{\ddagger}$ by ⩾16 kcal/mol. Such $\Delta\Delta G^{\ddagger}$ lowering must be supplied by a part of binding energy utilized for catalysis. Hay and Finke reported that this is plausible by assuming ⩾10 surface binding/hydrogen bonding sites of AdoCbl [70]. When all the substrate was consumed, the Co-C bond of the holoenzyme was regenerated. That is, the holoenzyme possesses a mechanism to open and close the catalytic center in response to the presence of substrate. This mechanism seems reasonable for generating a primary organic radical, a highly reactive species, only when necessary. Therefore, it can be considered that AdoCbl is a molecule to preserve the organic and cobalt radicals in the masked form in living organisms. The apoenzyme discloses the function of this coenzyme.

Structural requirements of AdoCbl for binding and catalysis have been determined with diol dehydrase via use of many of its analogs [11]. By mapping the groups on AdoCbl that are important for the Co-C bond cleavage, the mechanism of labilization of the Co-C bond of AdoCbl in enzymatic reactions has been postulated with diol dehydrase.

4.2. Labilization of the Co-C Bond by Interaction at the Adenosyl Group

4.2.1. Binding Site of Apodiol Dehydrase for the Adenosyl Group

No direct relationship between the chemical reactivity of the Co-C bond of analogs and their efficiency as coenzymes was observed [16]. Only small modifications were permitted without complete loss of activity. Such earlier observations strongly suggest the presence of the adenosyl group binding site in the apoenzyme. As shown in Fig. 5, an irreversible 1:1:1 ternary complex was formed between enzyme, Cbl^{II}, and 5′-deoxyadenosine when the enzyme was incubated first with the nucleoside and then with Cbl^{II} [71]. When this order of addition was reversed, no 5′-deoxyadenosine was bound to the enzyme–Cbl^{II} complex. These results indicated that the binding site for the adenosyl group of AdoCbl is present in the apoenzyme. The binding site was shown to be spatially limited and highly specific for adenine nucleosides.

4.2.2. Analogs of AdoCbl with a Modified Adenine Moiety

Analogs of AdoCbl modified in the adenine moiety of the adenosyl group were examined for coenzymic activity with diol dehydrase [16,72,73]

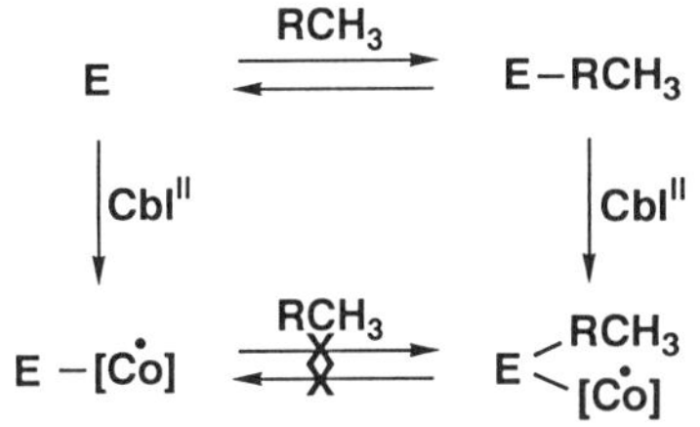

FIG. 5. Binding of 5′-deoxyadenosine and Cbl^{II} to apodiol dehydrase [71]. E, apoenzyme; [Ċo], Cbl^{II}; RCH_3, 5′-deoxyadenosine.

and glycerol dehydrase [74–77]. As shown in Fig. 6, 1-deaza, 3-deaza, 3-iso, and N^6-methyl analogs and nebularyl-Cbl were active as coenzymes, whereas 7-deaza, N^6-ethyl, N^6,N^6-dimethyl, 1-methyl, and 1,N^6-etheno derivatives of AdoCbl and inosyl-Cbl did not show detectable coenzymic activity. 7-Deaza and N^6,N^6-dimethyl analogs acted as strong competitive inhibitors. The formation of Cbl^{II} as intermediate in the catalytic reaction was spectroscopically observed with active analogs of the coenzyme but not with inactive analogs. These results indicated that the importance of the nitrogen atoms in the adenine

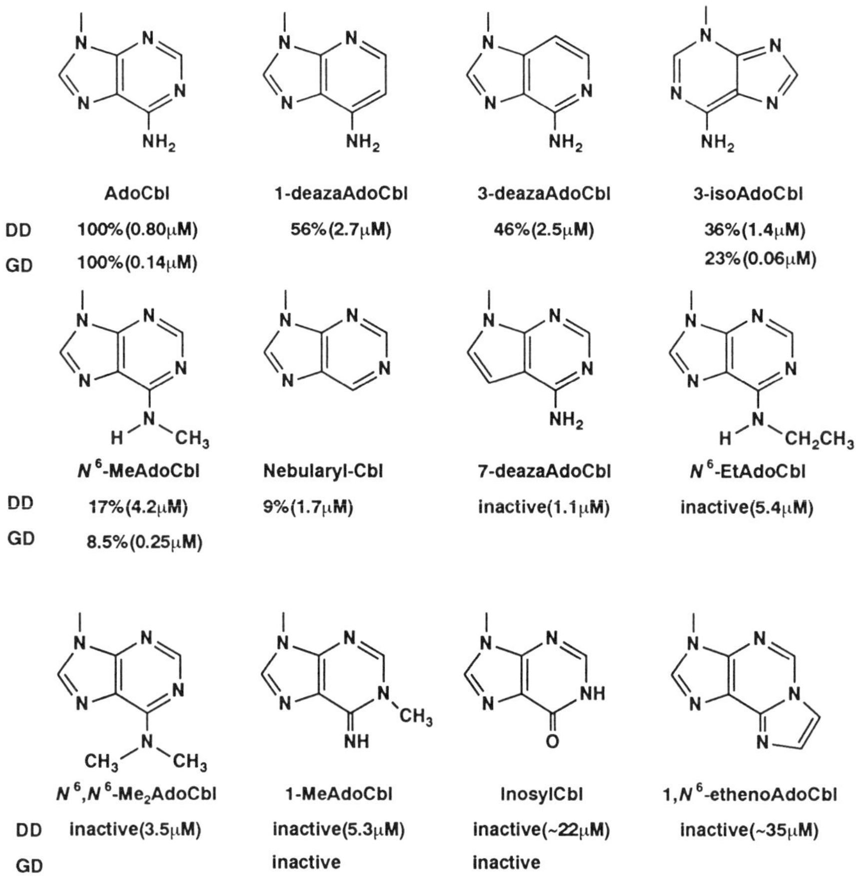

FIG. 6. Coenzymic activity of adenine-modified analogs and their affinity for diol dehydrase (DD) and glycerol dehydrase (GD) [16,72–77]. Relative activity, %; K_m or K_i in parentheses.

moiety for the manifestation of a catalytic function and for activation of the Co-C bond decreases in the following order: N-7 > 6-NH_2 > N-3 > N-1 [73].

4.2.3. *Analogs of AdoCbl with Modified Ribose Moiety*

Analogs of AdoCbl modified in the D-ribose moiety of the adenosyl group were examined for coenzymic activity with diol dehydrase [16,78,79] and glycerol dehydrase [76]. As shown in Fig. 7, adenine-arabinosylcobalamin (araAdoCbl), aristeromycyl-Cbl, 2′-dAdoCbl, and 3′-dAdoCbl were active coenzymes. 2′,3′-SecoAdoCbl and its dialdehyde derivative were totally inactive analogs. It was thus evident that the β-D-ribofuranose ring itself, possibly its rigid structure, is essential and that the importance of the functional groups of the ribose moiety decreases in the following order: β-D-ribofuranose ring ≫ 3′-OH > 2′-OH > -O- [79]. With active analogs, homolysis of the Co-C bond was observed spectroscopically, but not with 2′,3′-secoAdoCbl.

Adeninylethyl-Cbl (AdeEtCbl) [78] and 2′,3′-secoAdoCbl dialdehyde [79] underwent irreversible cleavage of the Co-C bond by interaction with apodiol dehydrase, although they were inactive analogs. The Co-C bond of AdeEtCbl was cleaved heterolytically forming Co^{III} and a carbanion or a carbanion-like species (3). Adeninylpropyl-Cbl and other longer chain homologs did not undergo the Co-C bond cleavage [16].

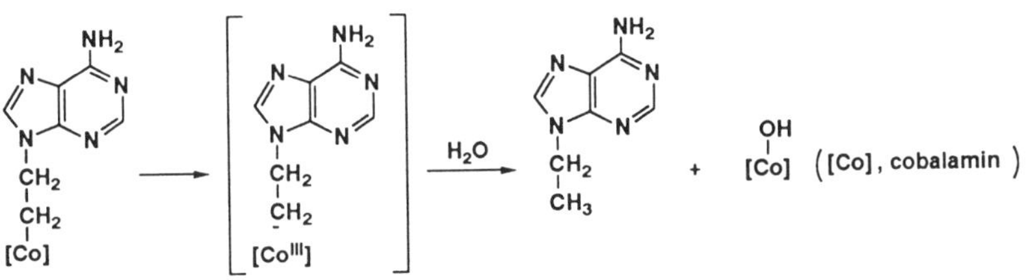

(3)

4.3. Labilization of the Co-C Bond by Interactions at the Cobalamin Moiety

4.3.1. *Importance of Interactions at the Cobalamin Moiety*

Upon binding of the adenosyl group of the coenzyme to diol dehydrase at 37°C ΔG° was estimated to be approximately −6.6 kcal/mol from the K_D value for 5′-deoxyadenosine [73], indicating that the interaction be-

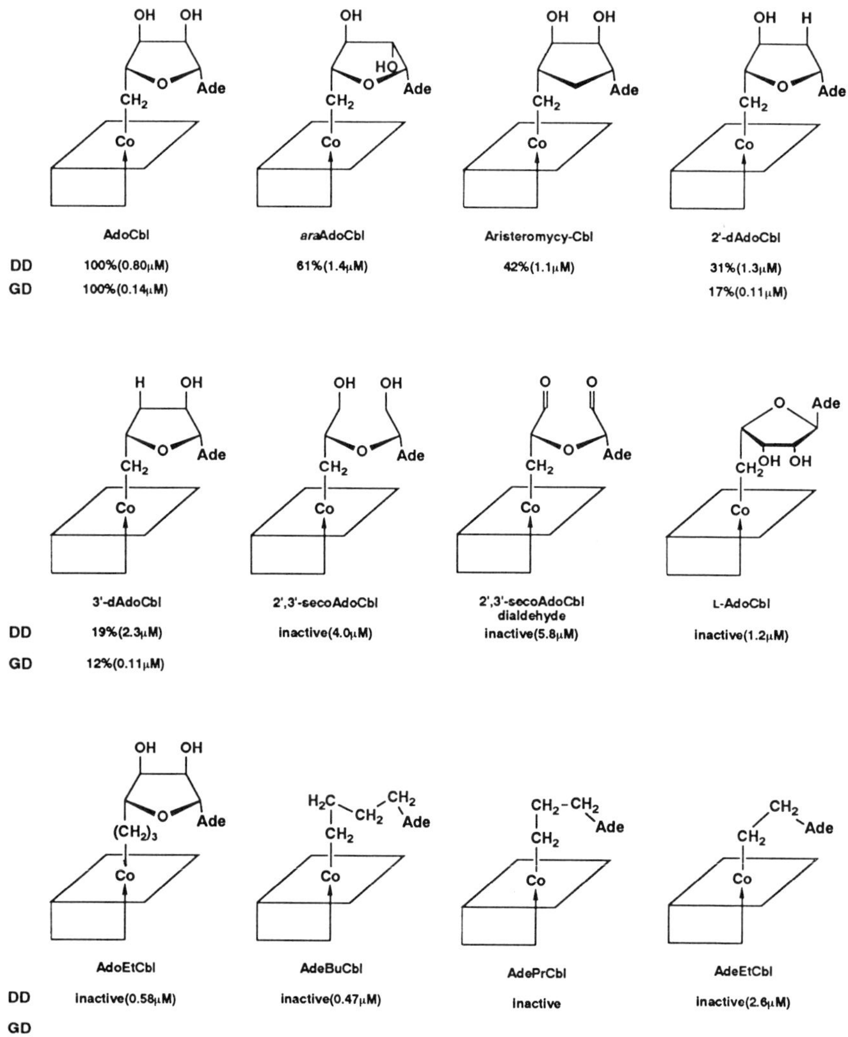

FIG. 7. Coenzymic activity of ribose-modified analogs and their affinity for diol dehydrase (DD) and glycerol dehydrase (GD) [16,76,78,79]. Relative activity, %; K_m or K_i in parentheses.

tween the adenosyl group and its binding site is not strong enough to cleave the Co-C bond of AdoCbl. Basal labilization of the Co-C bond was assumed to take place when the coenzyme was bound by the apoenzyme. Cleavage of the Co-C bond of sterically hindered alkylcobalamins bearing neither adenine nor functional groups, such as isobutyl-Cbl, neopentyl-Cbl, and cyclohexyl-Cbl, was markedly accelerated by interaction with apodiol dehydrase, although these cobalamins did not function as coenzymes [80]. This was due to labilization of their Co-C bond by binding to the active site of the enzyme. Such basal labilization was considered to be caused by a steric distortion of the corrin ring, which may be induced by specific tight interaction of the cobalamin moiety with the apoprotein.

4.3.2. Analogs of AdoCbl with Modified Peripheral Side Chains

The importance of the sterically unfavorable upward distortion of the corrin ring in labilization of the Co-C bond has been pointed out from model studies. Such a distortion could be brought about by the specific tight interaction with the enzyme at the peripheral amide side chains. As shown in Table 2, isomers of AdoCbl in which one of the three amide groups of the propionamide side chain of the corrin ring was converted to $-COO^-$, $-COOCH_3$, or $-CONHCH_3$ were examined for coenzymic activity with diol dehydrase [58]. With the *d*-site modifications, the relative activity decreased in the order $-COO^- \gg -CONHCH_3$, $-COOCH_3$, suggesting inhibition by steric effects, while at the *b*-site the activity decreased in the order $-CONHCH_3 \gg -COO^- > -COOCH_3$, suggesting the importance of hydrogen bond donation from coenzyme to apoenzyme. Three *b*-site analogs and the *d*-$CONHCH_3$ analog caused irreversible inactivation during catalysis. The 13-epimer showed 13% as active as AdoCbl [81]. Hogenkamp and coworkers demonstrated that the coenzyme activity of spin-labeled analogs of AdoCbl is also dependent on the position of the bulky spin label [82]. These indicated that the interactions of the coenzyme propionamide side chains with the apoenzyme facilitate the cleavage of the Co-C bond.

Changes in optical and circular dichroism (CD) spectra observed upon binding of several cobalamins to apodiol dehydrase [84] and apoethanolamine ammonia-lyase [85] may reflect the distortion-induced changes in the electronic structure of the corrin ring.

TABLE 2

Coenzymic Activity of Peripheral Side Chain–Modified Analogs and Their Affinity for Diol Dehydrase

Analog[a]	Relative activity (%)	K_m (μM)	Ref.
AdoCbl	(100)	0.3	58
AdoCbl (*d*-OH)	86	2.3	58
AdoCbl (*b*-OH)	11	4.5	58
AdoCbl (*e*-OH)	66	8.0	58
AdoCbl (*d*-OCH_3)	14	3.1	58
AdoCbl (*b*-OCH_3)	7	1.9	58
AdoCbl (*e*-OCH_3)	14	n.d.[b]	58
AdoCbl (*d*-$NHCH_3$)	16	2.1	58
AdoCbl (*b*-$NHCH_3$)	41	0.94	58
AdoCbl (*e*-$NHCH_3$)	43	1.5	58
AdoepiCbl	14	13	81
AdoCbl (*d*-TEMPO)	2.2	46	82
AdoCbl (*b*-TEMPO)	0	n.d.[b]	82
AdoCbl (*e*-TEMPO)	10	3.6	82
AdoepiCbl (*e*-TEMPO)	0	n.d.[b]	82

[a]AdoCbl (*b*, *d*, or *e*-OH), AdoCbl *b*-, *d*-, or *e*-carboxylic acid; AdoCbl (*b*, *d*, or *e*-OCH_3), AdoCbl *b*-, *d*-, or *e*-methyl ester; AdoCbl (*b*, *d*, or *e*-$NHCH_3$), AdoCbl *b*-, *d*-, or *e*-methylamide; AdoCbl (*b*, *d*, or *e*-TEMPO), AdoCbl *b*-, *d*-, or *e*-(2,2,6,6-tetramethylpiperidine-N-oxyl) amide; AdoepiCbl, 13-epimer of AdoCbl. The identification of isomers according to Anton et al. [83]. For the other analogs tested, see Ref. 11 and references therein.
[b]Not determined.

4.3.3. Analogs of AdoCbl with the Modified Nucleotide Loop Moiety

Among the analogs in which the D-ribose moiety of the nucleotide loop was replaced by an oligomethylene group, trimethylene and tetramethylene analogs were partially active as coenzyme in the diol dehydrase reaction (Fig. 8) [55]. Dimethylene and hexamethylene analogs were not active coenzymes but weak competitive inhibitors. This indicated that neither the α-D-ribofuranose ring nor the functional groups

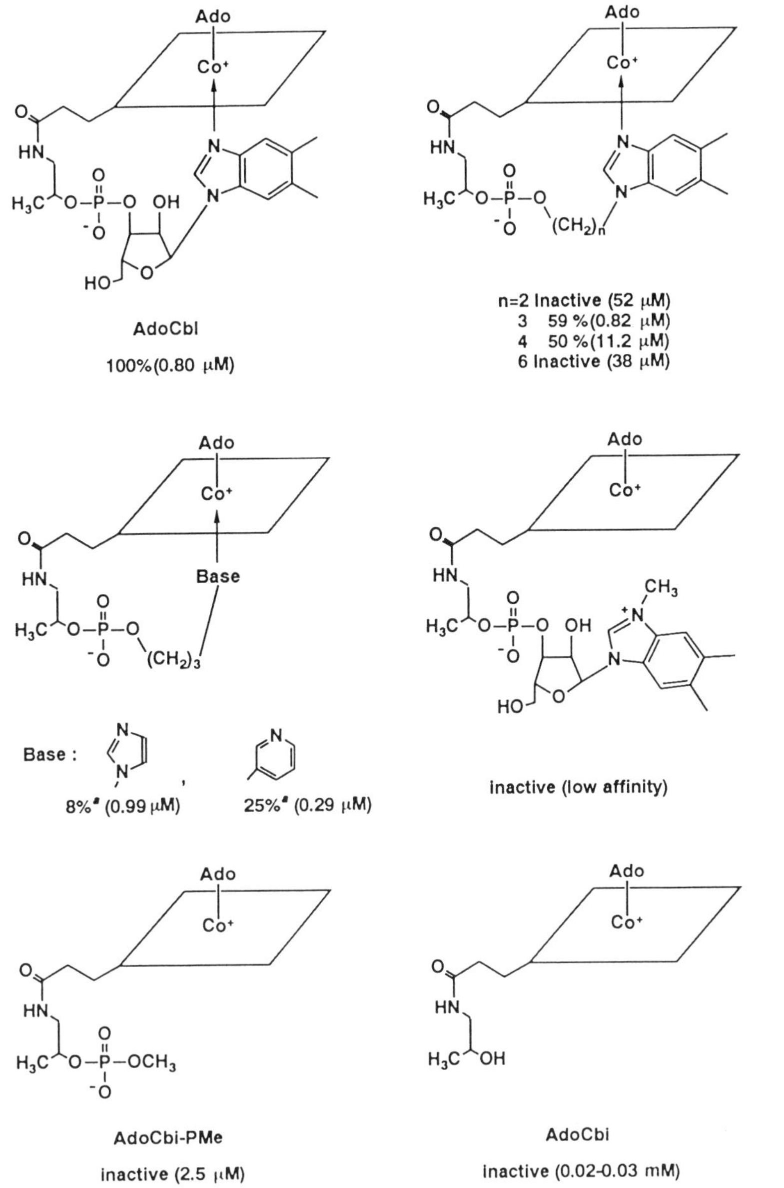

FIG. 8. Coenzymic activity of nucleotide loop–modified analogs and their affinity for diol dehydrase [55,56,86–88]. Relative activity, %; K_m or K_i in parentheses. [a]Calculated from the initial velocity. For the other analogs tested, see Ref. 11 and references therein.

of the ribose moiety are essential for coenzymic function. There was an optimum loop size of the analogs for catalysis and tight binding to the apoenzyme, which corresponds to the loop size of cobalamins. Therefore, the D-ribose moiety seems important as a spacer to keep 5,6-dimethylbenzimidazole in the proper position [55].

Adenosylcobamides containing other heterocyclic bases instead of 5,6-dimethylbenzimidazole possessed coenzymic activity as high as AdoCbl, although their affinity for the apoenzyme varied [11]. Trimethylene analogs containing imidazole [55] or pyridine [86] were partially active coenzymes in the diol dehydrase reaction, but the reaction with them as coenzymes was accompanied with concomitant inactivation during catalysis (see Sec. 5.1).

AdoCbi methyl phosphate did not show detectable coenzymic activity but behaved as a pseudocoenzyme [56] (see Sec. 5.1). Since AdoCbi neither functioned as coenzyme nor bound tightly to apoenzyme [87,88], it was concluded that the phosphodiester moiety of the nucleotide loop of AdoCbl is essential for tight binding to the apoenzyme and thus for subsequent activation of the Co-C bond and catalysis [56]. The analog of AdoCbl in which the N-3 position of 5,6-dimethylbenzimidazole is methylated and no longer coordinated to the cobalt atom was not active as coenzyme with diol and glycerol dehydrases and showed low affinity for the apoenzyme [88,89].

5. INACTIVATION AND REACTIVATION

5.1. Inactivation of the Holoenzyme

5.1.1. Suicide Inactivation by Glycerol and Substrate Analogs

Glycerol acted as both a good substrate and a potent inactivator for diol dehydrase (Table 1) [5,18]. Inactivation by glycerol was mechanism-based and involved irreversible dissociation of the Co-C bond forming 5′-deoxyadenosine and an alkyl-Cbl–like species. The cobalamin species formed remained very tightly bound to the enzyme, resulting in irreversible inactivation. Richards and coworkers demonstrated that the “S”-binding conformation is primarily responsible for the inactivation [18]. 1,2-Ethanediol and other substrates also induced gradual inactivation of diol dehydrase during catalysis, but to a lesser extent [18]. The holoenzyme was inactivated by glycolaldehyde [51] or chloroacetaldehyde [52]. The Cbl^{II} and 5′-deoxyadenosine were formed from the coenzyme upon inactivation. The mechanism of inactivation by these aldehydes or their hydrated forms seems to be related to the normal catalytic mechanism, although the details are not yet clear.

Glycerol dehydrase also underwent suicide inactivation by glycerol [30], but the pseudo–first-order rate constant (k_{inact}) was smaller than that of diol dehydrase [18,90].

5.1.2. Inactivation by O_2 in the Absence of Substrate

Aerobic incubation of holodiol dehydrase in the absence of substrate led to total inactivation [3,51]. The inactivation was accompanied by formation of OH-Cbl, suggesting irreversible cleavage of the activated Co-C bond of the holoenzyme by reaction with O_2. Substrates and analogs protected the reacting holoenzyme from O_2 inactivation [91].

The holoenzyme of glycerol dehydrase also underwent irreversible inactivation by O_2 [30,90]. The holoenzyme was protected from O_2 by the addition of K_2SO_3.

5.1.3. Inactivation by Coenzyme Analogs

Coenzyme analogs in which the D-ribose and 5,6-dimethylbenzimidazole moieties of the nucleotide loop were replaced by a trimethylene group and imidazole or pyridine, respectively, were partially active in the diol dehydrase reaction, but the reaction with these analogs as coenzyme was accompanied by concomitant inactivation during catalysis with k_{cat}/k_{inact} of 3.3×10^3 and 1.4×10^4, respectively [55,86]. Upon inactivation, 5′-deoxyadenosine and a Cbl^{II}-like species were formed without modification of the apoenzyme. This is a kind of suicide inactivation, and the dimethylbenzo moiety of AdoCbl seems to play an essential role in preventing the intermediate complexes from inactivation. Some of the other analogs also induced suicidal inactivation of diol dehydrase, but their efficiency of inactivation was much lower. For example, k_{cat}/k_{inact} for 3′-dAdoCbl, 2′-dAdoCbl, and AdoCbl *b*-carboxylic acid were 7.4×10^4 [79], 5.5×10^4 [79], and 1.3×10^4 [54], respectively, while the value for AdoCbl was 1.44×10^6 [58].

Although AdoCbi methyl phosphate did not show detectable coenzymic activity, the Co-C bond of this analog was irreversibly cleaved upon incubation with apodiol dehydrase in the presence of substrate [56]. The cleavage was not catalytic but stoichiometric, implying that the Co-C bond of the analog gets activated when the analog binds to the active site of the enzyme. 5′-Deoxyadenosine and Co(II)-corrinoid were formed upon the Co-C bond cleavage without modification of the apoenzyme. It was demonstrated that this analog acts as a pseudocoenzyme ($k_{cat}/k_{inact} \sim 0$) [56] (Eq. (4)).

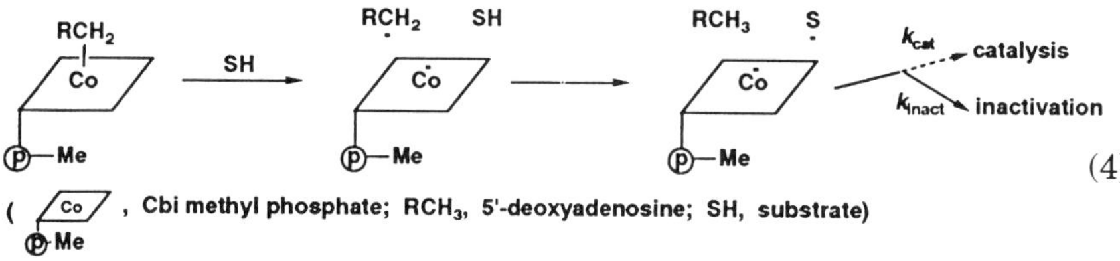

(4)

5.2. Protection of Free Radical Intermediates from Side Reactions

The reactions catalyzed by diol dehydrase or glycerol dehydrase are chemically difficult reactions ($k_e/k_n \sim \infty$). Such reactions could proceed with assistance of the high reactivity of a free radical. In the cases of diol dehydrase and glycerol dehydrase, the adenosyl radical derived from the coenzyme plays an essential role in the catalysis by abstracting a hydrogen atom from substrate forming a substrate radical. The substrate radical rearranges to a product radical by a hydroxyl group transfer. These free radical species must sustain their high reactivity at the active site and become extinct in the only way destined for the reaction. Once a radical intermediate is quenched or stabilized by side reactions, the catalytic turnover ceases, i.e., inactivation takes place during catalysis. Therefore, the radical species must be strictly protected from side reactions by the apoenzyme and other components of the intermediate complexes.

It was established that the nucleotide loop of AdoCbl plays a pivotal role in protecting the highly reactive free radical intermediates from unfruitful side reactions [55,56]. In this sense, it is evident that Cbl^{II} formed by homolysis of the Co-C bond is not simply a "spectator" but a "stabilizer" for organic free radical intermediates. Since the trimethylene analogs with a less bulky base than 5,6-dimethylbenzimidazole behaved as suicide coenzymes, the coordination of a bulky base to the cobalt atom seems essential for the protection through inducing a distortion of the corrin ring [55,86]. When substrates other than 1,2-propanediol are used as substrate, a substrate radical or a product radical may become extinct either by intramolecular reactions, by side reactions with nearby groups, or by escaping from the active site. Such undesirable side reactions of the reactive radical intermediates would lead to inactivation during catalysis. It should be reemphasized that a negative control of the reactivity of free radical intermediates by enzymes is obligatory for a continuous progress of a catalytic cycle. This effect of enzymes might be exerted by spatial isolation of a radical species from nearby groups.

5.3. Reactivation of Inactivated Enzymes

5.3.1. *In Situ Reactivation by a Coenzyme Exchange Mechanism*

Glycerol served as a suicide inactivator not only for diol dehydrase but also for glycerol dehydrase [30,90], although it was a growth substrate in the induction media for the bacteria [10,23,24]. Glycerol-inactivated enzymes in permeabilized cells (in situ system) underwent rapid reactivation by addition of ATP and Mg^{2+} (or Mn^{2+}) in the presence of an excess of free AdoCbl [92]. Such reactivation did not occur at all in cell-free extracts. The reactivating system seems inducible by glycerol.

In the presence of ATP and Mg^{2+}, enzyme-bound OH-Cbl, CN-Cbl, and MeCbl were rapidly displaced in situ by added AdoCbl forming the catalytically active holoenzyme [93]. In contrast, enzyme-bound adeninylbutyl-Cbl and adenosylethyl-Cbl were not displaceable by added coenzyme. It was therefore concluded that the relative affinity of the enzymes in situ for cobalamins with simple Coβ ligands is markedly lowered in the presence of ATP and Mg^{2+}, whereas that for cobalamins with adenine-containing ligands is not. These indicated that the inactivated dehydrases are reactivated in situ in the presence of ATP and Mg^{2+} by exchange of the modified coenzyme for intact AdoCbl (coenzyme exchange mechanism).

Holodiol dehydrase which was inactivated by O_2, inactivated during dehydration of 1,2-ethanediol, or inactivated during catalysis with araAdoCbl as coenzyme was also reactivated in situ by ATP, Mg^{2+}, and AdoCbl [93].

5.3.2. Reactivation by a Readenosylation Mechanism

Upon inactivation of the holoenzyme by substrates or coenzyme analogs, the adenosyl group of the enzyme-bound coenzyme was severed from cobalamin forming Cbl^{II} or OH-Cbl. It was demonstrated that holodiol dehydrase inactivated by O_2 in the absence of substrate and the enzyme–CN-Cbl complex, an inactive model complex, undergo reactivation upon incubation with cell-free extracts in the presence of ATP, Mg^{2+} (or Mn^{2+}), and the fully reduced form of riboflavin 5′-phosphate ($FMNH_2$) [94]. Since the cofactor requirement of the reactivation was identical to that of Co adenosylation of the free cobalamin, it seems likely that the inactivated holoenzyme is converted to the active form by readenosylation of the modified coenzyme (readenosylation mechanism).

6. PHYSIOLOGICAL ROLES

6.1. Distribution

Diol dehydrase and glycerol dehydrase were produced inducibly in selected genera of Enterobacteriaceae, e.g., *Klebsiella* and *Citrobacter*, grown anaerobically in complex media containing 1,2-propanediol and glycerol, respectively [10,23,24]. Of the species of Enterobacteriaceae tested, *K. oxytoca* ATCC 8724 [95] and *Salmonella typhimurium* [96] formed only diol dehydrase, while *C. freundii* NCIB 418 [24] formed only glycerol dehydrase. Beside Enterobacteriaceae, the formation of glycerol dehydrase was reported with *Lactobacillus* sp. [6,97] and of diol dehydrase with *Propionibacterium* [23,98], *Flavobacterium* [99], *Nocardia* [100], and *Lactobacillus* [101] species.

6.2. Regulation of the Apoenzyme Synthesis

Regulation of the synthesis of apodiol dehydrase was studied with *K. oxytoca* ATCC 8724, since this species produces only a single dehydrase (diol dehydrase) [95]. A high level of activity was detected in cells grown on 1,2-propanediol, 1,2-ethanediol, or 1,2-butanediol. The glycerol-grown cells also exhibited the activity, but to a much lesser degree. This may be due to low specificity for an inducer. Glucose-grown cells and citrate-grown cells did not possess any detectable activity. When the organism was cultivated in a glucose plus 1,2-propanediol medium, the level of enzyme activity remained very low until the glucose in the medium had been consumed. Thus, the synthesis of diol dehydrase seems regulated by catabolite repression [95]. Cyclic AMP overcame the catabolite repression of this enzyme, indicating that both an inducer and cyclic AMP are necessary for the diol dehydrase synthesis. Catabolite inactivation of the enzyme was unlikely, since the addition of glucose did not affect the rate of degradation of the already existing enzyme.

Glycerol dehydrase is also an inducible enzyme and formed during anaerobic growth on glycerol. It is a member of the *dha* system (see Sec. 6.3.) and under separate control from diol dehydrase [102,103]. The enzyme was induced by dihydroxyacetone, rather than glycerol directly, in a coordinated manner with other enzymes of the *dha* system, suggesting that their genes form the *dha* regulon. The synthesis of this enzyme in *K. pneumoniae* underwent catabolite repression, which was overcome by cyclic AMP in the glucose medium [104].

6.3. Metabolic Roles

Since 1,2-propanediol, 1,2-ethanediol, and glycerol are more reduced substrates than the corresponding carbohydrates, only a limited number of bacteria can ferment them without external electron acceptors under anaerobic conditions. Close relationship was observed between the capacities of bacteria to grow anaerobically on 1,2-diols or on glycerol without exogenous electron acceptors [23] and to produce specific dehydrases.

When 1,2-propanediol or 1,2-ethanediol was fermented anaerobically by cells of Enterobacteriaceae [105] or *Propionibacterium freudenreichii* [98], 1,2-diols were converted to the corresponding acids and alcohols. The activities of diol dehydrase, alcohol dehydrogenase, CoA-dependent aldehyde dehydrogenase, phosphotransacylase, and acetate

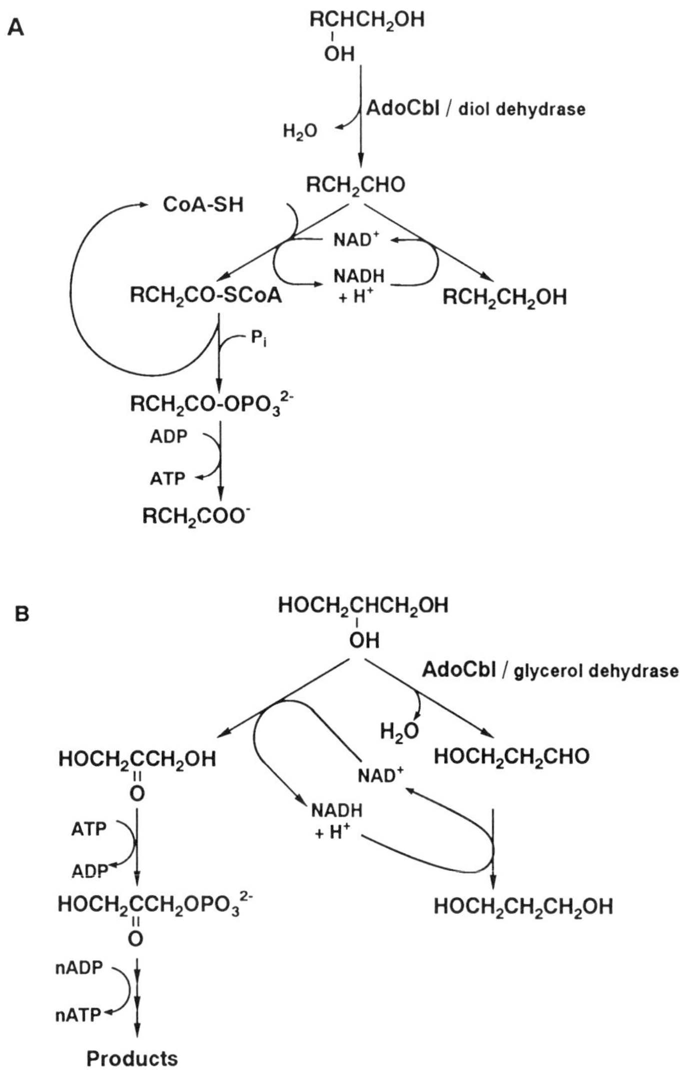

FIG. 9. Metabolic roles of diol dehydrase (A) [98,105] and glycerol dehydrase (B) [102,103] in anaerobic dissimilation of 1,2-diols and glycerol, respectively. R = CH_3 or H in A.

kinase were detected in cell-free extracts of the 1,2-diol-grown cells. Thus, it was concluded that 1,2-diols are fermented via aldehydes, acyl-CoA, and acyl phosphates [98,105], as described in Fig. 9A. This pathway was later found in *Clostridium glycolicum* [106] and *Salmonella typhimurium* [96] as well. The physiological role of diol dehydrase is therefore to produce both an electron acceptor and an intermediate for the incomplete oxidation of 1,2-diols. In *Nocardia* species, propionaldehyde was reported to be aerobically metabolized via propionate and succinate [100].

When glycerol was fermented by *Klebsiella*, *Citrobacter*, or *Lactobacillus* species, accumulation of 1,3-propanediol in the growth medium was observed [23,101]. Since glycerol is a more reduced substrate than the corresponding trioses by two hydrogen atoms, the conversion of glycerol to an electron acceptor seems essential for its anaerobic dissimilation [2]. Reoxidation of NADH to NAD with β-hydroxypropionaldehyde was catalyzed by 1,3-propanediol dehydrogenase [23,102,103]. Therefore, the metabolic role of glycerol dehydrase is to produce an electron acceptor, β-hydroxypropionaldehyde, in the glycerol fermentation. It was reported that four enzymes, i.e., glycerol dehydrogenase, dihydroxyacetone kinase, glycerol dehydrase, and 1,3-propanediol dehydrogenase, form a system for anaerobic dissimilation of glycerol (the "*dha* system") [102,103]. As illustrated in Fig. 9B, the first two enzymes participate in oxidation of glycerol and subsequent phosphorylation, whereas the latter two enzymes are for reoxidation of NADH to NAD^+. Organisms that do not possess the *dha* system, such as *E. coli*, require the addition of external electron acceptors, such as fumarate or nitrate, for anaerobic growth on glycerol [107]. Recently, the genes of the *Klebsiella* [108,109] or *Citrobacter* [110] *dha* regulon were cloned in *E. coli*. The recombinant *E. coli* strains were enabled to grow anaerobically on glycerol without added electron acceptors, and the formation of 1,3-propanediol was observed.

7. CONCLUDING REMARKS

In the three decades since the discovery of diol and glycerol dehydrases, overviews of enzymological properties, subunit structures, mechanisms of action, and metabolic roles have been revealed. One of the most important and unusual features is that the reactions catalyzed by these enzymes proceed by a free radical mechanism in common. A free radical species is formed by homolysis of the Co-C bond of the enzyme-bound AdoCbl. That is, the coenzyme functions as a radical-generating reagent for nature. Diol dehydrase of *Clostridium glycolicum*, another type of the isofunctional enzyme, is a noncobalamin enzyme and reported to contain a protein radical (see Chap. 7). It seems likely that the dehydrations of 1,2-diols and glycerol under physiological conditions are chemically difficult reactions and enabled to occur only by the assistance of the high reactivity of a free radical species.

In the free radical–mediated reactions, in general, the formation and sustenance of free radical species are important for continuous progress of a catalytic cycle. How does the apoenzyme labilize and provoke homolysis of the Co-C bond of the coenzyme? This question has been partly answered by mapping the groups of AdoCbl required for binding and catalysis, but it has not yet been fully solved. Another interesting question is how enzymes protect the free radical intermediates from side reactions. Undesired side reactions of the radical species lead to inactivation of the enzymes. Understanding how the enzymes activate the coenzyme and deal with highly reactive free radicals must await the elucidation of the three-dimensional structures of the active sites by X-ray crystallography.

ABBREVIATIONS

AdeEtCbl	adeninylethylcobalamin
AdoCbi	adenosylcobinamide
AdoCbl	adenosylcobalamin
araAdoCbl	adeninearabinosylcobalamin
secoAdoCbl	2′,3′-secoadenosylcobalamin
ADP	adenosine 5′-diphosphate
AMP	adenosine 5′-monophosphate
ATP	adenosine 5′-triphosphate
Cbi	cobinamide
Cbl	cobalamin

Cbl^{II}	cob(II)alamin
CD	circular dichroism
CN-Cbl	cyanocobalamin
OH-Cbl	hydroxocobalamin
*p*CMB	*p*-chloromercuribenzoate
CoA	coenzyme A
DEAE	diethylaminoethyl
dha	dihydroxyacetone
DNPH	2,4-dinitrophenylhydrazine
EPR	electron paramagnetic resonance
$FMNH_2$	riboflavin 5′-phosphate, fully reduced form
MBTH	3-methyl-2-benzothiazolinonehydrazone
MeCbl	methylcobalamin
M_r	relative molecular mass
NAD	nicotinamide adenine dinucleotide, oxidized form
NADH	nicotinamide adenine dinucleotide, reduced form
SDS	sodium dodecyl sulfate

REFERENCES

1. C. E. Voisenet, *Ann. Inst. Pasteur*, *32*, 476 (1918).
2. R. H. Abeles, A. M. Brownstein, and C. H. Randles, *Biochim. Biophys. Acta*, *41*, 531 (1960).
3. H. A. Lee, Jr., and R. H. Abeles, *J. Biol. Chem.*, *238*, 2367 (1963).
4. R. H. Abeles, and H. A. Lee, Jr., *J. Biol. Chem.*, *236*, 2347 (1961).
5. T. Toraya, T. Shirakashi, T. Kosuga, and S. Fukui, *Biochem. Biophys. Res. Commun.*, *69*, 475 (1976).
6. K. L. Smiley and M. Sobolov, *Arch. Biochem. Biophys.*, *97*, 538 (1962).
7. J. Pawelkiewicz and B. Zagalak, *Acta Biochim. Polon.*, *12*, 207 (1965).
8. Z. Schneider, E. G. Larsen, G. Jacobson, B. C. Johnson, and J. Pawelkiewicz, *J. Biol. Chem.*, *245*, 3388 (1970).
9. A. Stroinski, J. Pawelkiewicz, and B. C. Johnson, *Arch. Biochem. Biophys.*, *162*, 321 (1974).

10. T. Toraya and S. Fukui, *Eur. J. Biochem.*, *76*, 285 (1977).
11. T. Toraya and S. Fukui, in B_{12}, Vol. 2 (D. Dolphin, ed.), John Wiley and Sons, New York, 1982, p. 233.
12. R. H. Abeles, in *Vitamin* B_{12} (B. Zagalak and W. Friedrich, eds.), Walter de Gruyter, Berlin, 1979, p. 373.
13. R. H. Abeles and D. Dolphin, *Acc. Chem. Res.*, *9*, 114 (1976).
14. R. H. Abeles, *The Enzymes*, *5*, 481 (1971).
15. B. Zagalak, *Bull. Acad. Polon. Sci.*, *16*, 67 (1968).
16. T. Toraya, K. Ushio, S. Fukui, and H. P. C. Hogenkamp, *J. Biol. Chem.*, *252*, 963 (1977).
17. M. I. Yakusheva, A. A. Malahov, A. A. Poznanskaya, and V. A. Yakovlev, *Anal. Biochem.*, *60*, 293 (1974).
18. W. W. Bachovchin, R. G. Eagar, Jr., K. W. Moore, and J. H. Richards, *Biochemistry*, *16*, 1082 (1977).
19. R. H. Abeles, *Meth. Enzymol*, *9*, 686 (1966).
20. A. A. Poznanskaja, K. Tanizawa, K. Soda, T. Toraya, and S. Fukui, *Arch. Biochem. Biophys.*, *194*, 379 (1979).
21. D. E. McGee, and J. H. Richards, *Biochemistry*, *20*, 4293 (1981).
22. T. Toraya and S. Fukui, *J. Biol. Chem.*, *255*, 3520 (1980).
23. T. Toraya, S. Kuno, and S. Fukui, *J. Bacteriol.*, *141*, 1439 (1980).
24. R. G. Forage and M. A. Foster, *Biochim. Biophys. Acta*, *569*, 249 (1979).
25. M. K. Essenberg, P. A. Frey, and R. H. Abeles, *J. Am. Chem. Soc.*, *93*, 1242 (1971).
26. K. Tanizawa, N. Nakajima, T. Toraya, H. Tanaka, and K. Soda, *Z. Naturforsch.*, *42c*, 353 (1987).
27. T. Toraya, M. Uesaka, and S. Fukui, *Biochemistry*, *13*, 3895 (1974).
28. D. E. McGee, S. S. Carroll, M. W. Bond, and J. H. Richards, *Biochem. Biophys. Res. Commun.*, *108*, 547 (1982).
29. Z. Schneider, K. Pech, and J. Pawelkiewicz, *Bull. Acad. Pol. Sci.*, *14*, 7 (1966).
30. B. C. Johnson, A. Stroinski, and Z. Schneider, *Meth. Enzymol.*, *42*, 315 (1975).
31. T. Tobimatsu, T. Hara, M. Sakaguchi, Y. Kishimoto, H. Wada, M. Isoda, and T. Toraya (manuscript in preparation).

32. R. G. Eagar, Jr., W. W. Bachovchin, and J. H. Richards, *Biochemistry*, *14*, 5523 (1975).
33. K. W. Moore, and J. H. Richards, *Biochem. Biophys. Res. Commun.*, *87*, 1052 (1979).
34. B. Zagalak, P. A. Frey, G. L. Karabatsos, and R. H. Abeles, *J. Biol. Chem.*, *241*, 3028 (1966).
35. T. Yamane, T. Kato, S. Shimizu, and S. Fukui, *Arch. Biochem. Biophys.*, *113*, 362 (1966).
36. F. R. Jensen, and R. A. Neese, *Biochem. Biophys. Res. Commun.*, *62*, 816 (1975).
37. T. Toraya, Y. Sugimoto, Y. Tamao, S. Shimizu, and S. Fukui, *Biochemistry*, *10*, 3475 (1971).
38. S. Kuno, T. Toraya, and S. Fukui, *Arch. Biochem. Biophys.*, *210*, 474 (1981).
39. S. Kuno, T. Toraya, and S. Fukui, *Arch. Biochem. Biophys.*, *205*, 240 (1980).
40. S. Kuno, T. Toraya, and S. Fukui, *Arch. Biochem. Biophys.*, *211*, 722 (1981).
41. S. Kuno, S. Fukui, and T. Toraya, *Arch. Biochem. Biophys.*, *277*, 211 (1990).
42. B. M. Babior, *BioFactors.*, *1*, 21 (1988).
43. J. Rétey, A. Umani-Rouchi, J. Seibl, and D. Arigoni, *Experientia*, *22*, 502 (1966).
44. A. M. Brownstein and R. H. Abeles, *J. Biol. Chem.*, *236*, 1199 (1961).
45. J. Rétey, A. Umani-Rouchi, and D. Arigoni, *Experientia*, *22*, 72 (1966).
46. R. H. Abeles and B. Zagalak, *J. Biol. Chem.*, *241*, 1245 (1966).
47. P. A. Frey, M. K. Essenberg, and R. H. Abeles, *J. Biol. Chem.*, *242*, 5369 (1967).
48. P. A. Frey and R. H. Abeles, *J. Biol. Chem.*, *241*, 2732 (1966).
49. P. A. Frey, S. S. Kerwar, and R. H. Abeles, *Biochem. Biophys. Res. Commun.*, *29*, 873 (1967).
50. P. A. Frey, M. K. Essenberg, R. H. Abeles, and S. S. Kerwar, *J. Am. Chem. Soc.*, *92*, 4488 (1970).
51. O. W. Wagner, H. A. Lee, Jr., P. A. Frey, and R. H. Abeles, *J. Biol. Chem.*, *241*, 1751 (1966).

52. T. H. Finlay, J. Valinsky, K. Sato, and R. H. Abeles, *J. Biol. Chem.*, *247*, 4197 (1972).

53. W. W. Bachovchin, K. W. Moore, and J. H. Richards, *Biochemistry*, *17*, 2218 (1978).

54. T. Toraya and R. H. Abeles, *Arch. Biochem. Biophys.*, *203*, 174 (1980).

55. T. Toraya and A. Ishida, *J. Biol. Chem.*, *266*, 5430 (1991).

56. A. Ishida and T. Toraya, *Biochemistry*, *32*, 1535 (1993).

57. B. M. Babior, T. J. Carty, and R. H. Abeles, *J. Biol. Chem.*, *249*, 1689 (1974).

58. T. Toraya, E. Krodel, A. S. Mildvan, and R. H. Abeles, *Biochemistry*, *18*, 417 (1979).

59. R. H. Abeles and H. A. Lee, Jr., *Ann. N. Y. Acad. Sci.*, *112*, 695 (1964).

60. T. H. Finlay, J. Valinsky, A. S. Mildvan, and R. H. Abeles, *J. Biol. Chem.*, *248*, 1285 (1973).

61. J. E. Valinsky, R. H. Abeles, and A. S. Mildvan, *J. Biol. Chem.*, *249*, 2751 (1974).

62. J. E. Valinsky, R. H. Abeles, and J. A. Fee, *J. Am. Chem. Soc.*, *96*, 4709 (1974).

63. S. A. Cockle, H. A. O. Hill, R. J. P. Williams, S. P. Davies, and M. A. Foster, *J. Am. Chem. Soc.*, *94*, 275 (1972).

64. K. L. Schepler, W. R. Dunham, R. H. Sands, J. A. Fee, and R. H. Abeles, *Biochim. Biophys. Acta*, *397*, 510 (1975).

65. E.-I. Ochiai, *J. Inorg. Nucl. Chem.*, *37*, 351 (1975).

66. B. T. Golding, in B_{12}, Vol. 1 (D. Dolphin, ed.), John Wiley and Sons, New York, 1982, p. 543.

67. R. G. Finke, D. A. Schiraldi, and B. J. Mayer, *Coord. Chem. Rev.*, *54*, 1 (1984).

68. J. Halpern, S. H. Kim, and T. W. Leung, *J. Am. Chem. Soc.*, *106*, 8317 (1984).

69. B. P. Hay and R. G. Finke, *J. Am. Chem. Soc.*, *108*, 4820 (1986).

70. B. P. Hay and R. G. Finke, *J. Am. Chem. Soc.*, *109*, 8012 (1987).

71. T. Toraya, *Arch. Biochem. Biophys.*, *242*, 470 (1985).

72. K. Ushio, S. Fukui, and T. Toraya, *Biochim. Biophys. Acta*, *788*, 318 (1984).

73. T. Toraya, T. Matsumoto, M. Ichikawa, T. Itoh, T. Sugawara, and Y. Mizuno, *J. Biol. Chem.*, *261*, 9289 (1986).
74. J. Pawelkiewicz, and B. Zagalak, *Ann. N. Y. Acad. Sci.*, *112*, 703 (1964).
75. B. Zagalak and J. Pawelkiewicz, *Acta Biochim. Polon.*, *12*, 103 (1965).
76. B. Zagalak and J. Pawelkiewicz, *Acta Biochim. Polon.*, *12*, 219 (1965).
77. M. I. Yakusheva, A. A. Poznanskaya, T. A. Pospelova, I. P. Rudakova, A. M. Yurkevich, and V. A. Yakovlev, *Biochim. Biophys. Acta*, *484*, 216 (1977).
78. T. Toraya, N. Watanabe, M. Ichikawa, T. Matsumoto, K. Ushio, and S. Fukui, *J. Biol. Chem.*, *262*, 8544 (1987).
79. M. Ichikawa and T. Toraya, *Biochim. Biophys. Acta*, *952*, 191 (1988).
80. T. Toraya and A. Ishida, *Biochemistry*, *27*, 7677 (1988).
81. T. Toraya, T. Shirakashi, S. Fukui, and H. P. C. Hogenkamp, *Biochemistry*, *14*, 3949 (1975).
82. D. L. Anton, P. K. Tsai, and H. P. C. Hogenkamp, *J. Biol. Chem.*, *255*, 4507 (1980).
83. D. L. Anton, H. P. C. Hogenkamp, T. E. Walker, and N. A. Matwiyoff, *J. Am. Chem. Soc.*, *102*, 2215 (1980).
84. T. Toraya, N. Watanabe, K. Ushio, T. Matsumoto, and S. Fukui, *J. Biol. Chem.*, *258*, 9296 (1983).
85. J. S. Krouwer, B. Holmquist, R. S. Kipnes, and B. M. Babior, *Biochim. Biophys. Acta*, *612*, 153–159 (1980).
86. T. Toraya, S. Miyoshi, M. Mori, and K. Wada, *Biochim. Biophys. Acta* (in press).
87. G. Eberhard, H. Schlayer, H. Joseph, E. Fridrich, B. Utz, and O. Müller, *Biol. Chem. Hoppe-Seyler*, *369*, 1091 (1988).
88. A. Ishida, M. Ichikawa, K. Kobayashi, T. Hitomi, S. Kojima, and T. Toraya, *J. Nutr. Sci. Vitaminol.*, *39*, 115 (1993).
89. B. Zagalak and J. Pawelkiewicz, *Acta Biochim. Polon.*, *11*, 49 (1964).
90. A. A. Poznanskaya, M. I. Yakusheva, and V. A. Yakovlev, *Biochim. Biophys. Acta*, *484*, 236 (1977).
91. T. Toraya and S. Fukui, *Biochim. Biophys. Acta*, *284*, 536 (1972).

92. S. Honda, T. Toraya, and S. Fukui, *J. Bacteriol.*, *143*, 1458 (1980).
93. K. Ushio, S. Honda, T. Toraya, and S. Fukui, *J. Nutr. Sci. Vitaminol.*, *28*, 225 (1982).
94. T. Toraya and S. Fukui, manuscript in preparation.
95. T. Toraya, S. Honda, S. Kuno, and S. Fukui, *J. Bacteriol.*, *135*, 726 (1978).
96. N. Obradors, J. Badia, L. Baldoma, and J. Aguilar, *J. Bacteriol.*, *170*, 2159 (1988).
97. T. L. Talarico and W. J. Dobrogosz, *Appl. Environ. Microbiol.*, *56*, 1195 (1990).
98. N. Hosoi, K. Morimoto, C. Ozaki, Y. Kitamoto, and Y. Ichikawa, *J. Ferment. Tech.*, *56*, 566 (1978).
99. A. Willetts, *Biochim. Biophys. Acta*, *588*, 302 (1979).
100. J. A. M. D. Bont, J. P. V. Dijken, and K. G. V. Ginkel, *Biochim. Biophys. Acta*, *714*, 465 (1982).
101. H. Schutz and F. Radler, *Arch. Microbiol.*, *139*, 366 (1984).
102. R. G. Forage and M. A. Foster, *J. Bacteriol.*, *149*, 413 (1982).
103. R. G. Forage and E. C. C. Lin, *J. Bacteriol.*, *151*, 591 (1982).
104. A. Stroinski, in *Vitamin* B_{12} (B. Zagalak and W. Friedrich, eds.), Walter de Gruyter, Berlin, 1979, p. 1029.
105. T. Toraya, S. Honda, and S. Fukui, *J. Bacteriol.*, *139*, 39 (1979).
106. M. G. N. Hartmanis and T. C. Stadtman, *Arch. Biochem. Biophys.*, *245*, 144 (1986).
107. J. H. Quastel, M. Stephenson, and M. D. Whetham, *Biochem. J.*, *19*, 304 (1925).
108. G. A. Sprenger, B. A. Hammer, E. A. Johnson, and E. C. C. Lin, *J. Gen. Microbiol.*, *135*, 1255 (1989).
109. I. T. Tong, H. H. Liao, D. C. Cameron, *App. Environ. Microbiol.*, *57*, 3541 (1991).
110. R. Daniel and G. Gottschalk, *FEMS Microbiol. Lett.*, *100*, 281 (1992).

9

Adenosylcobalamin (Vitamin B_{12} Coenzyme)-Dependent Enzymes

Ei-Ichiro Ochiai

Department of Chemistry, Juniata College,
Huntingdon, Pennsylvania 16652, USA

1. SURVEY OF ADENOSYLCOBALAMIN–DEPENDENT ENZYMES

"Adenosylcobalamin" (often abbreviated as AdoCbl) is also called B_{12}-coenzyme, in which the cobalt atom located at the center of a corrin ligand is bound to the 5′-carbon atom of the ribose of the adenosyl group on the one side of the corrin ring and also to the nitrogen atom of a benzimidazole group on the other (Fig. 1). Another cobalamin coenzyme is methyl cobalamin, which is also involved in several biologically important processes such as methane and acetate formation. This chapter, however, deals only with adenosylcobalamin.

AdoCbl is a cofactor of enzymes which catalyze the 1,2 rearrangement (see below) and the ribonucleotide reductase reaction. The 1,2-rearrangement reactions can be represented by:

$$XYHC(1)-C(2)H_2Z \rightarrow YH_2C(1)-C(2)HXZ \quad (1)$$

The enzymatic 1,2-rearrangement reactions catalyzed by AdoCbl-dependent enzymes are illustrated by:

Diol dehydratase (or diol dehydrase):

$$(HO)CH_2CH_2(OH) \rightarrow CH_3CH(OH)_2\text{–(spontaneous)} \rightarrow CH_3CHO + H_2O \quad (2)$$

Ethanolamine ammonia lyase:

$$(NH_2)CH_2CH_2OH \rightarrow CH_3CH(OH)(NH_2)\text{–(spontaneous)} \rightarrow CH_3CHO + NH_3 \quad (3)$$

L-β-Lysine aminomutase:

$$(NH_2)CH_2CH_2(CH_2CH_2CH(NH_2)COOH) \rightarrow CH_3CH(NH_2)(CH_2CH_2CH(NH_2)COOH) \quad (4)$$

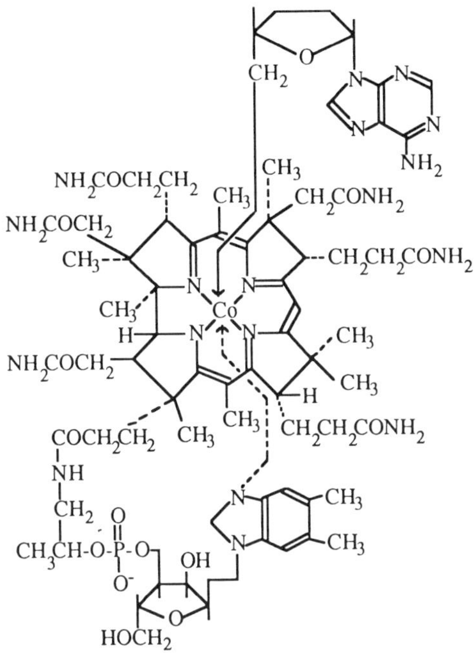

FIG. 1. Structure of adenosylcobalamin.

D-Ornithine aminomutase:

$$(NH_2)CH_2CH_2(CH_2CH(NH_2)COOH) \rightarrow CH_3CH(NH_2)(CH_2CH(NH_2)COOH) \quad (5)$$

Glutamate mutase:

$$((HOOC)(NH_2)CH))CH_2CH_2(COOH) \rightarrow CH_3CH(COOH)(CH(NH_2)(COOH)) \quad (6)$$

Methylmalonyl-CoA mutase:

$$(CoAS(O{=}C)(HOOC)CHCH_3 \rightarrow (HOOC)CH_2CH_2((C{=}O)SCoA) \quad (7)$$

α-Methyleneglutarate mutase:

$$(CH_2{=}C(COOH)){-}CH(COOH)CH_3 \rightarrow (HOOC)CH_2CH_2(C(COOH){=}CH_2) \quad (8)$$

The first two reactions [Eqs. (2) and (3)], however, may not necessarily be of the 1,2-rearrangement type in their mechanism, though they are popularly believed to be so. The enzyme diol dehydratase involved in reaction (2) is discussed in detail in Chap. 8 of this volume.

Ribonuleotide reductases catalyze the reduction of the OH on the C-2′ of the ribose of ribonucleotides (triphosphates) [reaction (9)]. The

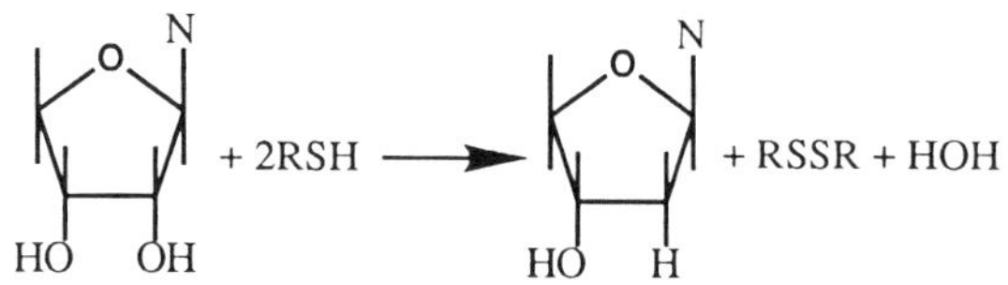

(9)

mechanism of this reaction may or may not involve 1,2 rearrangement as exemplified above. Ribonucleotide reductases dependent on iron and manganese are treated in Chaps. 4 and 5.

The characteristics and properties of the enzymes listed above are summarized in Ref. 1 and more fully discussed in the compendium B_{12} edited by Dolphin [2]. It is to be noted that sulfhydryl groups seem to be essential for these enzymatic activities, in addition to AdoCbl. Some enzymes, including glutamate mutase, diol dehydratase, 1-β-lysine aminomutase, and ornithine aminomutase, consist of a cobalamin-binding protein and a sulfhydryl protein. Each of the other enzymes such as methylmalonyl CoA mutase, ethanolamine ammonia lyase, α-methylglutarate mutase, and ribonucleotide reductase is a single protein which contains the essential sulfhydryl groups as well as a cobalamin binding site.

Despite intensive and extensive investigations, the mechanism of AdoCbl-dependent enzymatic reactions has eluded elucidation [3–6]. The first step of the enzymatic reactions involving AdoCbl, however, is believed to be the homolytic cleavage of the cobalt-to-carbon bond, which leads to the formation of a free radical (at C-5′ of adenosyl) and Co(II). These entities have been well documented to form in some enzymatic systems including diol dehydratase and ethanolamine ammonia lyase. They have been deteced and characterized by EPR. Only recently have EPR data been reported on the other enzyme systems including glutamate mutase and methylmalonyl CoA mutase. How this homolytic bond cleavage is induced is, however, yet to be determined.

This chapter will attempt to summarize the relevant data concerning the Co-C bond cleavage, discussing possible mechanisms of the bond cleavage, and then discuss relevant model study data and possible mechanisms of the enzymatic reactions.

2. MECHANISMS OF THE Co-C BOND CLEAVAGE

Here we will evaluate the various factors that might contribute to the Co-C bond cleavage both under protein-free conditions and under protein-bound conditions. It will be shown that contributions from the local environment (particularly the steric effects) alone do not appear to be sufficiently large to cause the Co-C bond cleavage. It seems necessary to invoke a global effect including the protein. A mechanism will be suggested based on these considerations.

2.1. Factors Influencing the Co-C Bond in AdoCbl and Its Model Compounds

The crucial first step of the AdoCbl-dependent enzymatic reactions is believed to be the homolytic cleavage of the Co-C (of the adenosyl group) bond. In order to obtain an insight into the mechanism of this bond splitting, the Co-C bond dissociation energies have been measured for a number of model compounds (as well as AdoCbl itself) under various conditions.

The compounds, generally denoted here as RCoLB, undergo the following reaction:

$$\mathrm{RCoLB} \rightarrow \mathrm{R}^{\bullet} + \mathrm{Co}^{\mathrm{II}}\mathrm{LB} \tag{10}$$

where R = alkyl group, L = in-plane ligand, and B = ligand *trans* to R group (R = adenosyl group, L = corrin, and B = benzimidazole in the case of AdoCbl). The bond dissociation energy (BDE) is simply the enthalpy change of this reaction, and hence BDE can be described by:

$$\mathrm{BDE} = \Delta H_f(\mathrm{R}^{\bullet}) + \Delta H_f(\mathrm{Co}^{\mathrm{II}}\mathrm{LB}) - \Delta H_f(\mathrm{RCoLB}) \tag{11}$$

With a fixed alkyl group (e.g., the adenosyl group), the BDE would be reduced by either lowering $\Delta H_f(\mathrm{Co}^{\mathrm{II}}\mathrm{LB})$ or raising $\Delta H_f(\mathrm{RCoLB})$ or both (see Fig. 2).

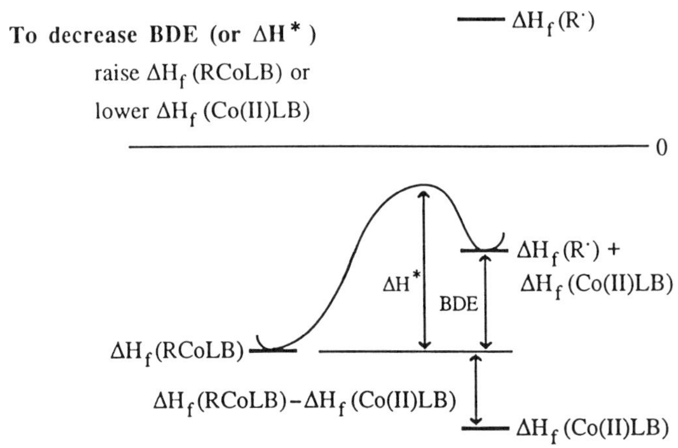

FIG. 2. Bond dissociation energy of R-Co in RCoLB [7].

The BDE values have been determined for a variety of R, L, and B combinations, and are summarized in Table 1. An interesting feature of this set of data is that BDE values of these compounds (i.e., those which are stable enough to be isolated) vary over a relatively narrow range despite the wide range of variations in R, L, and B. Statistically, they fall in the range of 97 ± 16 kJ/mol (the latter value is the standard deviation), although the average and the standard deviation values do not have the usual statistical significance here. This indicates that RCoLB systems tolerate significant changes in L, B, R, and steric effects (at least to a certain extent) and still maintain somewhat similar BDE values. In other words, this means that the BDE of RCoLB is not too sensitive to the variation in L and B. We will try to evaluate these factors.

The effect of B on the BDE value for a set of the same R and L can be gleaned from the data in Table 1. Among these bases (from H_2O to PPh_3), the variation in BDE values is within about 30 kJ/mol. This is, however, irrelevant for the enzymatic system, where B is fixed as a benzimidazole moiety. Only the distance between this ligand and the Co atom is perhaps of significance. The distance may be shortened to "push" the corrin ring (resulting in the rupture of the Co-C bond), or it may be lengthened. No definite evidence has been obtained either way in enzymatic systems.

TABLE 1

Cobalt-to-Carbon Bond Dissociation Energy (BDE)

B_{12} model compound	BDE(Co-C) (kJ/mol)	Ref.
$[CH_3CH_2CH_2\text{-Co(saloph)(pyr)}]$	105	3
$[(CH_3)_2CH\text{-Co(saloph)(pyr)}]$	84	3
$[(CH_3)_3CCH_2\text{-Co(saloph)(pyr)}]$	75	3
$[C_6H_5CH_2\text{-Co(saloph)(pyr)}]$	92	3
$[C_6H_5(CH_3)CH\text{-Co(DMG)}_2\text{(4-NH}_2\text{-pyr)}]$	88	3
[------------------------(4-CH_3-pyr)]	84	3
[------------------------(pyr)]	82, 84	3
[------------------------(4-CN-pyr)]	75	3
[------------------------(imidazole)]	88	3
[------------------------(PMe_2Ph)]	100	3
[------------------------($P(CH_2CH_2CN)_3$)]	84	3
[------------------------(P(*n*-Bu)$_3$)]	88	3
[------------------------($PEtPh_2$)]	80	3
[------------------------(PPh_3)]	71	3
$[C_6H_5CH_2\text{-Co(DMG)}_2(PMe_2Ph)]$	127	4
[-------------------(P(*n*-Bu)$_3$)]	121	4
[-------------------($PEtPh_2$)]	112	4
[-------------------(PPh_3)]	108	4
[-------------------(P(cyclo-$C_6H_{11})_3$)]	95	4
[-------------------(H_2O)]	100 ± 4	7
$[C_6H_5CH_2\text{-Co(DMG-BF}_2)_2(H_2O)]$	83 ± 11	7
$[C_6H_5CH_2\text{-Co(OEP)}(PMe_2Ph)]$	113	4
[-----------------(P(*n*-Bu)$_3$]	123	4
[-----------------($PEtPh_2$)]	109	4
[-----------------(PPh_3)]	100	4
[-----------------(P(cyclo-$C_6H_{11})_3$)]	124	4
AdoCbl	109, 126	5,16

It may be mentioned that Halpern and coworkers interpreted the effects of basicity of the B moiety on BDE as arising from the stabilizing effect on RCoLB [3]. But it seems more reasonable to ascribe it to the destabilizing effect on the product $Co^{II}LB$, as indicated in Fig. 2 and Table 2 below.

For R = $C_6C_5CH_2$, compounds of three different types of L [L = $(DMG)_2$, saloph, and OEP)] with various B's have been prepared and their BDE values determined. The average BDE value is 114 kJ/mol for [RCo(OEP)B], 111 kJ/mol for [RCo$(DMG)_2$B], and 92 kJ/mol for [RCo-(saloph)B]; that is, 20 or so kJ/mol variation. The difference in BDE between [RCo(acacen)(pyr)] and [RCo(F_6-acacen)(pyr)] was determined to be 8 kJ/mol with R = CH_3 and 4 kJ/mol with R = C_2H_5 [5]. The BDE difference between [$C_6H_5CH_2Co(DMG)_2(H_2O)$] and [$C_6H_5CH_2$Co(DMG-$BF_2)_2H_2O$)] has been estimated to be 16 kJ/mol [7], where (DMG-$BF_2)_2$ is a $(DMG)_2$ in which the bridging H^+ ions are replaced by BF_2^+ ions. These data indicate that the BDE of RCoLB is not very sensitive to the variation in L, at least with the types of compounds studied. An implication of this would be that an electronic perturbation in the corrin ring that might occur as a result of its conformational change in the enzymatic reactions may not affect the Co-C bond very much. An estimate of this effect (which may be termed as "cis-electronic effect") may be 20 or so kJ/mol at most.

The steric effects are the most difficult to evaluate. For one thing we have no data as to the extent of an increase of steric interaction between the adenosyl group and the corrin group in the enzymatic reaction systems, as required by the popularly held mechanism, i.e., mechanochemical hypothesis [3,8]. The static steric effects that exist in model compounds can be estimated as follows, using a set of BDE data with [RCo(saloph)(pyr)] [3]: $\Delta H_f(R^\cdot)$ can readily be estimated from the bond energy of R-H, and the "$\Delta H_f(Co^{II}LB) - \Delta H_f(RCoLB)$" can then be estimated using Eq. (11). This latter value should reflect the true energy to break the R-Co bond. These values are plotted against C-H bond energy in Fig. 3. The R-Co bond energy may reasonably be expected to be linearly related to R-H bond energy, were it not for steric effects. Assuming that the steric effect is negligible with R = CH_3, CH_3CH_2, and $C_6H_5CH_2$, a straight line is drawn through these points. The points with R = $(CH_3)_3CCH_2$ and R = $(CH_3)_2CH$ lie significantly higher than the line. The discrepancies between the points and line may then be regarded as arising from steric effects. Therefore, the static steric

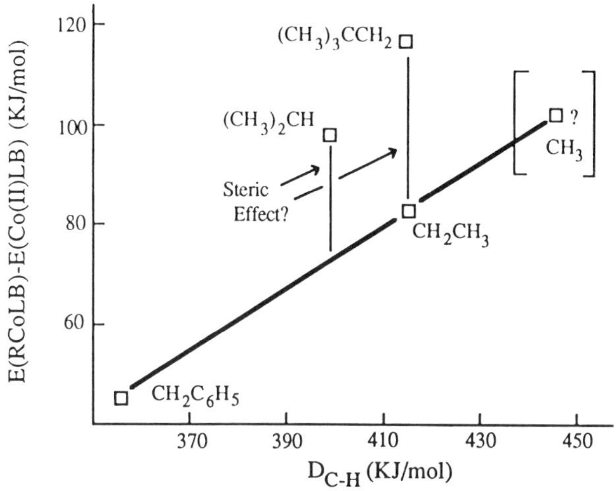

FIG. 3. Estimate of steric effects [7].

effects in these kinds of compounds may be estimated to be about 30 or so kJ/mol at most [7]. This value certainly would not represent the steric effect that might occur due to a conformation change in the corrin ring during enzymatic reactions. However, an implication of this estimate is that it would require a very substantial degree of conformational change in the corrin ring to break the Co-C bond, if the cleavage were to occur only through the steric interaction of this kind.

Table 2 summarizes the extents of various effects on BDE as discussed above. The local effects on the BDE (Co-C) appear to amount to about 50–60 kJ/mol at most, i.e., about 50% of the typical BDE value, unless a very large conformational change in the corrin ring does actually take place in the right direction to increase the steric interaction. However, no experimental evidence has been obtained for a very significant conformational change during enzymatic reactions, as discussed below.

2.2. Spectral Data on Enzymatic Reaction Systems

Spectral data on AdoCbl and AdoCbl-dependent enzymatic reaction systems have been reviewed in Ref. 2. Important spectral data on the following reaction scheme can be summarized as follows:

TABLE 2

Possible Effects on the BDE and Estimates of Their Extents

Nature of effects	On RCo(III) state [estimate of extent]	On Co(II) state [estimate of extent]
Steric	Steric hindrace can destabilize this state raising E(RCoLB) and thus reduces BDE (probably 30–40 kJ/mol at most; see text)	Insignificant
Electronic		
(1)	Effects of trans ligand: lengthen Co-C bond distance. (not significant)	Destabilizes Co(II) and increases BDE (30–40 kJ/mol at most)
(2)	Effects of cis ligand: A certain degree of π conjugation seems necessary to stabilize the Co-C bond. Disruption of this system (through conformational change) could reduce BDE (? 20 kJ/mol at most)	π^*-Conjugate system could stabilize Co(II) state. Disruption of this effect could increase BDE [?].

$$E + RCoLB \rightarrow E(RCoLB)\ (+\ S) \rightarrow E(R^{\cdot}, Co^{II}LB, S) \quad (12)$$

where E = enzyme and S = substrate. That the R-Co bond splits homolytically and results in the formation of $Co^{II}LB$ (often called "Co(II)balamin") and $R^{\cdot}$ has been most directly confirmed by EPR spectroscopy [9]. The EPR spectrum that was obtained upon addition of amino-2,1-propanol to an ethanolamine ammonia lyase preparation shows a kind of doublet [9]. This has been interpreted to be due to a coupled species of Co(II)balamin plus an organic free radical [9]. The distance between the Co(II) center and a free radical was estimated to be about 1000 pm (10 Å). A similar doublet-like spectrum was also observed with a ribonucleotide reductase reaction system [9].

Until quite recently, no definitive data had been obtained regarding whether the cobalt-to-carbon bond in AdoCbl splits homolytically in the reactions of mutases [Eqs. (4)–8)]. Retey and coworkers [10] reported recently that an EPR spectrum was observed when the substrate was added to a methylmalonyl-CoA mutase–AdoCbl holoenzyme preparation. The spectrum is somewhat different from those found with ethanolamine ammonia lyase and ribonucleotide reductase mentioned above, and has not been interpreted adequately. However, it appears that Co(II)balamin does form in this reaction too.

Buckel and his coworkers [11] demonstrated (1) that a preparation of 2-methyleneglutarate mutase from *Clostridium barkeri* showed an EPR spectrum indistinguishable from that of Co(II)balamin, implying that a portion of AdoCbl in the holoenzyme existed in that form, but (2) that a different spectrum appeared when the substrate was added to the preparation. The difference spectrum appeared similar to that observed with methylmalonyl-CoA mutase by Retey (see above).

These spectra are not like the doublet spectrum observed with the ethanolamine ammonia lyase system or the ribonucleotide reductase system [9]. They are more like a singlet, but asymmetric, and there is a sign that the low-*g* side may be split. In other words, these spectra of the mutase systems may be interpreted as resulting from the interaction between a free radical and Co(II)balamin, but with a combination of different relative positions of the moieties involved, as compared to the system of ethanolamine ammonia lyase.

EPR spectra have also been obtained for glutamate mutase from *Clostridium cochlearium* [12], but whether or not a free radical and Co(II)balamin would form upon addition of a substrate in this system as well has not yet been determined by EPR [12].

The optical spectrum of AdoCbl does not change appreciably when bound to an enzyme protein. The circular dichroismic spectrum of AdoCbl changes, but not too significantly, upon binding to ethanolamine ammonia lyase [13]. The intermediate formed after the cleavage of the Co-C bond, i.e., Co(II)balamin bound to diol dehydratase, is indistinguishable from that of free Co(II)balamin [14]. These findings indicate that the conformational change in the corrin ring, if it does take place during the enzymatic reactions, is not very significant. It must be pointed out, however, that the optical spectra are rather insensitive to conformational changes in the case of AdoCbl or cobalamin, as they are due to π–π^* transitions of the corrin ring. However, the conformational changes in the Co(II)balamin that occur upon addition of a substrate appeared to be somewhat appreciable in the EPR spectra [15].

2.3. Energetics of the "E+[RCoLB]+S" System

The BDE value of AdoCbl has been determined to be approximately 110 kJ/mol [5] or 130 kJ/mol [16]. The homolytic cleavage of this Co-C bond in enzymatic reaction systems is almost instantaneous; therefore, perhaps something like at least 100 kJ/mol (which would cause an approximately 1000-fold rate increase of the Co-C bond cleavage) of energy needs to be supplied somehow as a result of the additon of E (enzyme protein) and then S to E(AdoCbl). The question is: where does this energy come from?

Kinetic studies yielded the following results [14]:

$$\text{E(diol dehydratase)} + \text{AdoCbl} \rightarrow \text{E(AdoCbl)}: \quad K_m = 0.3\text{–}0.9 \text{ mM } (25°\text{C}) \tag{13}$$

$$\text{E(AdoCbl)} + \text{S} \rightarrow \text{E(AdoCbl, S)}: K_m = 0.03\text{–}0.18 \text{ mM} \tag{14}$$

K_m values can be converted to $\Delta G°$ values; taking typical values, $\Delta G°$ (for E + AdoCbl) = -36 kJ/mol (K_m = 0.5 mM) and $\Delta G°$ (for E(AdoCbl) + S) = -23 kJ/mol (K_m = 0.1 mM).

AdoCbl is known to bind to the protein via hydrogen bonds through the amide groups and other groups on the corrin ring [13,14] (and likely also through those on the benzimidazole connecting portion) [17]. Let us suppose that all the groups on AdoCbl that are potentially capable of forming a hydrogen bond to an appropriate protein residue are utilized in binding to the enzyme protein [7]. There are 16 such sites on AdoCbl: 6 $CONH_2$ on the corrin ring, 2 OH/4N on adenosine, and 1 O^-(P)/2 OH/1 CONH on the benzimidazole connecting portion. Typical H-bonding energies are known to be -25 kJ/mol for O−H--O, and -16 kJ/mol for N−H--O. Taking a conservative "-12 kJ/mol" as the average of the H-bonding interactions involved in this case, the binding energy $\Delta H°$ between the coenzyme and the protein can be estimated to be about -190 kJ/mol [7].

Yet the $\Delta G°$ value for this binding is experimentally determined to be about -36 kJ/mol (see above). Since $\Delta S°$ for this type of binding may be estimated to be about -40 eu [18], $T\Delta S°$ = -50 kJ/mol. Therefore,

ΔH^o (observed) would be -86 kJ/mol. The difference, about 100 kJ/mol [$= -190 - (-86)$], between this observed value and the estimated value is unaccounted for. This energy, or at least a major portion of it, could be used to modify the conformations of the protein or the coenzyme or both in such a way that the Co-C bond will be weakened but unbroken at this stage. Perhaps this could be manifested as an elongation of the Co-C bond. This state may be termed as "ready state". The binding of a substrate would supply the rest of energy to complete the rupture of the Co-C bond. All together a little over 100 kJ/mol of the coenzyme/substrate binding energy may have been expended to break the bond [7]. This amount, it is to be noted, is approximately the same as the BDE (Co-C) of AdoCbl.

2.4. A Possible Mechanism

The discussion above suggests the following mechanism for the Co-C bond cleavage in the enzymatic reaction systems. Binding of the coenzyme to an apoenzyme causes a conformational change in the protein, which would stretch the Co-C bond (perhaps as much as 100 pm (1 Å), yet leaving it still bound (ready state). When a substrate binds, a further conformational change in the protein forces the Co-C bond to split homolytically. That is, the adenosyl portion of the coenzyme may be pulled away from the Co atom by the protein [perhaps as much as 700 pm (7 Å) or more, as deduced from an EPR spectrum of the diol dehydratase system] [9] (see Fig. 4).

This mechanism assumes a significant change in the protein conformation, and since the coenzyme is strongly bound to the protein, the conformation of the corrin ring would also be changed in the process. Whether this is sufficiently large that the steric interaction alone causes the bond splitting cannot be determined. However, the optical spectrum of the Co(II)balamin formed at this stage is known to be indistinguishable from that of the free form of Co(II)balamin, at least in some enzymatic reaction systems [14]. This suggests that the conformational change in the corrin ring at this state is not large enough to be detected by the optical spectral means and hence that the steric effect (as suggested by the mechanochemical hypothesis) may not be the major cause of the bond cleavage.

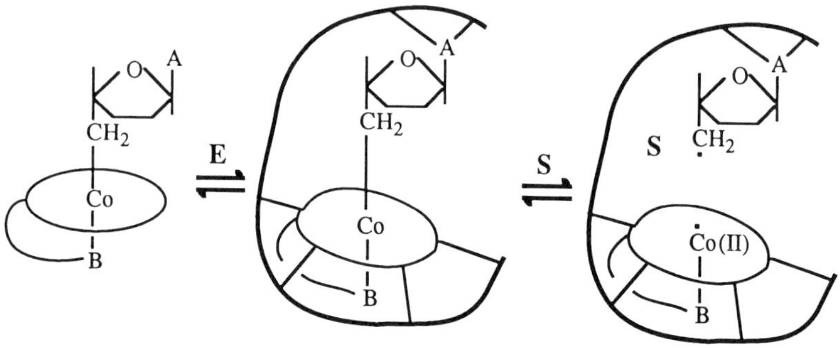

FIG. 4. Possible model for the Co-C(Ado) bond cleavage in enzymatic reactions [7].

A recent study reports a rate enhancement of about 10-fold of R-Co bond cleavage of R-cobalamin (where R = *i*-butyl or neopentyl) upon binding to an apoenzyme (diol dehydratase) [19]. The authors [19] attribute this enhancement (6 kJ/mol in terms of energy) to an increased steric hindrance due to a conformational change in the corrin ring upon its binding to the protein. Since in this case the R group does not have residues that are capable of hydrogen-bonding, the pulling-away effect as assumed above cannot be expected and the enhancement could be solely steric. However, this effect seems to be rather small for such sterically bulky groups as *i*-Bu or neopentyl. The rate enhancement effect on a less bulky or less hindered group is expected to be even smaller.

That in the ready state (protein-bound state) the Co-C bond is weakened significantly has also been demonstrated by an accelerated conversion of adenine-ethylcobalamin to hydroxycobalamin when it is bound to diol dehydratase [20]. Here the R group has a base, which is capable of hydrogen-bonding.

In these cases, the conformational change accompanying the second stage (binding of a substrate) was not examined. However, the extent of this conformational change of the corrin ring would be smaller than that in the first stage (binding of the coenzyme to the protein) because the energy that would become available upon binding of a substrate is expected to be much smaller.

3. MECHANISMS OF THE SUBSEQUENT REACTIONS

3.1. Hydrogen Transfer Between the AdoCbl Moiety, the Substrate, and the Protein

The second step is believed to be the abstraction of an H atom from a substrate by the adenosyl radical formed [21]. This has been demonstrated by use of labeled (either D or T) substrates and/or AdoCbl (at the 5′ position) for diol dehydratase [22], methylmalonyl-CoA mutase [23,24], ethanolamine ammonia lyase [25], L-β-lysine aminomutase [26], and others [21]. That is:

$$
\begin{aligned}
&[\mathrm{Co{-}CH_2(5'\text{-}adenosyl)}] \rightarrow [\mathrm{Co^{II}\ {}^{\bullet}CH_2(5'\text{-}Ado)}]\\
&[\mathrm{Co^{II}\ {}^{\bullet}CH_2(5'\text{-}Ado)}] + \text{S-T (substrate)} \rightarrow [\mathrm{Co^{II}\ TCH_2(5'\text{-}Ado)}] + \mathrm{S^{\bullet}}\\
&[\mathrm{Co^{II}\ TCH_2(5'\text{-}Ado)}] + \mathrm{S^{\bullet}} \rightarrow [\mathrm{Co^{II}\ TCH_2(5'\text{-}Ado)}] + \mathrm{P^{\bullet}}\\
&[\mathrm{Co^{II}\ TCH_2(5'\text{-}Ado)}] + \mathrm{P^{\bullet}} \rightarrow\\
&\qquad [\mathrm{Co{-}CHT(5'\text{-}Ado)}] + \text{P-H (rearranged product)}
\end{aligned}
\tag{15}
$$

or

$$[\mathrm{Co{-}CHT(5'\text{-}Ado)}] + \text{S-H} \rightarrow [\mathrm{CoCHH(5'\text{-}Ado)}] + \text{P-T} \tag{16}$$

It was noted in the study on diol dehydratase [22] that product formation was faster than tritium transfer by a factor of about 250, i.e., $k_{cat}/k_H = 250$. Assuming that T transfer competes with H transfer from the three equivalent hydrogens on the C-5′(Ado), one estimates a tritium isotope effect $k_H/k_T = 125$. Based on the $k_H/k_D = 10–12$ for the overall reaction, k_H/k_T can be predicted to be 35. Therefore, the observed tritium isotope effect is unusually large (by a factor of approximately 4).

In contrast, the k_H/k_T value for the hydrogen abstraction step [the second step in Eq. (15)] in the ethanolamine ammonia lyase reaction was determined to be 4.7. This suggests that this is not the rate-determining step [27]. After the migration of the amino group [the third step in Eq. (15)], the product radical abstracts back an H atom from C-5′(Ado). The k_H/k_T value of this step was found to be very large, i.e., 160 [27]. The k_H/k_D for the same step was determined to be 22. These values are thus consistent with the overall deuterium isotope effect value (of $k_H/k_D = 7.4$), implying that the last step of Eq. (15) is rate determining. The k_H/k_T of 160, however, is much larger (by a factor of about 9) than the expected value.

It has also been demonstrated that about 90% of tritium was accounted for in the products or in the recovered AdoCbl, and that the rest (10%) of the tritium was found in the exchangeable pool [28]. These results prompted a hypothesis that there might be a second source of hydrogen [28]. The presence of an alternative hydrogen for the product radical to abstract would make the loss of tritium from the labeled AdoCbl appear slower than otherwise, thus making the isotope effect much larger than expected. The hydrogen is assumed to be attached to an oxygen or a sulfur (i.e., sulfhydryl) [29].

3.2. Mechanisms of 1,2 Rearrangement

3.2.1. Ethanolamine Ammonia Lyase

Ethanolamine ammonia lyase will produce acetaldehyde and ammonia from ethanolamine in the following manner [see also Eq. (3)]:

$$(NH_2)CH_2CH_2(OH) \rightarrow (NH_2)CH_2(C^{\bullet})H(OH) \rightarrow$$
$$CH_3(C^{\bullet})(OH)(NH_2) \rightarrow CH_3CH(OH)(NH_2) \rightarrow CH_3CH{=}O + NH_3 \quad (17)$$

The migrating group has been established to be the NH_2 rather than the OH group. The question is whether and what role the enzyme protein and the Co(II)balamin entity play in this rearrangement reaction.

Bansal *et al.* [30] determined the rate constant for the following reaction to be 9×10^5 (s^{-1}):

$$^{\bullet}CH(OH)CH_2(OH)^+ \rightarrow {}^{\bullet}CH_2CHO + H^+ + H_2O \quad (18)$$

This is a spontaneous reaction, and the rate is larger by a factor of a few orders of magnitude than that of the diol dehydratase reaction (see Chap. 8). Therefore, it can be hypothesized that the rearrangment reaction step is just spontaneous, requiring no extra entity.

A similar spontaneous reaction rate has not been determined in the case of the ethanolamine free radical. However, it appears reasonable to assume that the amine group migration may be spontaneous in the ethanolamine ammonia lyase reaction, too. No definitive data have been obtained to indicate otherwise, i.e., that these rearrangements require the participation of the protein or the Co(II)balamin entity (however, see Ref. 31 and Sec. 3.2.3).

3.2.2. *Methylmalonyl-CoA Mutase, 2-Methyleneglutarate Mutase, and Glutamate Mutase: Model Studies*

The mechanism up to the homolytic cleavage of the cobalt-to-carbon bond of AdoCbl seems to be common to all AdoCbl-dependent enzymatic reactions. Some spectroscopic data to this effect have been discussed above for these enzymes.

The crucial step, then, is the 1,2-rearrangement reaction. Would it occur spontaneously in these reactions, as in the case of diol dehydratase and ethanolamine ammonia lyase as discussed above? Is the cobalamin moiety involved? Radical or nonradical mechanism?

Wollowitz and Halpern [32] treated the bromide (I, below) with Bu_3SnH in the presence of azobis(isobutyronitrile) (AIBN) (a radical initiator). It resulted in the formation of a small amount (2–9%) of the rearranged product (II):

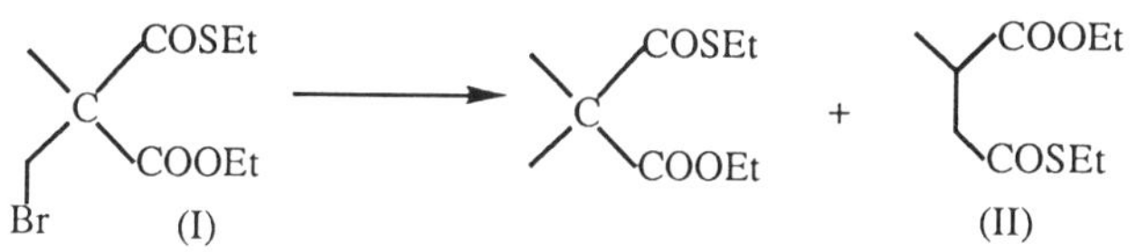

(19)

Another bromide (III), however, did not yield any rearranged product when it was treated with Bu_3SnH. That is, a truly free radical reaction did not seem to lead to the formation of the 1,2-rearranged reaction product [21]. On the other hand, the same bromide (III) yielded a rearranged product (IV) in a good yield (60%) in the presence of Co(I)balamin [21,33].

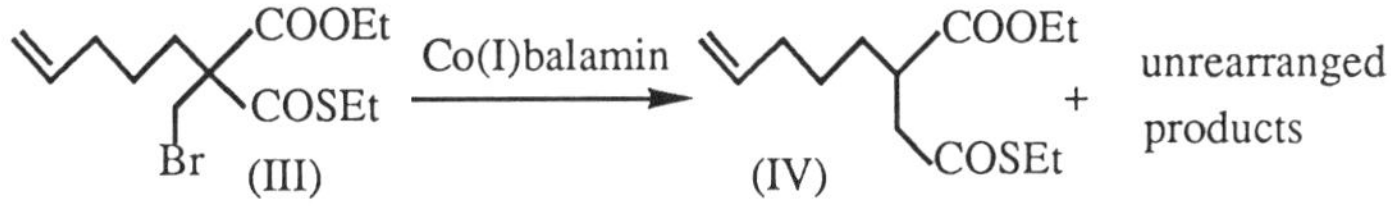

(20)

One of the unrearranged products (in the sense of 1,2 rearrangement) is a radical cyclization product bound to the cobalt atom of cobalamin. The yield of this product was about 11%. Dowd [21] interpreted these results as follows: "A free radical is produced at the outset of the reaction of (III) with Co(I)balamin. This can be partitioned so that a small part undergoes cyclization and trapping with Co(I)balamin or Co(II)balamin. The major part accepts a second electron or reforms the carbon–cobalt bond and goes on to rearrange to (IV) by a nonradical pathway". The first portion of this contention seems to be valid in view of the EPR spectral and other spectral data mentioned earlier (Sec. 2.2) as well. He did not elaborate on the nonradical pathway.

Suzuki and Barker [34] showed that no exchange with exogenous labeled ammonia or α-ketoglutarate could be detected in the glutamate mutase reaction [Eq. (6)]. They interpreted this result to indicate that no free radical intermediate was involved and argued that a possible role of enzyme-bound intermediates could not be ruled out. A possibility has been suggested that a Schiff base formation may facilitate the rearrangement (methylaspartate to glutamate), though there is no indication that a pyridoxal moiety is present in the enzyme [34].

Accordingly, Dowd and coworkers [35] constructed a model system to simulate the reverse reaction of glutamate mutase using a keto form instead of an imine form:

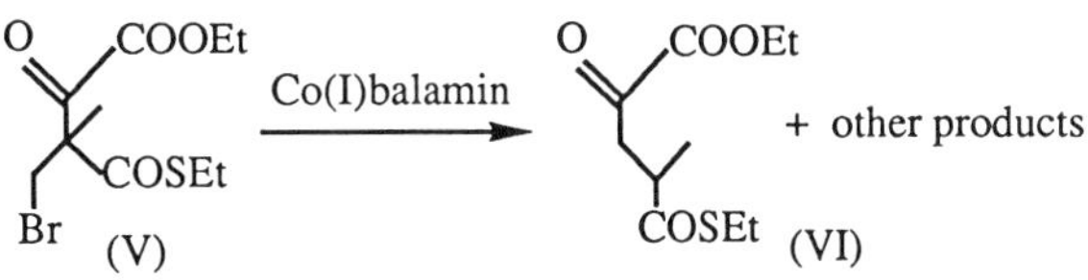

(21)

In this case, too, the yield of the rearranged product (VI) was quite high (70%). However, a control reaction with Bu_3SnH (with AIBN), too, did produce the rearranged product (VI) in a good yield. A Schiff base analog of (V), phenylimine (C = N−Ph instead of C = O), also produced the corresponding rearrangement product when treated with either Bu_3SnH or Co(I)balamin [36]. These results have shown that a ketimine derivative, if formed, can rearrange either free radically or in the presence of Co(I)balamin, and consequently these results are inconclusive regarding the mechanism of the enzymatic rearrangement.

3.2.3. Possible Mechanisms

The mechanism of the 1,2-shift reaction has not been well understood. It is usually considered in terms of a bridged intermediate [1]. If a bridged species is indeed the intermediate (or transition state), then it can be concluded that (1) a carbonium ion would provide the most facile rearrangement and (2) a free radical mechanism can be feasible only in the case of compounds of a special type, such as those in which the migrating group is an aromatic ring system. The second point can be understood in terms of a stabilizing effect of the conjugated double-bond system on the bridged intermediate (see Fig. 5). The facile free radical rearrangement of a phenyl ketimine derivative mentioned above may be cited as an example of this effect. This, however, does not necessarily endorse a free radical mechanism (without any intervention of the protein and/or cobalamin) for the enzymatic reactions.

As a matter of fact, most model studies mentioned in the previous section suggest what is termed a "nonradical pathway" [21]. Possibilities of participation of the carbonium ion form as a substrate intermediate have been proposed in the past [1], but these mechanisms were based on heterolytic cleavage of the cobalt-to-carbon bond of AdoCbl.

Let us suppose, as has been demonstrated, that the carbon-to-carbon bond of AdoCbl splits homolytically and the adenosyl radical abstracts a hydrogen from the substrate, resulting in a substrate free radical intermediate. Then the next step is either (1) a formation of a carbonium ion of the substrate, which requires the abstraction of an

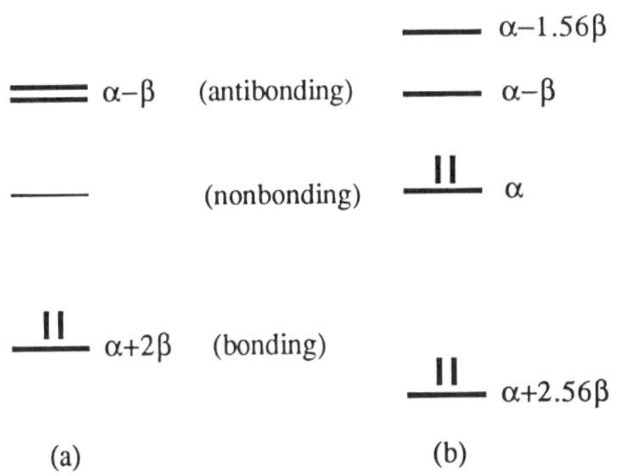

FIG. 5. Energetics of the bridged species (transition state) for a 1,2 rearrangement: (a) the case of a carbonium ion reaction (without participation of Co(II)), and (b) the case of Co(II)-assisted reaction. Each vertical short bar represents an electron. The simple Hückel calculations were based on the assumption that all the atoms involved have the same Coulomb integral and that all the exchange integrals between the adjacent atoms have the same value. (Modified from Ref. 15.)

electron from the substrate intermediate (by what?), or (2) a rearrangement of the free radical by participation of the Co(II)balamin entity. Hypothesis (1) is highly unlikely in view of the data obtained so far. Dowd et al. [37] formulated hypothesis (2) as a recombination of the free radical with the Co(II) and a subsequent rearrangement. This does not say anything about the mechanism of the rearrangement itself, except implying that the cobalt is participating in the reaction.

We proposed a mechanism involving the Co(II) species in 1975 [1,15]. We will present a slightly modified version of it here to conclude this chapter. It is merely a model, but it seems to adequately explain most of the observations mentioned above. The energetics of the bridge intermediate is sketched in Fig. 5a, which depicts the situation with a carbonium ion. The figure suggests that a carbonium ion would lead to a relatively stable intermediate, since a pair of electrons participate in the bridge formation by occupying the bonding level. In the cases of free radical and anion species, one or two electrons have to occupy the antibonding levels, thus making the formation of a bridged intermediate unfavorable. The proposed mechanism, as sketched in Fig. 6, assumes that the bridged species interacts with the unpaired electron of the Co(II) species. The energetics of this system (third species in Fig. 6) is shown in Fig. 5b. The four electrons involved [three from the free radical and one from the Co(II)] would occupy the lowest two levels, rendering the bridged species relatively favorable. This could facilitate the 1,2 rearrangement.

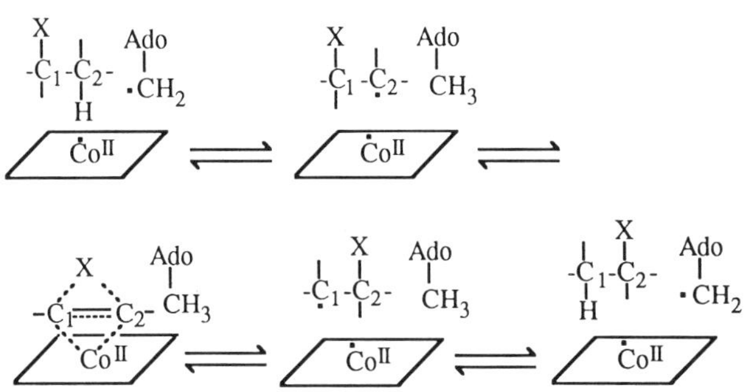

FIG. 6. Possible mechanism of Co(II)-assisted 1,2 rearrangement. (Modified from Ref. 15.)

This mechanism may be called Co(II)-assisted 1,2 rearrangement. This is possible with Co(II) because it has an unpaired electron. It may not be possible with other transition metals such as Fe since the iron might have to be Fe(I), which is not easy to obtain. Perhaps this is the reason that nature has chosen cobalt for this purpose. It is also possible that a mechanism similar to that suggested in Fig. 6 may be used in the case of diol dehydratase and ethanolamine ammonia lyase as well.

3.3. Ribonucleotide Reductase

The first step of ribonucleotide reductase is believed to be the homolytic splitting of the cobalt-to-carbon bond in AdoCbl. The rest of the reaction is not yet known. Readers are referred to Sec. 4.2.1 and 4.2.3 in Chap. 1 and to Chap. 4.

4. CONCLUSION

The mechanisms of AdoCbl-dependent enzymatic reactions including that of ribonucleotide reductase are far from being well understood. This chapter tried to discuss what's known and what's unknown, and then to present some new ideas, in the hope of enhancing our understanding of this interesting group of enzymes.

ACKNOWLEDGMENT

The author thanks Dr. P. Dowd and Dr. W. Buckel for providing preprints and reprints of their recent work.

ABBREVIATIONS

acacen	bis(acetylacetone)ethylenediimine
AdoCbl	adenosylcobalamin, B_{12}-coenzyme
AIBN	azobis(isobutyronitrile)

BDE	bond dissociation energy
CD	circular dichroism
CoA	coenzyme A
Co(II)balamin	Co(II) form of cobalamin
DMG	dimethylglyoxime
E	enzyme protein
EPR	electron paramagnetic resonance or ESR (electron spin resonance)
i-Bu	isobutyl
OEP	octaethylporphyrin
pyr	pyridine
S	substrate
saloph	bis(salicylaldehyde)phenylenediimine

REFERENCES

1. E.-I. Ochiai, *Bioinorganic Chemistry: An Introduction*, Allyn and Bacon, Boston, 1977, Chap. 12.
2. D. Dolphin (ed.), B_{12}, Vols. 1 and 2, John Wiley and Sons, New York, 1982.
3. J. Halpern, *Science*, *227*, 869 (1985).
4. J. Halpern, S.-H. Kim, and T. W. Leung, *J. Am. Chem. Soc.*, *106*, 8317 (1984).
5. M. K. Geno and J. Halpern, *J. Am. Chem. Soc.*, *109*, 1238 (1987).
6. (a) P. J. Toscano and L. G. Marzilli, *Prog. Inorg. Chem.*, *31*, 105 (1984); (b) L. G. Marzilli, W. O. Parker, J.-P. Charland, M. F. Summers, L. Randaccio, N. Bresciani-Pahor, and E. Zangrado, in *Frontiers in Bioinorganic Chemistry* (A. V. Xavier, ed.), VCH, Weinheim, 1986, p. 467 ff.
7. E.-I. Ochiai and C. E. Loope, *J. Inorg. Chem.*, *36*, 194 (1989).
8. J. H. Grate and G. N. Schrauzer, *J. Am. Chem. Sco.*, *101*, 4601 (1979).
9. J. R. Pilbrow, in B_{12}, Vol. 1 (D. Dolphin, ed.), John Wiley and Sons, New York, 1982, p. 431 ff.
10. Y. Zhao, P. Such, and J. Retey, *Angew. Chem. Int. Engl. Ed.*, *31*, 215 (1992).

11. C. Michel, S. P. J. Albracht, and W. Buckel, *Eur. J. Biochem.*, *205*, 767 (1992).
12. U. Leutbecher, R. Boecher, D. Linder, and W. Buckel, *Eur. J. Biochem.*, *205*, 759 (1992).
13. B. M. Babior, in B_{12}, Vol. 2 (D. Dolphin, ed.), John Wiley and Sons, New York, 1982, p. 263 ff.
14. T. Toraya and S. Fukui, in B_{12}, Vol. 2 (D. Dolphin, ed.), John Wiley and Sons, New York, 1982, p. 233 ff.
15. E.-I. Ochiai, *J. Inorg. Nucl. Chem.*, *37*, 351 (1975).
16. R. G. Finke and B. P. Hay, *Inorg. Chem.*, *23*, 3041 (1984).
17. T. Toraya and A. Ishida, *Vitamin*, *62*, 221 (1988).
18. W. R. Jencks, *Adv. Enzymol.*, *43*, 219 (1975).
19. T. Toraya and A. Ishida, *Biochemistry*, *27*, 7677 (1988).
20. T. Toraya, N. Watanabe, M. Ichikawa, T. Matsumoto, K. Ushio, and S. Fukui, *J. Biol. Chem.*, *262*, 8544 (1987).
21. For a recent review, see P. Dowd, in *Selective Hydrocarbon Activation: Principles and Progress* (J. J. Davies, P. L. Watson, A. Greenberg, and J. F. Liebman, eds.), VCH, Weinheim, p. 265 ff.
22. (a) R. H. Abeles and B. Zagalak, *J. Biol. Chem.*, *241*, 1245 (1966); (b) P. A. Frey and R. H. Abeles, *J. Biol. Chem.*, *241*, 2732 (1966); (c) P. A. Frey, M. K. Essenberg, and R. H. Abeles, *J. Biol. Chem.*, *242*, 5369 (1967); (d) M. K. Essenberg, P. A. Frey, and R. H. Abeles, *J. Am. Chem. Soc.*, *93*, 1242 (1971).
23. J. Retey and D. Arigoni, *Experientia*, *24*, 783 (1966).
24. W. W. Miller and J. H. Richards, *J. Am. Chem. Soc.*, *91*, 1498 (1969).
25. B. M. Babior, *Biochem. Biophys. Acta*, *167*, 456 (1968).
26. J. Retey, F. Kunz, T. C. Stadtman, and D. Arigoni, *Experientia*, *25*, 801 (1969).
27. D. Q. Weisblat and B. M. Babior, *J. Biol. Chem.*, *246*, 6064 (1971).
28. R. J. O'Brien, J. A. Fox, M. G. Kopczynski, and B. M. Babior, *J. Biol. Chem.*, *260*, 16131 (1985).
29. B. M. Babior, *Biofactors*, *1*, 21 (1988).
30. K. M. Bansal, M. Graetsel, A. Henglein, and E. Janata, *J. Phys. Chem.*, *75*, 455 (1973).
31. R. G. Finke, D. A. Schiraldi, and B. J. Mayer, *Coord. Chem. Rev.*, *54*, 1 (1984).

32. S. Wollowitz and J. Halpern, *J. Am. Chem. Soc.*, *110*, 3113 (1988).
33. G. Choi, S.-C. Choi, A. Galan, B. Wilk, and P. Dowd, *Proc. Natl. Acad. Sci. USA*, *87*, 3174 (1990).
34. F. Suzuki and H. A. Barker, *J. Biol. Chem.*, *241*, 878 (1966).
35. P. Dowd, S.-C. Choi, F. Duah, and C. Kaufman, *Tetrahedron*, *44*, 2137 (1988).
36. P. Dowd and S.-C. Choi, *J. Am. Chem. Soc.*, *111*, 2313 (1989).
37. P. Dowd, G. Choi, B. Wilk, S.-C. Choi, S. Zhang, and R. E. Shepherd, in *Chemical Aspects of Enzymes Biotechnology* (T. O. Baldwin, F. M. Raushel, and A. I. Scott, eds.), Plenum Press, New York, 1990, p 235 ff.

10
S-Adenosylmethionine-Dependent Radical Formation in Anaerobic Systems

Kenny K. Wong and John W. Kozarich

Department of Chemistry and Biochemistry, University of Maryland, College Park, Maryland 20742, USA

1. INTRODUCTION

In this chapter we will review some recent and exciting developments in the formation of radical-containing enzymes that are mediated through the cleavage of S-adenosylmethionine (SAM). The three enzyme systems, pyruvate formate lyase (PFL) and its activating enzyme (AE), lysine 2,3-aminomutase (LAM), and a SAM-dependent anaerobic ribonucleotide reductase (ARR) each use a 5′-deoxyadenosyl radical derived from SAM as a B_{12} surrogate. In the case of LAM the radical cofactor functions in a catalytic manner with respect to substrate and is responsible for the generation of substrate radical intermediates. For PFL and ARR, the radical cofactor is utilized stoichiometrically with respect to the enzyme to generate an enzyme radical localized on a glycine residue. These systems define a new role for SAM in enzymatic catalysis and will certainly be the subject of continued intensive research in the future.

2. PYRUVATE FORMATE LYASE–ACTIVATING ENZYME

2.1. The Pyruvate Formate Lyase System

Pyruvate formate lyase (PFL, EC. 2.3.1.54) of *Escherichia coli* is a central enzyme in the anaerobic metabolism of pyruvate. It reversibly converts pyruvate to acetyl-CoA and formate. During anaerobiosis, in the absence of a terminal electron acceptor, substrate level phosphorylation from glycolysis and the PFL system is the major source of ATP (Fig. 1). Historically, the activity of PFL was thought to constitute the "phosphoroclastic" reaction involving the formation of acetyl phosphate from pyruvate and inorganic phosphate (Fig. 2A) [1,2]. However, measure-

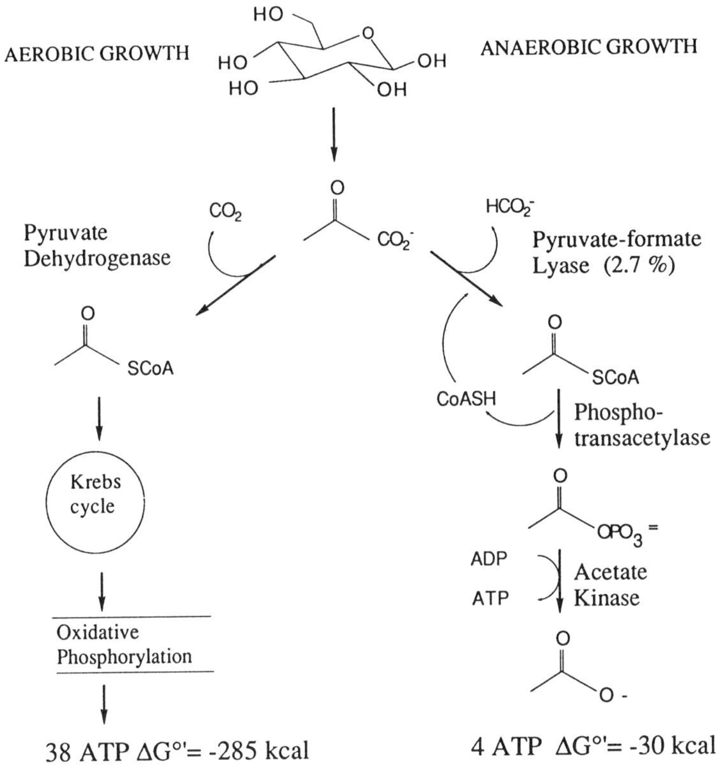

FIG. 1. ATP synthesis under aerobic and anaerobic growth in *Escherichia coli*. Under aerobic growth, pyruvate dehydrogenase oxidatively converts pyruvate to acetyl-CoA and carbon dioxide. Acetyl-CoA then can proceed through the Krebs cycle and oxidative phosphorylation to generate 38 mol of ATPs per mol of glucose. When *E. coli* switches to anaerobic growth, pyruvate dehydrogenase is rapidly inhibited and is replaced by pyruvate formate lyase. It is expressed to 2.7% of the soluble protein. The conversion of pyruvate to acetyl-CoA is a nonoxidative process and followed by two enzymes, phosphotransacetylase and acetate kinase to generate 1 mol of ATP per mol of pyruvate with an overall yield 4 ATPs per mol of glucose.

ments of the amount of acetyl phosphate formed relative to pyruvate consumption did not support this hypothesis. The eventual discovery that the lyase reaction was CoA-dependent led to the reformulation of the reaction (Fig. 2B) [3,4]. Earlier studies indicated that the lyase reaction was highly sensitive to oxygen; however, anaerobic incubation

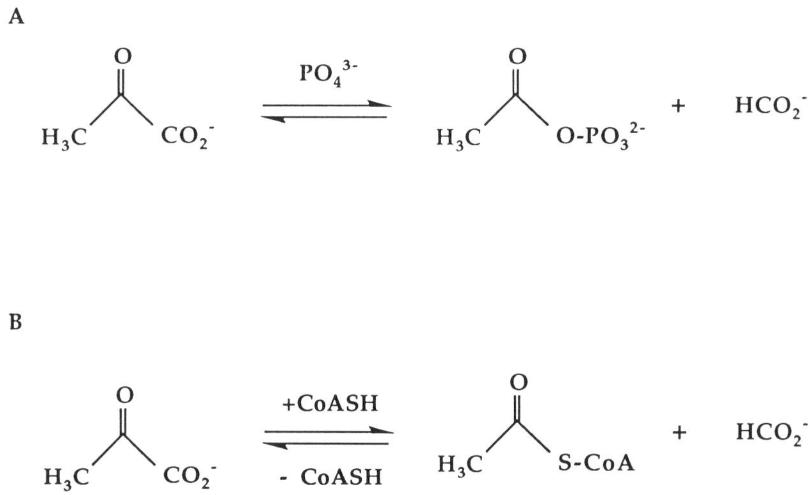

FIG. 2. (A) The phosphoroclastic reaction. The activity of PFL was believed to catalyze the formation of acetyl phosphate from pyruvate and inorganic phosphate. (B) The actual reaction catalyzed by PFL that forms acetyl-CoA and formate.

of aerobically isolated crude extracts resulted in the recovery of PFL activity. Furthermore, the recovery of PFL activity was found to be dependent on the presence of S-adenosylmethionine [5]. This suggested that the PFL may exist in two interconvertible forms. Subsequent work by Knappe and coworkers resulted in the isolation of four protein fractions that were required for PFL activity [6,7]. Two of the protein fractions were later identified as flavodoxin and flavodoxin reductase. The other two fractions were the lyase itself (170000 D, dimer) and an activating enzyme (30000 D). PFL can be isolated aerobically from anaerobically grown extract in the inactive form. The conversion to active PFL can be accomplished by incubation with the activating enzyme in the presence of Fe(II), SAM, reduced flavodoxin, and flavodoxin reductase. This clearly established that PFL existed in two interconvertible forms (Fig. 3): an inactive form, isolated under aerobic conditions, that is devoid of enzymatic activity (PFL_i) and an active form, stable under anaerobic conditions, that has full enzymatic activity (PFL_a). The activation of PFL after aerobic purification greatly simplified the isolation of this unusual enzyme.

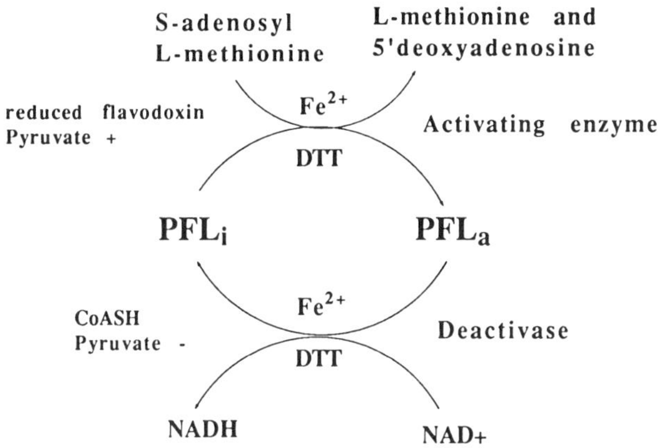

FIG. 3. The complex posttranslational regulation of PFL by activating enzyme and deactivase. Pyruvate is a positive allosteric effector for the activation reaction and a negative allosteric effector for the deactivation reaction. Both reactions require Fe(II) and strict anaerobic conditions.

Recently a PFL deactivase has been discovered that can convert PFL_a into PFL_i [8]. The deactivase was also found to be an NAD-dependent alcohol/aldehyde dehydrogenase. Thus, the switching on and off of PFL activity is tightly regulated by these two enzymes and clearly reflects the importance of PFL in the metabolic flux of pyruvate during anaerobiosis (Fig. 3). In addition to the intricate level of post-translational regulation, the genetic regulation of PFL expression is equally complex. The PFL operon is composed of an open reading frame, *orf*, of unknown function and *pfl*, the gene for PFL which is located immediately downstream of *orf* [9]. Downstream of *pfl* is *act*, encoding for PFL-activating enzyme. Fnr protein and more recently the ArcA gene product have been shown to affect the levels of transcription of both *pfl* and *orf* [10,11]. In-depth studies on the transcription of the operon have suggested that the transcription of *pfl* occurs at seven different promoters some of which are nested within *orf*. All of the promoters appear to be under the control of the dual Fnr and ArcA protein regulatory complex. The reason for such an elaborate transcriptional mechanism is not known. PFL-activating enzyme, on the other hand, is transcribed under its own promoter and is expressed constitutively under all growth conditions.

The concentrations of PFL and PFL-activating enzyme are approximately equal under aerobic conditions. However, upon switching to conditions of anaerobic growth, PFL is rapidly expressed and converted to the active radical form [7]. The rate of pyruvate consumption by the PFL system is faster than the rate of glycolysis, thus no buildup of pyruvate is observed under anaerobic growth. When the medium is depleted with glucose, PFL is converted to the inactive form by the deactivase system. Presumably, a switch to aerobic growth would also trigger the deactivase system to rapidly convert PFL to the inactive form. The inhibition of pyruvate dehydrogenase under anaerobic conditions and the tight regulation of PFL activity ensure that pyruvate consumption continues to supply the energetic needs of the cell.

2.2. The Glycyl Radical of Pyruvate Formate Lyase

The catalytic mechanism of PFL_a has been elusive. Studies designed to detect covalent modifications in the active protein by SAM were unsuccessful. Furthermore, PFL activity had been shown to require only PFL_a and was not due to a protein complex between PFL and the activating enzyme. This suggested that the activating enzyme effected a post-translational modification in PFL that was responsible for catalytic activity.

Then, in 1988, Knappe and coworkers made the highly significant observation that PFL_a exhibited an electron paramagnetic resonance (EPR) signal with a *g* value of 2.0037 and a visible absorbance at 340 nm [12,13]. This organic radical displayed a doublet with unresolved hyperfine splitting. In contrast, PFL_i was found to be EPR-silent and exhibited no visible absorbance features. The principal doublet appeared to derive from a solvent-exchangeable hydrogen coupled to the unpaired electron as observed by the collapse of the doublet to a singlet when incubated in D_2O. The doublet could be recovered with subsequent incubation in H_2O [13]. In a series of EPR experiments with labeled amino acids incorporated into PFL, Knappe and coworkers observed no perturbation in the EPR spectrum with labeled tyrosine, tryptophan, phenylalanine, histidine, and cysteine, thus excluding these amino acids as the site of the radical [13]. Partial labeling of the protein with ^{13}C from cells grown in ^{13}C-enriched glucose yielded unresolved hyperfine splitting due to the coupling of the ^{13}C nucleus to the unpaired electron. From these studies it was concluded that the radical resided on the backbone of an amino acid.

Recently, Knappe and coworkers tentatively identified the radical to be localized on the α carbon of glycine-734 [14]. This assignment was initially based on the observation that dioxygen irreversibly fragmented PFL into a 3-kD fragment that was not amenable to N-terminal analysis and suggested that the N-terminal amino acid had been chemically modified by dioxygen. Proteolytic digestion and amino acid sequencing of this fragment indicated that the 3-kD piece was derived from the 27-amino-acid, C-terminal end of PFL. The amino acid residue at the N-terminal end of this fragment appeared to be an oxalyl moiety which was derived from glycine-734 after reaction with dioxygen (Fig. 4A). This fragmentation reaction is reminiscent of the mechanism of the α-amidating enzyme which is responsible for cleavage of the C-terminal

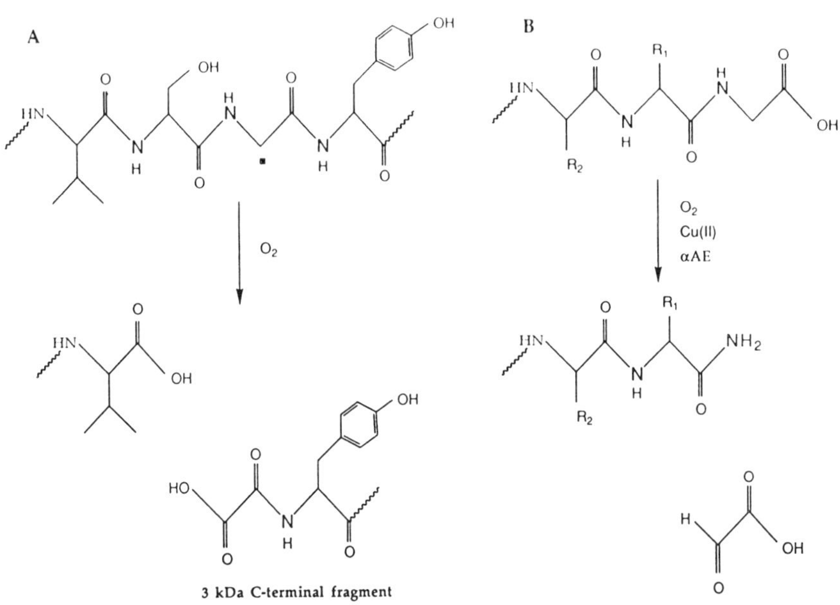

FIG. 4. (A) Oxygen-mediated fragmentation of active PFL (PFL_a). PFL_a rapidly reacts with oxygen and fragments into two pieces consistent of a 167 kD protein having a valine at the C-terminal end and a 3 kD fragment with an oxalyl moiety at the N terminus. (B) The reaction catalyzed by the copper monooxygenase, α-amidating enzyme.

glycine residue of a number of prohormones to form glyoxalate and the mature hormone with a C-terminal amide [15,16]. The α-amidating reaction requires Cu(II) and dioxygen and has been proposed to proceed via a glycyl radical intermediate (Fig. 4B) [17].

This important chemical observation aided in a systematic labeling of the protein with specific ^{13}C amino acids that resulted in the identification of the radical site. When the protein was labeled with [^{13}C-C-2]glycine, a perturbation of the EPR signal was observed suggestive of a strong coupling between the ^{13}C nucleus and the unpaired electron. Line broadening of the EPR signal was observed when the protein was labeled with [^{13}C-C-1]glycine or dual-labeled with [^{13}C-C-1]glycine/[^{13}C-C-1]serine. Similar line broadening was also observed with a uniformly labeled ^{15}N protein. These observations taken together provided very strong evidence that the resting radical on PFL_a resides on the α carbon of glycine-734.

The notion of a reactive α-carbon center radical at the active site of PFL may at first glance seem astonishing. However, the α-glycyl radical is known in organic chemistry and is particularly stable relative to other amino acid radicals. This is based on the captodative hypothesis of radical stabilization [18,19]. This concept is based on the mutual reinforcement of substituent effects adjacent to the radical center. One substituent is electron withdrawing (captor, the amide carbonyl of glycine-734) and the other electron releasing (donor, the amino group of glycine-734). The combined effect increases the resonance stability of the radical (Fig. 5A). Moreover, the geometric and steric requirements for optimization of the effect render the glycyl radical more stable than the corresponding radical in other amino acids [20]. This is due to the absence of nonbonding interactions between the side chain of glycine and the carbonyl oxygen of the adjacent amide which would contribute to the destabilization of the radical (Fig. 5B). The captodative effect on radical stabilization has experimental validation and would explain the unusual stability of the PFL radical [20].

With experimental and theoretical validation of a glycyl radical at the active site of PFL, the generality of a glycyl radical motif in the catalysis of other enzymatic reactions is worth considering. We have conducted an exhaustive homology search through the available protein and DNA data bases on PFL and the results reveal that several proteins have striking homologies to the region of PFL containing the glycine

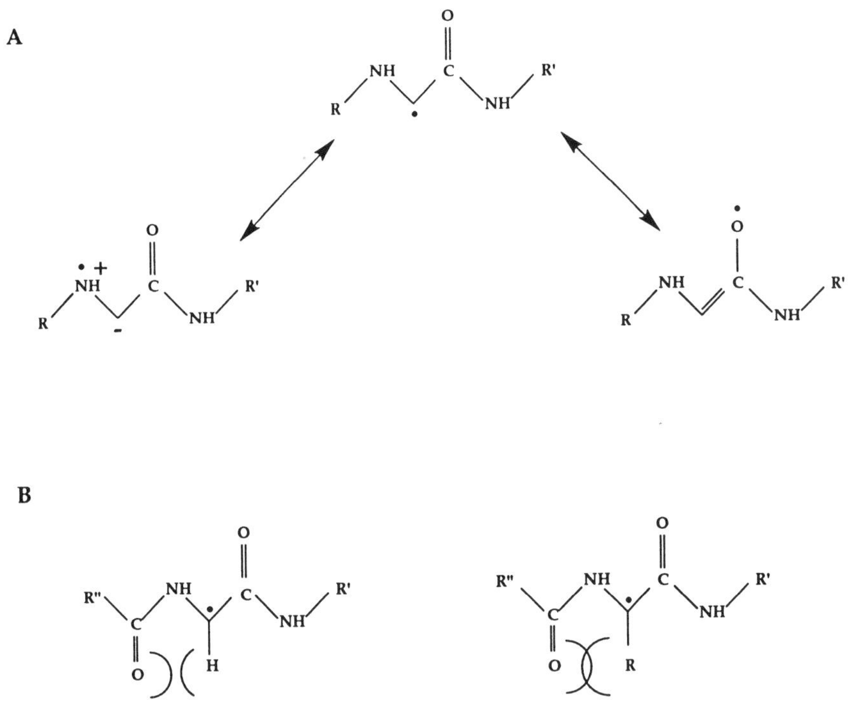

FIG. 5. The captodative effect. (A) Radicals can be stabilized through the positioning of electron donating and withdrawing groups adjacent to the radical site. (B) In particular, α-glycyl radical on the backbone of proteins can be stabilized via this principle due to the adjacent electron donor, the amide, and the electron acceptor, the carbonyl group. Other α-amino acid radicals are less stable due to nonbonding interactions.

radical (Fig. 6) [21]. Two hypothetical bacteriophage proteins from T4 and T5 and a partial protein sequence from *Serratia liquifaciens* show sequence identity with PFL spanning amino acids 722–740 of the PFL sequence [22–24]. The sequence, PEKYPQLTIRVSGYAVRFN, includes glycine-734. Extensive homologies also extend in both the N-terminal and C-terminal directions from this 19-amino-acid domain and encompass amino acid positions 699–759 of the PFL sequence. These striking homologies suggest that these proteins may have a functional relationship to PFL. The protein from *S. liquifaciens* was only a partial sequence and may be a PFL; however, the genetic organization would be

```
                  |680      |690        |700      |710     720|
PFL               DDEVRKTHLAGLMDGYFHHEASI--EGGQHLNVNVMNREMLLDAME
T4-HP             FVDPEIFYGLHAREIEIEPQPTVKIEGGQHLNVNVLRHETLEDAVK
T5-HP             HFDPESF-----REVKVEVAPQI--EGGQHLNVNVLSRDQLLDAQK
Serratia-HP       EGAVSDLGQIEYREVPLEMQPTVRVEGGQHLNVNVLRRETLEDAVK


                           |730 *    |740    |750
PFL               NPEKYPQLTIRVSGYAVRFNSLTKEQQDVITRTFTQSM
T4-HP             HPEKYPQLTIRVSGYAVRFNSLTPEQRDVIARTFTESL
T5-HP             HPEKYPQLTIRVSGYAVRFNASTREQNDVISRTFTQAM
Serratia-HP       HPEKYPQLTIRVSGYAVRFNSLTPEQRDVIARTFTESL
```

FIG. 6. Homologies between PFL and three proteins of unknown functions at the active radical site. The starred residue is glycine-734 of PFL implicated to be the radical site. The bold amino acids are identical residues. A, alanine; C, cysteine; D, aspartate; E, glutamate; F, phenylalanine; G, glycine; H, histidine; I, isoleucine; K, lysine; L, leucine; M, methionine; N, asparagine; P, proline; Q, glutamine; R, arginine; S, serine; T, threonine; V, valine; W, tryptophan; Y, tyrosine.

quite different from that of the *E. coli* counterpart because immediately 3′ downstream of this gene is an extracellular phospholipase gene [22], not a corresponding activating enzyme. The T5 protein was found to be situated downstream of the tRNA gene cluster suggesting that it is expressed early in the phage life cycle although it is not certain this *orf* codes for a real protein. The T4 protein, 63.1, was found situated in the early expressed genetic region between the *denV* (endonuclease V) and *tk* (thymidine kinase) genes. The genetic organization suggests that T4 63.1 gene encodes for a functional protein. The T4 and T5 bacteriophage proteins are most certainly not PFLs because of their sizes (~14000 D) and their lack of any cysteine residues [23,24]. It is conceivable, however, that they are part of a multimeric PFL in which these proteins constitute only one subunit. It is also tempting to speculate that they may represent other glycyl radical–containing enzymes of yet unknown activities which are important in the phage life cycle under anaerobic conditions. Since little research has focused on phage infections under anaerobic conditions, these proteins may be important tools for studying this question. Since *E. coli* is a host for both T4 and T5 bacteriophages, these striking homologies may also suggest a close functional relationship between these proteins, perhaps a protein–protein interaction that might modify the activity of the lyase reaction. We therefore recently cloned and expressed the T4 63.1 protein in *E. coli* to address the functional relationship of this protein with PFL. Studies with purified protein are in progress.

2.3. Chemical Mechanism of Pyruvate Formate Lyase

The catalytic mechanism of PFL is believed to be a homolytic process, although a heterolytic mechanism has not been rigorously ruled out. A heterolytic mechanism involving a thiolate attack to form a tetrahedral intermediate followed by elimination and protonation of the formate dianion would yield the expected products (Fig. 7A). However, the subsequent elimination of the formate dianion is energetically unfavorable. Likewise the decarboxylation of the pyruvate forming carbon dioxide and the acetyl anion is also energetically unfavorable in the absence of any cofactor assistance (i.e., thiamine). Even if this were feasible with the assistance of a yet undiscovered cofactor, the enzyme would still be required to reduce CO_2 and to attach an enzymic thiol to the acetate anion intermediate to give the expected products (Fig. 7B).

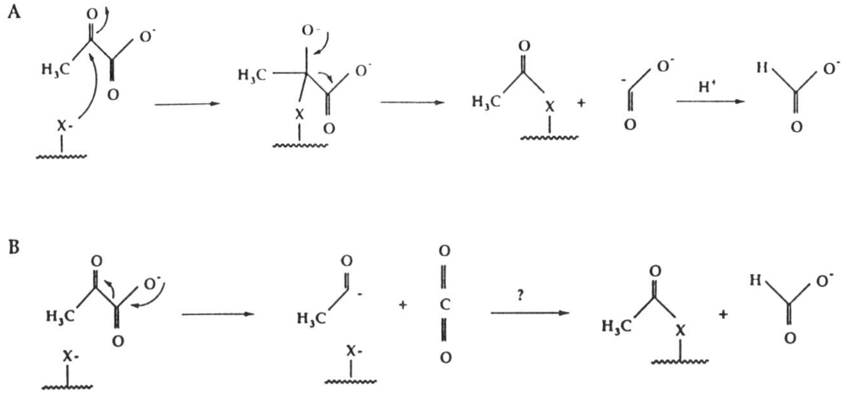

FIG. 7. Two hypothetical heterolytic mechanisms of C-1–C-2 of pyruvate to form acetyl-CoA and formate.

This exercise leads to the conclusion that a homolytic mechanism for the PFL reaction based on a functional role for the glycyl radical is worthy of consideration.

The nonoxidative cleavage of pyruvate via a homolytic mechanism by PFL is not without experimental support. The mechanism is supported by the observation of a paramagnetic species in the active form of PFL and not in the inactive form [13]. Furthermore, there is a correlation between the increase in the EPR signal of the organic radical and the increase in catalytic activity during activation implicating the participation of the radical in catalysis. Additional evidence for a homolytic process includes the observation of a primary deuterium kinetic isotope effect on the rates of pyruvate formation using deuterioformate and on the rate of hypophosphite inactivation using deuteriohypophosphite [25]. A radical intermediate has been observed during the inactivation of acetyl-PFL with hypophosphite [13]. Another inactivator, bisulfite anion, known to participate in radical chemistry, inactivates the radical form of PFL [26]. Finally, there is a chemical precedent for homolytic C-C bond cleavage with pyruvate esters via the Fenton-type Minisci reaction [27–29].

The strong circumstantial evidence and chemical precedent in support of a homolytic mechanism led us and others to propose the following mechanistic scenario for the cleavage of pyruvate (Fig. 8) [25,30]: Upon binding of pyruvate the first chemical step involves a thiyl radical attack of an active site cysteine residue (C418 or C419) on C-2 of pyruvate to form a radical tetrahedral intermediate. The thiyl radical site would be generated by a transient quenching of the glycine radical. A thiolate attack at C-2 of pyruvate followed by radical abstraction from the oxy anion is also feasible. The collapse of the tetrahedral radical intermediate would lead to the homolytic cleavage of C-1 and C-2 of the pyruvate forming acetyl enzyme and formyl radical anion. Formyl radical anion would then abstract the hydrogen from the transiently quenched glycine-734 yielding formate anion and the glycyl radical. This mechanism is modeled after the Minisci reaction for the cleavage of α-ketoesters by Fenton's reagent [27–29]. The formation of the formate radical anion has been observed previously and the ΔG for hydrogen abstraction has been measured to be a moderate 87 kcal/mol [31,32]. This first half-reaction has been well studied by isotope exchange using

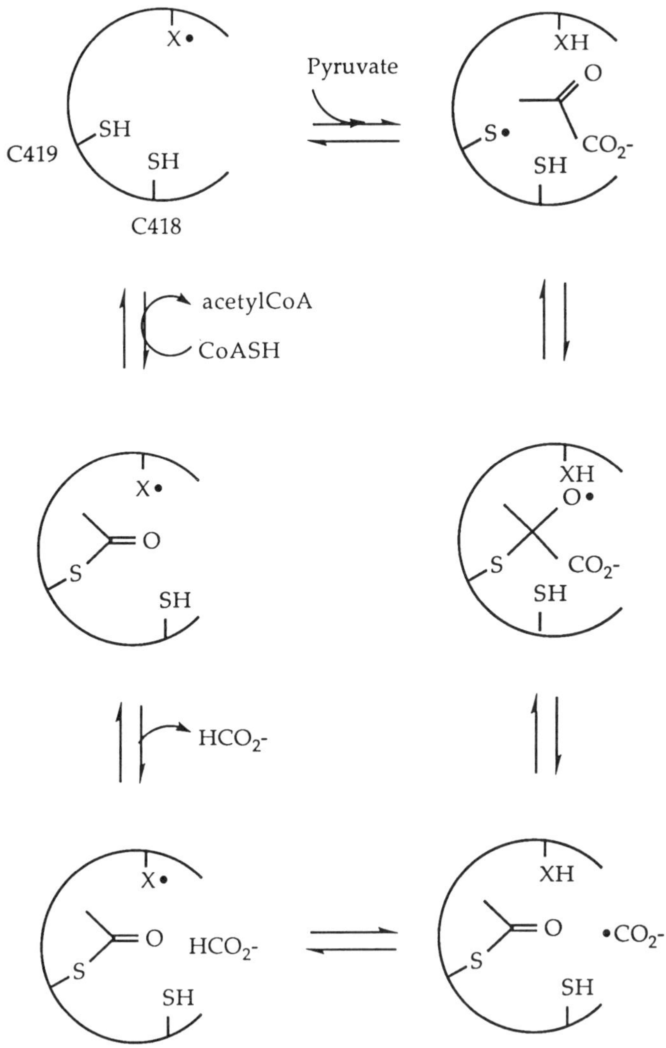

FIG. 8. Proposed homolytic mechanism of pyruvate dissimilation catalyzed by pyruvate formate lyase.

^{14}C-formate [33]. The second half-reaction would simply be a thioester exchange of the acetyl enzyme with CoA.

This overall mechanism is consistent with the observation of ping-pong kinetics and the isolation of an acetyl enzyme intermediate. The kinetic competence of the acetyl enzyme intermediate has not been explored, however, due to its lability [30]. The thiols implicated are cysteines-418 and 419 based on protein labeling studies [30]. Evidence suggests that cysteine-419 is the site of acetylation; however, site-directed mutagenesis of cysteine-418 to serine resulted in the loss of both half-reactions [10,34]. This indicates that cysteine-418 is also important in catalysis even though the C418S mutant can be converted to the radical form [10]. A possible explanation may be that cysteine-418 is the catalytic thiol involved in C-C bond cleavage and cysteine-419 is the site of acetylation involved in thiol–ester exchange. Using ^{32}P-radiolabeled CoA we have investigated the poorly studied thioester exchange half-reaction [34]. It appears that thioester exchange requires the active radical form of the enzyme indicating one of two possibilities: either the thioester exchange reaction also proceeds via a radical mechanism or the inactive enzyme is not in a correct conformational state for binding and heterolytic exchange of the acetyl enzyme with CoA.

2.4. Kinetic Studies on Formation of the Glycyl Radical

The radical on PFL is stable for days at 4°C and does not appear to be dependent on a metal ion, although the dependence of a metal ion on catalysis has not been rigorously ruled out due to the presence of Fe(II) in the activation mixture. The recent findings that this radical resides on glycine-734 places PFL in a separate class from that of the tyrosyl and tryptophanyl radical–containing enzymes [14]. The genesis of the glycyl radical on PFL is complex and requires AE, Fe(II), SAM, reduced flavodoxin, and pyruvate under strict anaerobic conditions. In vitro, reduced flavodoxin and pyruvate can be replaced by photoreduced 5-deazariboflavin and oxamate, respectively [7]. The catalytic competence of oxamate in PFL activation suggests that pyruvate functions as a positive allosteric effector during the activation. There is a strict requirement for SAM and Fe(II). Earlier studies showed that SAH could not support activation but did inhibit the reaction. The same observations were also observed for Co(II) [6,35]. The stoichiometry of SAM cleavage has been determined to be approximately 1:1 [36]. In the absence of PFL, a significant rate of SAM cleavage was not observed, indicating that glycyl radical formation is tightly coupled to reductive cleavage of SAM. This would make sense, since the biosynthesis of SAM has a high energetic price tag, especially under anaerobic conditions where ATP is at a premium.

We investigated the kinetics of PFL activation using a series of SAM analogs to gain insight into the binding interactions of SAM to PFL-activating enzyme. We found that only two substrate analogs were capable of supporting protein radical formation (**2** and **3**) but at very low rates (Fig. 9) [37]. Inhibition studies with other SAM analogs indicated that binding interactions with the activating enzyme can tolerate modifications in the methionine moiety, suggesting that the predominant binding interactions occur at the adenosylsulfonium moiety. Interestingly, catalysis was very sensitive to modifications of the methionine moiety. Good inhibitors such as **4** and **3** were not a substrate or a very poor substrate, respectively. The kinetics of glycyl radical formation followed saturation kinetics with respect to the substrates SAM and PFL_i. The K_m for SAM is approximately 3 μM. The K_m for PFL is 1 μM and is consistent with the level of PFL_i in the *E. coli* cell under aerobic conditions [9].

We also examined other metal ions in addition to Co(II) for catalytic activity and inhibition. No metal ions tested were capable of supporting activation, but Cu(II), Zn(II), and Co(II) were found to be very potent inhibitors of the activation [38]. Modest inhibition was observed for Hg(II) and Cd(II). With the availability of purified recombinant PFL-activating enzyme we demonstrated that AE bound to 1 mol of Fe(II) per mol of protein as determined by atomic absorption analysis. In addition, Cu(II) and Co(II) reconstituted enzyme bound with similar stoichiometry and were equally active under standard activation methods that contained 0.1 mM Fe(II). These observations suggest that the Fe(II) can freely exchange with solution Fe(II) during activation. The ability to reconstitute spectroscopically active metals into the activating enzyme will allow the characterization of the metal binding site by EPR spectroscopy.

We have begun biochemical characterization of the recombinant

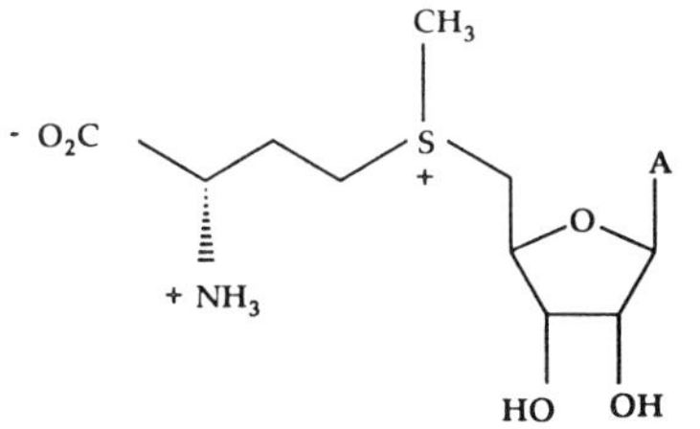

1

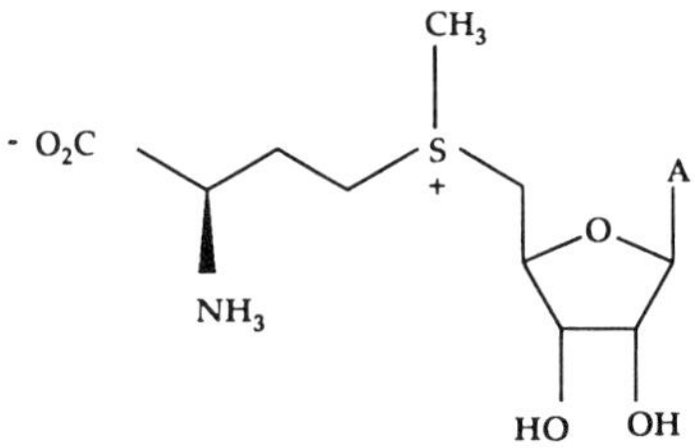

2

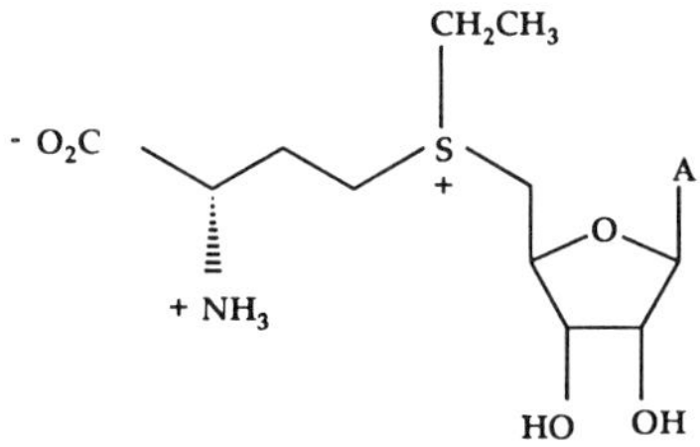

3

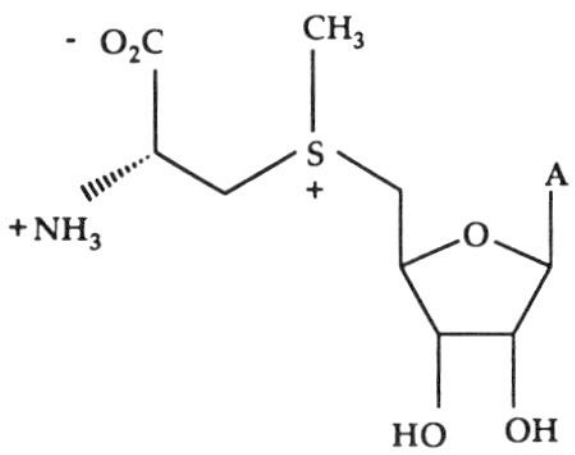

4

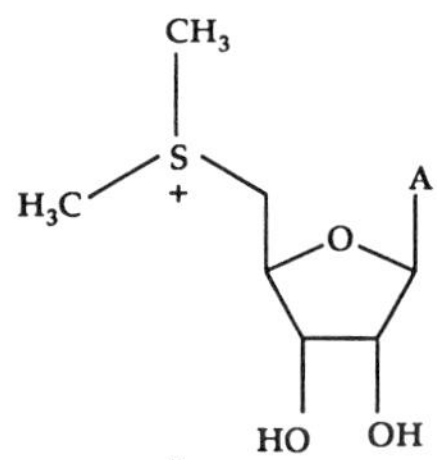

5

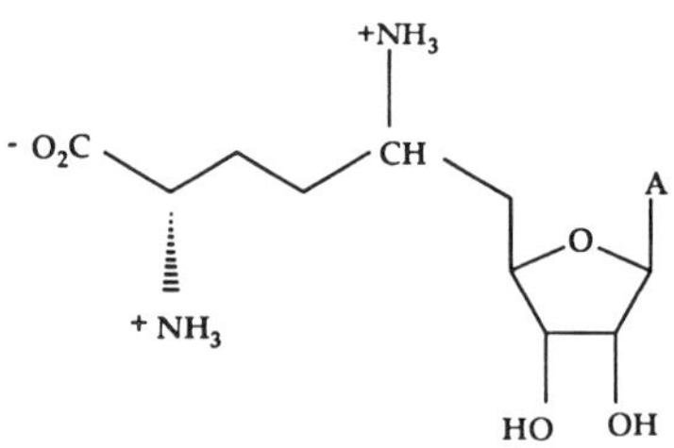

6

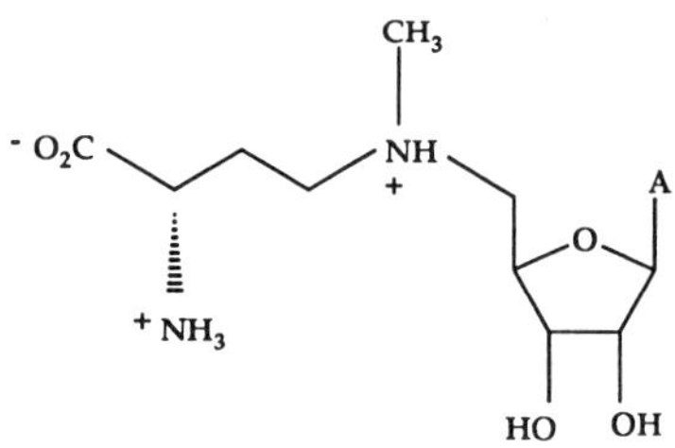

7

FIG. 9. Some substrates and inhibitors of the PFL activation reaction. 1, S-adenosyl-L-methionine; 2, S-adenosyl-D-methionine; 3, S-adenosyl-L-ethionine are substrates for the activation of PFL. 4, S-methyl-S-adenosyl-L-cysteine; 5, dimethylthioadenosine; 6, sinefungin; 7, 4-N,N-methyladenosyl-2-aminobutyrate are inhibitors of the activation of PFL.

activating enzyme [38]. PFL-activating enzyme is a monomer of approximately 30000 D consistent with protein of 28000 D predicted from the DNA sequence and consistent with the wild-type activating enzyme [7]. The N-terminal formylmethionine is removed in the matured protein. It contains six cysteines, four of which are clustered at the N-terminal end of the protein and have been suggested to be the metal binding region [39]. It is unclear whether or not the protein has a covalent chromophore associated with it [7,9]. In our studies with active recombinant PFL-activating enzyme no visible absorbance features consistent with a chromophore were observed [38]. Thus we believe that the enzyme does not have a covalently attached chromophore.

2.5. Mechanism of Protein Radical Formation

The activation of PFL requires stoichiometric reduction of SAM to L-methionine and 5′-deoxyadenosine [36]. Thus SAM functions as a redox cofactor rather than an alkylating agent. This net two-electron reduction of SAM is unusual, but other examples are known as discussed later. One of the two electrons needed for this reaction is probably derived from the oxidation of glycine-734 of PFL_i to the active glycyl radical form. The second electron probably comes from Fe(II), although there is no direct evidence that Fe(II) undergoes redox chemistry during the formation of the PFL radical. Fe(II) is needed in excess (i.e., 0.1 mM for half-maximal activity) with a constant consumption of reduced flavin. It has been proposed that reduced flavin aids in the regeneration of Fe(II) [12]. It is likely, then, that Fe(II) participates in protein radical formation.

By analogy to B_{12} chemistry, the mechanism for PFL activation can be envisioned to occur by an initial nucleophilic attack on the 5′ carbon of SAM by Fe(II) forming the product methionine and a Fe(IV)-adenosyl intermediate [12] (Fig. 10). Although there is no biochemical precedent for this intermediate, it would be analogous to the Co(III)-adenosyl intermediate formed from Co(I) and ATP [40]. Homolytic cleavage of the iron–carbon bond can then be proposed to form Fe(III) and 5′-deoxyadenosyl radical, which can subsequently abstract, directly or indirectly, a hydrogen from glycine-734 to form PFL_a. Fe(II) can be regenerated via reduced flavin or the reducing environment.

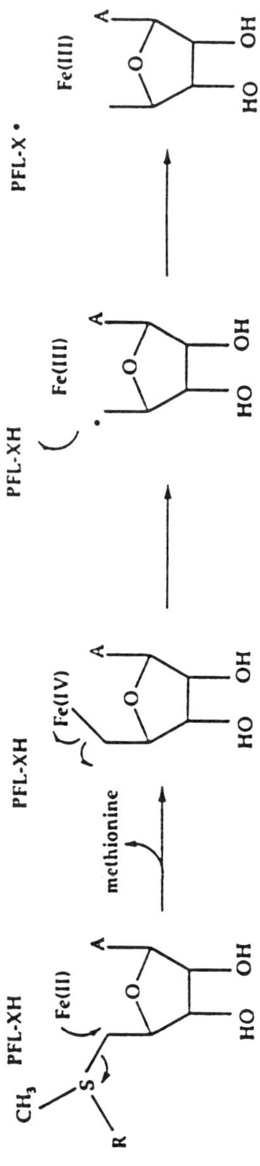

FIG. 10. Proposed mechanism for the activation of PFL involving an Fe(IV)-adenosyl intermediate.

Protein radical formation need not require a direct hydrogen transfer from glycine-734 but can involve first a hydrogen abstraction from a distant hetero atom (i.e., a cysteine residue) followed by long-range electron transfer mediated by appropriately placed amino acid residues of increasing negative potential to glycine-734. Such a mechanism has been proposed for ribonucleotide reductase where the resting tyrosyl radical is situated far from the active site [41]. Thus ribonucleotide reduction must involve some type of long-range electron transfer in order to deliver the radical to the active site for catalytic turnover.

To address the question of direct or indirect hydrogen abstraction, we have expressed PFL in a glycine-requiring strain of *E. coli*, which allowed for the incorporation of dideuterated glycine into the protein [42]. Preliminary results indicated a nominal primary kinetic isotope effect on the rate of activation. This suggested that carbon hydrogen cleavage of glycine-734 may be at least partially rate determining during radical formation. Moreover, the incorporation of *R*- and *S*-deuterated glycines into PFL has revealed that an isotope effect on activation is observed only for the *S* isomer suggesting that the hydrogen abstraction from glycine-734 is stereospecific [42]. We are also analyzing the product, 5′-deoxyadenosine, for deuterium incorporation in order to address the question of direct hydrogen transfer.

2.6. Quenching of the Glycyl Radical

The quenching of the glycyl radical in PFL can be achieved by two procedures. The first route is a chemical inactivation. A series of inactivators are known that inactivate PFL by quenching the radical. Most notable among these is dioxygen, which rapidly inactivates PFL with a half-life in the order of 4 sec at 4°C [12]. As mentioned above, oxygen inactivation has been shown to lead to the fragmentation of PFL at glycine-734 resulting in the formation of a 3-kD peptide. The resulting 164-kD protein dimer is inactive and cannot be reactivated. However, evidence also suggests that PFL inactivated by oxygen can be reactivated, supporting other routes of inactivation that may not result in fragmentation [25]. Indeed, PFL under strict anaerobic conditions does lose activity over time, which may support other chemical routes of inactivation. Under in vitro conditions, inactivation could be effected by exogenous thiols such as dithiothreitol (DTT) which can reductively quench the glycyl radical, thereby regenerating intact PFL_i.

Substrate analogs have also been shown to be potent mechanism-based inactivators of PFL. Hypophosphite is a mechanism-based inactivator, the mode of action of which is consistent with its recognition as a formate analog. The inactivation occurs with both free enzyme and acetyl enzyme. The rate of inactivation is faster with acetyl enzyme consistent with its binding as a formate analog [25]. Inactivation of acetyl enzyme by hypophosphite has been demonstrated to go through a radical intermediate having a structure similar to 2-hydroxyethylphosphonate, but attached to cysteine-418 at the phosphorus atom (Fig. 11) [13,30]. While the evidence for this mode of inactivation appears convincing, it does not explain the inactivation of the free enzyme by hypophosphite. Clearly, additional work is required to resolve this

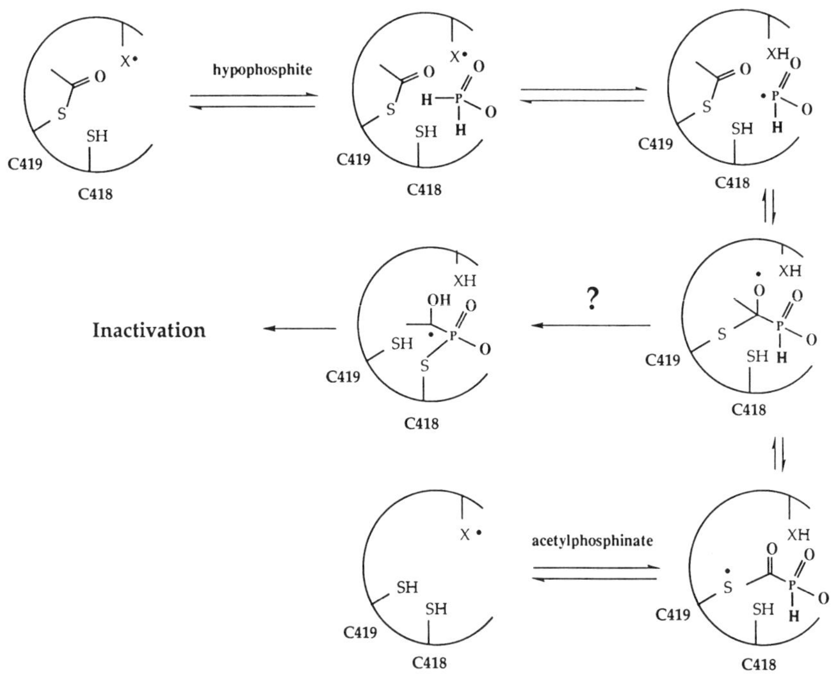

FIG. 11. Proposed mechanism of inactivation of PFL by the inactivators, hypophosphite and acetylphosphinate.

question. A primary kinetic isotope effect on phosphorus–hydrogen bond cleavage is observed suggesting that hydrogen abstraction is at least partially rate limiting [25]. A reasonable mechanism, then, involves an initial hydrogen abstraction from hypophosphite forming hypophosphoryl radical, which then can modify a side chain in the free enzyme or, by proximity, can react with the acetyl group of acetylated PFL. A pyruvate analog, acetylphosphinate, is believed to inactivate PFL by a similar pathway as that for hypophosphite inactivation, but via a C-P bond cleavage [43]. The homolytic cleavage of acetylphosphinate results in the formation of a hypophosphoryl radical anion and acetyl enzyme. This intermediate would be similar to the intermediate formed from hypophosphite and acetyl enzyme following hydrogen abstraction on hypophosphite (Fig. 11). The reaction pathway can partition between inactivation and hypophosphite release, forming the acetyl enzyme that is protected from further inactivation unless formate or CoA is added to deacetylate the enzyme. Quantitation suggests that the partition ratio is ~5, i.e., five turnovers are required for complete inactivation.

The deactivating system, an NAD-dependent alcohol/aldehyde dehydrogenase, constitutes a reversible enzymatic pathway for quenching the radical [8]. The deactivation does not result in protein fragmentation as observed with dioxygen and PFL_a can be regenerated with the activating enzyme. The reaction also requires Fe(II) and CoA as cofactors. The K_m values for Fe(II), NAD, and CoA are 0.1 mM, 1.7 mM, and 1 μM, respectively [44]. The reaction can be inhibited by desulfoCoA, NADH, and pyruvate. The K_i values for NADH and desulfoCoA were 0.15 mM and 15 μM, respectively. Oxamate cannot substitute for pyruvate in the inhibition of deactivation suggesting that the role of pyruvate is catalytic and not allosteric. This was supported by the observation that deacetylation of the acetyl enzyme by the addition of formate abolished inhibition of deactivation suggesting that the free enzyme is the substrate for the deactivase. The mechanism of quenching of the glycyl radical with the involvement of Fe(II), NAD, and CoA is puzzling. The formation of NADH requires the two-electron oxidation of active PFL, which is already oxidized by one electron from the activation. How the deactivase quenches the radical by an oxidative mechanism is intriguing and deserves more study. At any rate, the PFL inactivated by the deactivase can be reactivated once again, thus this deactivase system may serve to protect PFL from irreversible oxygen damage upon switching to aerobic growth conditions.

Aside from the uniqueness of the reaction it catalyzes, the deactivase is also unique in its molecular structure. The protein has a subunit molecular weight of 96 kD, but a native molecular weight of 2000 kD. Ultracentrifugation and crosslinking analysis suggested that the protein formed a rod-like supramolecular structure. Furthermore, the addition of substrates changed the susceptibility to proteolysis indicating a conformational change after binding of substrates. These observations are very suggestive that the rod-like oligomeric structure of the deactivase is related to the formation of spirosomes that is observed in facultative bacteria under conditions of anaerobic glycolysis [45].

3. LYSINE 2,3-AMINOMUTASE

3.1. General Properties of the Enzyme

Lysine 2,3-aminomutase (LAM) catalyzes the conversion of lysine to β-lysine (Fig. 12) [46]. In *Clostridium* the enzyme plays a key role in the formation of β-lysine, which is facilely converted to acetyl-CoA, thus permitting lysine to serve as a source of carbon and nitrogen [47]. In addition, the enzyme may play a role in the production of β-lysine-containing antibiotics in *Nocardia* and *Streptomyces* [48]. The initial purification of the enzyme from *Clostridium* SB_4 revealed that the enzyme has a hexameric structure (MW 280–290 kD; 47 kD subunits) and was activated by S-adenosylmethionine, iron, and pyridoxal phosphate (PLP) [49]. In addition, no protons derived from solvent are incorporated into lysine or β-lysine. Thus, the mechanistic sequence

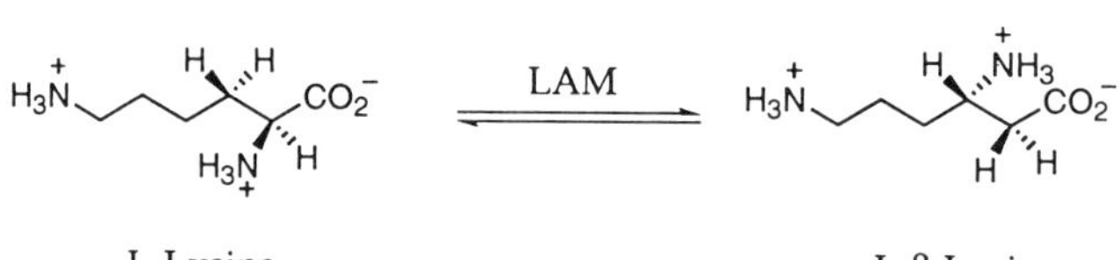

FIG. 12. Reaction catalyzed by lysine 2,3-aminomutase (LAM).

bears a striking similarity to the adenosylcobalamin-dependent skeletal rearrangements [50]; however, LAM does not require adenosylcobalamin or B_{12} derivative for activity. Moreover, the requirement for SAM suggests an unusual role for this cofactor as a B_{12} surrogate.

Frey and coworkers recently made significant progress on the purification and molecular properties of LAM [51]. The presence of inorganic sulfide and cobalt has been established in addition to iron and PLP. The variable amounts of copper and zinc reported in preparations result in a direct decrease in cobalt content and in overall enzymatic activity. The most completely constituted and active preparations contain up to 5 or more g-atoms of cobalt per mol of hexamer and minimal amounts of zinc and copper. The meticulous exclusion of oxygen is also essential for maximal enzyme activity. Under rigorously anaerobic conditions, LAM contains approximately 12 g-atoms of iron and of sulfur per mole of hexamer. Recent EPR data suggest that the iron and sulfur are configured as three Fe_4S_4 clusters [52]. Finally, purification in the presence of PLP and lysine affords up to 5.5 g-atoms of PLP per hexamer [53]. Thus, fully active LAM may be inferred to contain one PLP and one Co(II) for each subunit in the hexamer and three Fe_4S_4 clusters.

3.2. Role of S-Adenosylmethionine as a B_{12} Surrogate

Even when isolated under scrupulously anaerobic conditions LAM must always be activated with SAM. When isolated under aerobic conditions LAM requires a reductive incubation with dithionite and glutathione or dihydrolipoate prior to SAM activation [49]. While 50–75% of maximal activity can be achieved by SAM activation alone with anaerobically purified LAM, best results are also obtained with the initial reduction step.

The precise mechanistic role of SAM has been beautifully elucidated by the Frey group. An earlier stereochemical analysis by Aberhart and coworkers had shown that hydrogen transfer and amino group migration proceed with inversion of configuration with the 3-pro-R hydrogen of lysine being transferred to the 2-pro-R position of β-lysine and with an intramolecular migration of the amino group [54]. Additional work by this group using a mixture of unlabeled lysine and

[3,3-$^2H_2$]lysine suggested that hydrogen transfer was intermolecular due to the formation of monodeuterio-β-lysine. In a thorough reinvestigation the Frey group analyzed the pattern of deuterium labeling in β-lysine produced from different ratios of [3,3-$^2H_2$]lysine/lysine [55]. Their findings demonstrated that the conversion of lysine to β-lysine proceeds with both intramolecular and intermolecular transfer. This duality of mechanism is characteristic of adenosylcobalamin-mediated rearrangements [50]. In these reactions, a 5′-adenosyl radical or its equivalent abstracts a hydrogen from substrate to afford a 5′-deoxyadenosine and a substrate radical. The substrate hydrogen is torsionally randomized on the 5′-methyl group with the two adenosyl

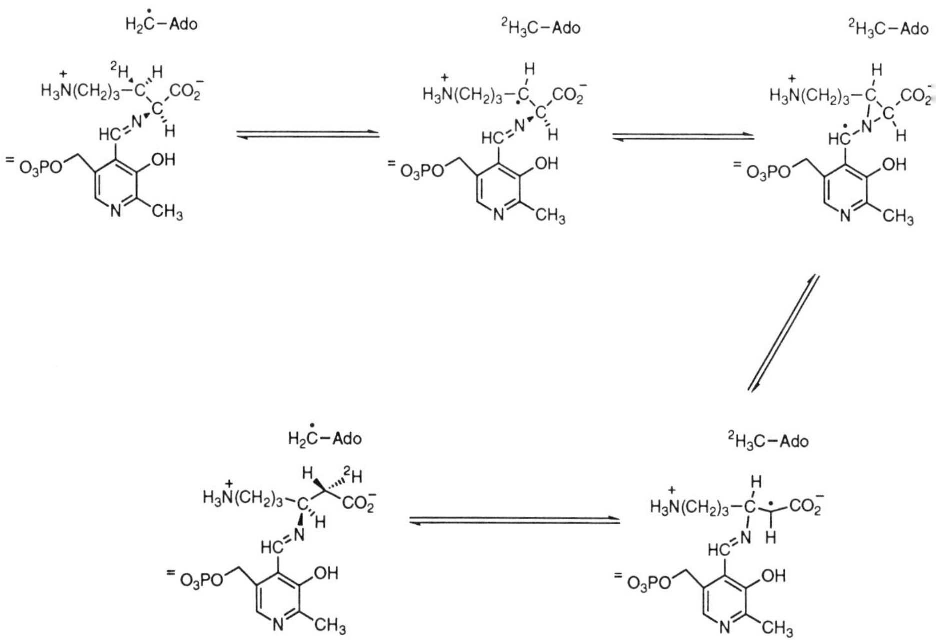

I

FIG. 13. Proposed mechanism of lysine 2,3-aminomutase (LAM) showing the role of the 5′-deoxyadenosyl radical in stereospecific deuterium transfer and the role of pyridoxal in the stabilization of the substrate radical intermediate during rearrangement [48]. The species designated I is the predominant radical species detected by EPR at steady state [48,58,59].

hydrogens initially present in the cofactor. The rearranged product radical then abstracts a hydrogen from the methyl group of the 5′-deoxyadenosine to generate the product and the 5′-adenosyl radical which reforms adenosylcobalamin. Thus, the initial substrate hydrogen can be transferred to the product to effect an intramolecular transfer or, if a cofactor hydrogen is delivered to the product, be stored in the cofactor at the end of the reaction cycle. The latter creates the possibility of intermolecular hydrogen transfer.

Several important experiments support the conclusion that the adenosyl moiety of SAM mediates hydrogen transfer in a manner analogous to that of adenosylcobalamin (Fig. 13) [48,55,56]. LAM has been shown to incorporate tritium from [5′-$^3H_2$]SAM into both lysine and β-lysine. In addition, the ratio of tritium in lysine and β-lysine is equivalent to the ratio of lysine and β-lysine at chemical equilibrium. The amount of tritium incorporated into lysine and β-lysine depends on the amount of enzyme present and not on the amount of labeled SAM when SAM is in excess over the enzyme. This establishes that the activation of LAM by SAM is stoichiometric and enzyme-bound SAM does not exchange with free SAM. Finally, when [3-^{3}H]lysine is used as the substrate, tritiated SAM has been reisolated from the reaction mixture [57]. Taken together, these data support the mechanistic proposal in Fig. 13 and solidify the role of SAM as a B_{12} surrogate in the aminomutase reaction.

3.3. Substrate Radicals as Intermediates

LAM has proven to be an ideal system for the detection of substrate/product radicals by EPR spectroscopy as elegantly demonstrated by Reed and Frey [48,58,59]. Initial work showed that reaction mixtures that contained reduced LAM, SAM, and L-lysine reproducibly afforded at 77 K an EPR signal centered at $g = 2.001$ [58]. The signal was dependent on the concentration and activity of the enzyme and the presence of L-lysine and not the D isomer. In addition, it was not a byproduct of the activation protocol. The g value, the hyperfine structure and narrow signal widths, and the lack of observable EPR signals for high-spin Co(II) and for the iron-sulfur centers at 77 K suggested that the signal was derived from an organic radical species of lysine. A time course of the EPR spectrum during the approach to chemical equilibrium between lysine and β-lysine further suggested that the EPR spectrum was inhomogeneous due to the presence of multiple substrate radical species. At equilibrium (steady state) one of these species predominated.

In order to refine the structure of the dominant radical species at equilibrium, experiments with isotopically labeled substrates were performed [58]. The EPR spectrum obtained with L-[3,3,4,4,5,5,6,6-^{2}H]lysine showed considerable enhancement of resolution arising from the loss of longer range couplings to the radical center by protons at carbons 3–6 upon substitution of deuterium at these positions. However, the observation that the unpaired spin was coupled to these lysine protons provided strong evidence that the organic radical is located on the carbon skeleton of lysine. Subsequent EPR experiments with [2-^{13}C]lysine, [2-^{2}H]lysine, and [2-^{15}N]lysine have firmly established the steady-state radical as a β-lysine imine radical (species I, Fig. 13) with the unpaired electron on the α carbon. Moreover, the bond linking carbon-3 to nitrogen is approximately coplanar with the p orbital on C-2 containing the unpaired electron [59]. Overall, these studies highlight the power of EPR spectroscopy in the study of LAM.

3.4. Role of Pyridoxal Phosphate

The role of pyridoxal phosphate in the aminomutase remains a subject of some speculation; however, an intriguing possibility has been forwarded by Frey and Reed [48]. By analogy to the well-known cyclopropylcarbinyl radical rearrangement, imine radicals could undergo a similar process, thus suggesting a route by which PLP could facilitate a 1,2-amino group migration. A model for the LAM reaction is the conversion of the benzaldimine of α-methyl-β-bromoalanine ethyl ester to a mixture of the benzaldimine derivatives of α-methylalanine ethyl ester and α-methyl-β-alanine ethyl ester in the presence of a free radical chain initiator [60]. The rearranged product, α-methyl-β-alanine ethyl ester, is more stable, being derived from the more stable α radical. The reaction may be viewed as an azacyclopropylcarbinyl rearrangement and, in the case of LAM, the pyridoxylidene group of PLP resonantly stabilizes the transiently formed azacyclopropylcarbinyl radical (Fig. 13). However, this attractive mechanism remains to be unambiguously proven.

Actually, little evidence exists for the direct participation of PLP in the mechanism of LAM [48]. While the enzyme is activated and stabilized by the presence of PLP, attempts to reduce PLP onto the enzyme or the substrate with $NaBH_4$ have been unsuccessful. Of course, this is also consistent with a lack of accessibility of the imine to the reducing agent or its presence as the less reactive neutral imine. The EPR data obtained in ^2H_2O suggests that the nitrogen in the predominant radical species is not bonded to more than one hydrogen atom and thus constitutes the best evidence to date for the occurrence of an imine between the substrate and PLP at the active site.

3.5. Role of Cobalt and the Iron-Sulfur Clusters

Much speculation exists concerning the function of cobalt and the iron-sulfur clusters in LAM. The activity of the enzyme clearly depends on both cofactors although no experiment directly implicates either in catalysis [48]. It seems reasonable that the cleavage of SAM by a reduced form of the enzyme may result in the formation of a catalytically functional adenosyl-Co(III) species with a weak cobalt–carbon bond similar to adenosylcobalamin. The corresponding weak bond in adenosylcobalamin is known to readily undergo homolysis to Co(II) and the 5′-deoxyadenosyl radical, thus providing the catalytic cofactor. The ability of SAM to undergo heterolysis with the formation of methionine and the 5′-deoxyadenosyl cation equivalent suggests that SAM may be cleaved by a nucleophile en route to the formation of the Co(III)-adenosyl species.

Two possible nucleophiles that could be generated in the reduced form of the enzyme are Co(I) and a reduced Fe_4-S_4 cluster [48,52]. No evidence is currently available for the presence of Co(I) in the reduced form of the enzyme and the reduction potentials of known cobalt species are lower than that provided by the reduction conditions. However, the nucleophilicity of Co(I) renders it a viable candidate. The possibility of a reduced Fe-S cluster would obviate the necessity for Co(I). The sulfurs of these reduced clusters are rich in negative charge and would be expected to be highly reactive nucleophiles. The adenosyl-FeS cluster formed by alkylation of a bridging sulfur by the adenosyl moiety in SAM could undergo sulfur–carbon bond homolysis to afford the one-electron oxidized cluster and the 5′-deoxyadenosyl radical. The adenosyl radical would then combine with Co(II) to form the Co(III)-adenosyl species. A wealth of additional experimental information will be required to fully delineate this intriguing process. It is clear, however, that the mechanism of adenosyl radical formation in LAM presents a fundamentally distinct motif from that of the PFL-activating enzyme.

4. ANAEROBIC RIBONUCLEOTIDE REDUCTASE

Ribonucleotide reductase (EC 1.17.4.1) is a multiclass enzyme responsible for the synthesis of deoxyribonucleotides from ribonucleotides and plays a central role in DNA biosynthesis [61,62]. This ubiquitous enzyme system is divided into four classes. Class I consists of the ferric/tyrosyl radical–containing enzymes; class II, the poorly characterized manganese-containing enzymes; class III, the B_{12}-containing enzymes; and class IV, the recently discovered SAM-dependent anaerobic enzyme (ARR). The class I and II reductases use ribonucleoside diphosphates as substrates while the class III and IV enzymes use ribonucleoside triphosphates as substrates. The class IV reductase is present only in anaerobically grown extracts of *E. coli* and is extremely sensitive to oxygen [63]. The anaerobic reductase has been purified by gel filtration fractionation followed by dATP affinity chromatography from a strain of *E. coli* devoid of the aerobic reductase. The purified protein consists of two major protein bands: a 90 kD α band and a 70 to 80 kD β band comprising three closely migrating subbands (β_1–β_3). Initial characterization showed that the activity was dependent on the presence of SAM, a loosely bound metal ion, and a protein fraction. Immunological studies with mono- and polyclonal antibodies raised against the B1 subunit of the aerobic class I reductase suggests a relationship between the B1 subunit and the α and β subunits of the anaerobic reductase. Recently, the α and β_3 bands observed on SDS-PAGE have been identified to be related to the ATP-dependent protease, clpB, and copurified with the reductase [64]. A regulatory role of the clpB proteins associated with the anaerobic ribonucleotide reductase has been proposed. More recently, Reichard and coworkers isolated the *nrdD* gene coding for the anaerobic reductase [65]. Sequence alignment with PFL revealed a region of homology at the active site glycine-734 of PFL and the *sunY* gene product of T4. The consensus sequence was RVXSG. Although not as highly homologous as the proteins from T4, T5, and the *Serratia* open reading frames, it is significant to suggest that the anaerobic reductase may also contain a glycyl radical at position 684. Indeed, circumstantial evidence in support of a glycyl radical in the anaerobic reductase has been obtained. Oxygen-inactivated enzyme exhibited an increase in the amount of the smaller 74 kD protein, indicative of a oxygen-dependent fragmentation process of the 77 kD protein at glycine-684 similar to what has been observed with PFL (vide supra).

With the availability of recombinant ARR, preliminary characterization of the enzyme has been reported [66]. The enzyme appears to contain an iron-sulfur center. This is based on the UV-Vis spectrum and EPR analysis of the unactivated enzyme. The EPR signal is centered at $g = 2.01$ and is temperature-sensitive. Further analysis of iron and inorganic sulfide content revealed a 1:1 stoichiometry. The inability to saturate the EPR signal at 14 K indicated that the center is a $(3Fe\text{-}4S)^+$ cluster. Upon activation with NADPH and photoreduced 5-deazariboflavin, the EPR signal of the iron-sulfur center disappeared and a new signal emerged with a g value of 2.0033. The EPR signal is a doublet reminiscent of the doublet observed in active PFL, but contains no minor hyperfine splittings. The doublet is not sensitive to solvent isotopic composition as it does not change in the presence of ^2H_2O. The doublet observed in PFL has been assigned to the coupling of the α hydrogen on glycine-734 to the radical [14]. The insensitivity of the doublet in ARR to solvent isotopic composition may reflect a difference in the microenvironment of the glycyl radical surrounding glycine-684. This is not surprising since the sequence containing the active site glycine is not identical to PFL. The replacement of serine with cysteine adjacent to glycine-684 may contribute to the observed EPR characteristics of the ARR glycyl radical.

The similarities of the reaction of the class III B_{12}-dependent enzyme and the class IV SAM-dependent enzyme indicates that SAM may be a B_{12} surrogate for the generation of 5′-deoxyadenosyl radical formation. Although a metal-dependent reductive cleavage of SAM has not yet been demonstrated in ARR, it is possible that the activation of the class IV reductase may be very similar to the activation of PFL. This poses the interesting question of whether PFL-activating enzyme may activate other anaerobic radical-containing enzymes. In addition, the cleavage of SAM may be mediated by the Fe-S cluster in a manner similar to LAM.

5. CONCLUDING REMARKS

In the last decade there has been an emergence of evidence that important radical processes in anaerobic systems involve the reductive cleavage of SAM. For many years, SAM-dependent radical formation appeared to be unique to the PFL system, but mounting evidence suggests that this may be a common occurrence in facultative bacteria and even some lower eukaryotes. Experiments indicate that the 5′-deoxyadenosyl radical can be generated via the reductive cleavage of SAM and can serve as a B_{12} surrogate in a number of enzyme systems. An obvious question is, does the SAM-dependent system predate the B_{12} system in the formation of the 5′-deoxyadenosyl radical [48]? The SAM-dependent radical systems in the lysine 2,3-aminomutase, PFL, and anaerobic ribonucleotide reductase may constitute a primitive cofactor from an ancient world where life existed in a reducing environment. As life evolved toward an oxidizing environment, the aerobic counterparts for PFL (i.e., pyruvate dehydrogenase) and the aerobic reductase (i.e., the B_{12}-dependent reductase) evolved. Alternatively, they may have evolved at the same time with a different chemical solution for the same problem under different environments.

Finally, another interesting lesson is the observation that pyridoxal phosphate (PLP) and SAM, which are traditionally thought of as molecules that perform two-electron chemistries, may be capable of performing one-electron chemistries under anaerobic conditions. Thiamine pyrophosphate has been shown to be required for PFL activity in *Streptococcus faecalis*, which may indicate that this cofactor also may perform one-electron chemistry [67]. Clearly, anaerobic metabolism in facultative organisms embodies a wealth of stimulating bioorganic and bioinorganic chemistry that we have only just begun to discover.

ABBREVIATIONS

act	gene for pyruvate formate lyase–activating enzyme
ADP	adenosine 5′-diphosphosphate
AE	activating enzyme
ArcA	an anaerobic transcriptional regulatory protein
ARR	SAM-dependent anaerobic ribonucleotide reductase
ATP	adenosine 5′-triphosphate
CoA/CoASH	coenzyme A
DTT	dithiothreitol
EPR	electron paramagnetic resonance
Fnr protein	an anaerobic transcriptional regulatory protein
HP	hypothetical protein
LAM	lysine 2,3-aminomutase
NAD	nicotinamide adenine dinucleotide
NADH	nicotinamide adenine dinucleotide (reduced)
NADPH	nicotinamide adenine dinucleotide phosphate (reduced)
orf	open reading frame
PFL	pyruvate formate lyase
pfl	gene for PFL
PFL_a	active form of pyruvate formate lyase
PFL_i	inactive form of pyruvate formate lyase
PLP	pyridoxal phosphate
SAH	S-adenosylhomocysteine
SAM	S-adenosylmethionine
SDS-PAGE	sodium dodecyl sulfate–polyacrylamide gel electrophoresis
XH	radical site of pyruvate formate lyase
UV	ultraviolet
Vis	visible

REFERENCES

1. G. Kalnitsky and C. H. Werkman, *Arch. Biochem. Biophys.*, *2*, 113–124 (1943).
2. M. F. Utter and C. H. Werkman, *Arch. Biochem. Biophys.*, *5*, 413–422 (1944).

3. H. Chantrenne and F. Lipmann, *J. Biol. Chem.*, *187*, 757–767 (1950).
4. E. R. Stadtman, G. D. Novelli, and F. Lipmann, *J. Biol. Chem.*, *191*, 365–376 (1951).
5. F. Lipmann and L. C. Tuttle, *J. Biol. Chem.*, *154*, 725–726 (1944).
6. J. Knappe, J. Schacht, W. Mockel, Th. Hoperner, H. Vetter, Jr. and R. Edenharden, *Eur. J. Biochem.*, *11*, 316–327 (1969).
7. H. Conradt, M. Holmann-Berger, H. P. Holmann, H. P. Blaschkowski, and J. Knappe, *Arch. Biochem. Biophys.*, *228*, 133–142 (1988).
8. D. Kessler, I. Leibrecht, and J. Knappe, *FEBS Lett.*, *281*, 59–63 (1991).
9. J. Knappe and G. Sawers, *FEMS Microbiol. Rev.*, *75*, 383–398 (1990).
10. G. Sawers and A. Bock, *J. Bacteriol.*, *171*, 2485–2498 (1989).
11. G. Sawers and B. Suppman, *J. Bacteriol.*, *174*, 3474–3478 (1992).
12. J. Knappe, F. A. Neugebauer, H. P. Blaschkowski, and M. Ganzler, *Proc. Natl. Acad. Sci. USA*, *81*, 1332–1335 (1984).
13. V. Unkrig, F. A. Neugebauer, and J. Knappe, *Eur. J. Biochem.*, *184*, 723–728 (1989).
14. A. F. V. Wagner, M. Frey, F. A. Neugebauer, W. Schafer, and J. Knappe, *Proc. Natl. Acad. Sci. USA*, *89*, 996–1000 (1992).
15. B. A. Epper, C. B.-R. Green, T. A. Campell, D. A. Stoffers, H. T. Kentmann, K. E. Mains, and L. H. Ouafik, *J. Biol. Chem.*, *267*, 4008–4015 (1992).
16. E. J. Husten and B. A. Eipper, *J. Biol. Chem.*, *266*, 17004–17010 (1991).
17. D. J. Merkler, R. Kulathila, A. P. Consalro, S. D. Young, and D. E. Ash, *Biochemistry*, *31*, 7282–7288 (1992).
18. H. G. Vieche, R. Merenyi, L. Stella, and Z. Janousek, *Angew. Chem., Int. Ed. Engl.*, *18*, 917–932 (1979).
19. H. G. Vieche, Z. Janousek, R. Merenyi, and L. Stella, *Acc. Chem. Res.*, *18*, 148–154 (1985).
20. V. Burgess, C. J. Easton, and M. P. Hay, *J. Am. Chem. Soc.*, *111*, 1047–1052 (1989).
21. K. K. Wong and J. W. Kozarich, (submitted).
22. M. Givkov, L. Olsen, and S. Molin, *J. Bacteriol.*, *170*, 5855–5862 (1988).

23. K. Valerie, J. Stevens, M. Lynch, E. E. Henderson, and J. K. deReil, *Nucl. Acid Res.*, *14*, 8637–8654 (1988).
24. V. N. Ksenzenko, M. G. Shlyapnikov, V. G. Azbarov, O. Garcia, V. M. Kryukov, and A. A. Bayer, *Nucl. Acid Res.*, *13*, 5480–5481 (1987).
25. E. J. Brush, K. A. Lipsett, and J. W. Kozarich, *Biochemistry*, *27*, 2217–2222 (1988).
26. C. Parast, K. K. Wong, and J. W. Kozarich, (submitted).
27. R. Bernardi, T. Caronna, R. Galli, F. Minisci, and M. Perchinunn, *Tetrahedron Lett*, *9*, 645–648 (1973).
28. G. Heinisch and G. Lotsch, *Angew. Chem., Int. Ed. Engl.*, *24*, 692–693 (1985).
29. G. Heinisch and G. Lotsch, *Tetrahedron*, *41*, 1199–1205 (1985).
30. W. Plaga, R. Frank, and J. Knappe, *Eur. J. Biochem.*, *178*, 445–450 (1988).
31. D. C. McKean, J. L. Ducan, and L. Batt, *Spectrochim. Acta, Part A*, *29*, 1037–1049 (1973).
32. P. S. Surdhar, S. P. Mezyk, and D. A. Armstrong, *J. Phys. Chem.*, *93*, 3360–3363 (1989).
33. H. Nakayama, G. G. Midwinter, and L. O. Krampitz, *Arch. Biochem. Biophys.*, *143*, 526–534 (1971).
34. S. A. Lewisch, K. K. Wong, and J. W. Kozarich, unpublished results.
35. T. Chase, Jr., and J. C. Rabinowitz, *J. Bacteriol.*, *96*, 1065–1078 (1968).
36. J. Knappe and T. Schmitt, *Biochem. Biophys. Res. Commun.*, *71*, 1110–1117 (1976).
37. K. K. Wong, B. W. Murray, and J. W. Kozarich, (submitted).
38. K. K. Wong, B. W. Murray, S. A. Lewisch, M. K. Baxter, T. W. Ridky, L. Ullisi-DeMario, and J. W. Kozarich, (submitted).
39. W. Rodel, W. Plaga, R. Frank, and J. Knappe, *Eur. J. Biochem.*, *177*, 153–158 (1988).
40. R. H. Abeles and D. Dolphin, *Acc. Chem. Res.*, *9*, 114–120 (1976).
41. S. S. Mao, T. P. Holler, G. X. Yu, J. M. Bollinger, Jr., S. Brooker, M. I. Jonston, and J. Stubbe, *Biochemistry*, *31*, 9733–9743 (1992).
42. B. W. Murray, K. K. Wong, and J. W. Kozarich, unpublished results.

43. L. Ullissi-DeMario, E. J. Brush, and J. W. Kozarich, *J. Am. Chem. Soc.*, *113*, 4341–4342 (1991).
44. D. Kessler, W. Herth, and J. Knappe, *J. Bacteriol.*, *170*, 5855–5862 (1992).
45. S. Matayoshi, H. Oda, and G. Sarwar, *J. Gen. Microbiol.*, *135*, 525–529 (1989).
46. R.N. Costilow, O. M. Rochovansky, and H. A. Barker, *J. Biol. Chem.*, *241*, 1573–1580 (1966).
47. T. C. Stadtman, *Adv. Enzymol. Relat. Areas Mol. Biol.*, *38*, 413–448 (1973).
48. P. A. Frey and G. H. Reed, *Adv. Enzymol.*, *66*, 1–39 (1993); and references cited therein.
49. T. P. Chirpich, V. Zappia, R. N. Costilow, and H. A. Barker, *J. Biol. Chem.*, *245*, 1778–1789 (1970).
50. R. H. Abeles and D. Dolphin, *Acc. Chem. Res.*, *9*, 114–121 (1976).
51. R. M. Petrovich, F. J. Ruzicka, G. H. Reed, and P. A. Frey, *J. Biol. Chem.*, *266*, 7656–7660 (1991).
52. R. M. Petrovich, F. J. Ruzicka, G. H. Reed, and P. A. Frey, *Biochemistry*, *31*, 10774–10781 (1992).
53. K. B. Song and P. A. Frey, *J. Biol. Chem.*, *266*, 7651–7655 (1991).
54. D. J. Aberhart, S. J. Gould, H.-J. Lin, T. K. Thiruvengadam, and B. H. Weiller, *J. Am. Chem. Soc.*, *105*, 5461–5470 (1983).
55. J. Baraniak, M. L. Moss, and P. A. Frey, *J. Biol. Chem.*, *264*, 1357–1360 (1989).
56. M. L. Moss and P. A. Frey, *J. Biol. Chem.*, *265*, 18112–18115 (1990).
57. J. L. Kilgore and D. J. Aberhart, *J. Chem. Soc. Perkin Trans.*, *I*, 79–84 (1991).
58. M. D. Ballinger, G. H. Reed, and P. A. Frey, *Biochemistry*, *31*, 949–953 (1992).
59. M. D. Ballinger, G. H. Reed, and P. A. Frey, *Biochemistry*, *31*, 10782–10789 (1992).
60. O. Han and P. A. Frey, *J. Am. Chem. Soc.*, *112*, 8982–8983 (1990).
61. P. Reichard, *Ann. Rev. Biochem.*, *57*, 349–374 (1988).
62. J. Stubbe, *Ann. Rev. Biochem.*, *58*, 257–285 (1989).

63. M. Fontecave, R. Eliason, and P. Reichard, *Proc. Natl. Acad. Sci. USA*, *86*, 2147–2151 (1989).
64. E. Potis, X. Sun, H. Jörnvall, M. Krook, and P. Reichard, *Biochem. Biophys. Res. Commun.*, *180*, 1222–1226 (1991).
65. X. Sun, J. Harden, M. Krook, H. Jörnvall, B.-M. Sjöberg, and P. Reichard, *Proc. Natl. Acad. Sci. USA*, *90*, 577–581 (1993).
66. E. Mulliez, M. Fontecave, J. Gaillaid, and P. Reichard, *J. Biol. Chem.*, *268*, 2296–2299 (1993).
67. P. G. Lindmark, P. Paolella, and N. P. Wood, *J. Biol. Chem.*, *244*, 3605–3612 (1969).

11
The Free Radical-Coupled Copper Active Site of Galactose Oxidase

James W. Whittaker

Department of Chemistry, Carnegie Mellon University,
4400 Fifth Avenue, Pittsburgh, Pennsylvania
15213-3890, USA

1. INTRODUCTION

1.1. The Catalytic Chemistry of Galactose Oxidase

Galactose oxidase is a fungal metalloenzyme that catalyzes the oxidation of primary alcohols (RCH_2OH) to the corresponding aldehydes, coupling this reaction to the reduction of dioxygen to hydrogen peroxide [1] [Eq. (1)]:

$$RCH_2OH + O_2 \rightarrow RCHO + H_2O_2 \qquad (1)$$

Galactose oxidase exhibits a surprisingly low specificity for the primary alcohol but is completely regioselective: no secondary alcohols are known to be substrates for this enzyme [1,2]. Aldehydes themselves have been shown to serve as effective substrates for catalysis, being slowly converted to the corresponding carboxylic acid [3]. The selectivity of galactose oxidase for primary hydroxyl groups is particularly interesting to synthetic chemists lacking efficient reagents for selective oxidation of terminal hydroxyls in the presence of secondary hydroxyl groups. Conventional oxidants such as chromium trioxide typically exhibit low discrimination between these two classes, and electrochemical methods are restricted by the kinetic stability of the alcohols, leading to large overvoltage requirements in alcohol oxidation and which can result in complex product mixtures. The possibility of applying galactose oxidase in semisynthetic strategies has already received attention from synthetic chemists [4]. Detailed studies on the active site of galactose oxidase are providing insights into alcohol oxidation catalysis and defining the essential features required for small-molecule catalytic complexes as a basis for the preparation of new reagents for alcohol oxidation. The synthesis of results from biochemical and spectroscopic studies which are described in this chapter represent a step toward this goal.

1.2. General Features of Redox Catalysis

Redox catalysts mediate the transfer of charge from a relatively electron-rich donor (substrate) to a relatively electron-deficient acceptor. In biochemical redox catalysis, metal ions often serve to activate coordinated substrates and to participate as redox-active sites for charge accumulation. Metalloenzymes with active sites containing Cu, Fe, Mo, W, and V are known to catalyze redox transformations of a wide variety of molecules (N_2, O_2, H_2, CO, and complex organics). In general, the number of metal ions involved in the reaction matches the number of electrons transferred. For example, superoxide dismutases [5–7], catalyzing one-electron redox disproportionation of superoxide radicals, universally contain a single isolated redox-active metal center of Fe, Mn, or Cu. The active sites of O_2 carrier proteins like hemerythrin [8] and hemocyanin [9], which bind O_2 as peroxide, are formed respectively from binuclear complexes of Fe and Cu, and Mn catalase, which catalyzes the disproportionation of hydrogen peroxide (two-electron redox chemistry) and contains a binuclear Mn cluster as its active site [10]. In cytochrome oxidase [11], four redox-active metal ions (two heme Fe and two Cu) are involved in the four-electron reduction of O_2 to H_2O, and four Mn ions are bound in the water-splitting active site in photosynthetic reaction centers of green plants, where the reverse reaction of O_2 formation occurs [12]. Many further examples of this sort could be cited to elaborate this basic rule. The reaction catalyzed by galactose oxidase [Eq. (1)] involves two-electron redox chemistry in both the alcohol oxidation and O_2 reduction half-reactions, yet the enzyme contains a single Cu ion in its structure [13]. This apparent exception to the general rule of biological redox catalysis has been shown to derive from a unique structural feature of the active site, which incorporates a free radical into the redox unit for alcohol oxidation chemistry [14]. The trail leading to the recognition of this unique catalytic structure is described below.

1.3. Galactose Oxidase and Free Radical Enzymology

Until very recently, free radicals in biology were relegated to obscure and deleterious roles, and considered mainly in the light of radiation damage mechanisms, toxicity and mutagenic effects of aromatic hydrocarbons, and aging [15]. More recently, there has been an emerging recognition of the central role played by free radicals in biochemistry, particularly through the identification of a stable free radical site in ribonucleotide reductase [16], and mounting evidence for free radical pathways in enzyme redox mechanisms. Indeed, the observation that all known ribonucleotide reductases of both eukaryotic and microbial origin, key enzymes in the biosynthesis of DNA required for cellular replication, all function by controlling free radical chemistry emphasizes that free radicals and biology have worked together for a long time and that free radicals have an essential role in biochemistry and redox catalysis. The recognition of this essential role for free radicals in biochemistry is leading to the emergence of a new field of free radical enzymology [17,18], and it is one of the aims of this chapter to illustrate the tools and techniques for exploring free radical species in catalytic active sites.

2. HISTORICAL SURVEY

2.1. Early Work

Galactose oxidase, first isolated [19] in 1957 and studied extensively in the laboratory of Horecker [13,20], is an exoenzyme secreted by a mycelial wood rot fungus originally identified as *Polyporus circinatus* Fr., subsequently [21] reidentified as *Dactylium dendroides*. Basic biochemical procedures for purification and characterization of the new enzyme were developed in these earliest studies [13,20], which also describe conditions for preparing crystals of the protein [13,22] like those which have recently provided a solution of the protein structure at atomic resolution [23]. Development of an enzyme assay based on a peroxidase-coupled reaction of the hydrogen peroxide produced by alcohol oxidation [20] permitted exploration of the basic substrate specificity requirements of the enzyme. Oligosaccharides containing a galactose residue at the nonreducing terminus were found to be the most efficient substrates supporting the highest turnover rates. This preference for a nonreducing galactose unit in the substrate led to the identification of the enzyme as galactose oxidase. However, the enzyme exhibits a rather low specificity for its organic substrate, and a wide range of alcohols, many potentially arising in the degradation of wood, were found to serve as substrates for the enzyme. These early studies also applied protein chemistry to defining the basic structural features of the purified protein, including the presence of a single Cu ion and no other detectable metal ions or dissociable cofactors [13].

2.2. Refinements

Subsequent studies on the enzyme have refined the cultivation conditions under which the fungus expresses galactose oxidase [24–26] and led to a modified purification procedure [27,28], resulting in increased yields of galactose oxidase essential for molecular studies of the purified protein. A refinement of the protein chemistry defined a molecular mass of 68000 D for the single polypeptide chain [26]. A number of studies [29–32] have focused on the well-developed electron paramagnetic resonance (EPR) signal arising from the Cu^{2+} active site metal ion, which was reported to quantitate to 0.7 $\pm$ 0.3 spin/mol active site [33], identified as arising from the homogeneous, active enzyme. This cupric EPR signal did not change on anaerobic addition of substrate, which was interpreted as evidence that Cu^{+} does not participate in the enzyme mechanism [32]. A careful analysis of the fine structure in the EPR spectrum permitted identification of two to three nitrogen ligands to the metal, presumed to represent histidine imidazole ligands [29]. The intense Cu^{2+} EPR signal of these samples also permitted studies of exogenous ligand interactions with the active site metal ion, which could be converted to complexes with small anions such as azide, fluoride, and cyanide. Effects of ligand interactions on the optical absorption and circular dichroism (CD) spectra for galactose oxidase have been reported in a series of publications from the Bioinorganic Graduate Research Group of SUNY–Buffalo [34–37].

2.3. Controversy

Hamilton et al. reported that the catalytic activity of the enzyme is variable and reaches its highest level in the presence of oxidants [38,39]. This redox activation process demonstrated that the enzyme could be deliberately and systematically interconverted between active and inactive forms in a redox titration, using ferri-/ferrocyanide solutions and measuring catalytic activity in an oxygen electrode, monitoring O_2 uptake. The midpoint potential for this activation process was found to be 0.41 V vs. NHE. The interconversion of active and inactive forms of galactose oxidase could be accounted for with a two-state model in which the catalytic activity of an inactive, reduced enzyme can be restored by oxidation [38]. Parallel EPR experiments performed on enzyme samples in ferri-/ferrocyanide solution showed that the intensity of the EPR signal varied inversely with activity, and that in the active oxidized form of the enzyme the Cu^{2+} EPR signal was essentially eliminated, suggesting that the EPR signal arises from inactive enzyme. These observations formed the basis of a plausible proposal that the metal ion is converted to a trivalent oxidation state in the active enzyme [39], consistent with the reported stabilization of formally trivalent Cu in peptide complexes [40] but ruled out by subsequent studies [14,41]. A variety of oxidants have been found to be effective for activation of the enzyme, including inorganic (Mn^{3+}-EDTA, Na_2IrCl_6) and organic (porphyrexide) reagents. In each case, the activated enzyme exhibited the minimum EPR signal intensity. While these results are consistent with an oxidized Cu complex, the alternative explanation proposed by the Buffalo group [32,33,37] was that paramagnetic redox agents were tenaciously bound to the oxidized protein in the proximity of the Cu^{2+} ion, the magnetic interactions between sites accounting for the absence of a detectable EPR signal in these complexes. These studies on the redox-activated enzyme were complicated by the observation that prolonged exposure to oxidants led to irreversible inactivation [42]. An additional sensitivity of galactose oxidase to inactivation by trace amounts of reductants provides an additional challenge.

Electrochemical studies of galactose oxidase have been motivated by the recognition of the potential applications of electrode-coupled enzymatic reactions in the design of galactose sensors for clinical applications. These studies [43] defined a redox potential for galactose oxidase activation of 0.40 V vs. NHE, consistent with Hamilton's [38] and our own [14] ferricyanide redox titrations. This step was (incorrectly) identified as a Cu^{+}/Cu^{2+} redox couple based on the color changes associated with the activation process, the appearance of the intense green color of the active enzyme being interpreted as resulting from oxidation from a weakly absorbing Cu^{+} to a strongly absorbing Cu^{2+} form. Coulometric measurements have made it possible to define this process as a one-electron redox step [43].

Kinetic and mechanistic studies on galactose oxidase have examined the requirements for the catalytic reaction and the inhibitory effects of anions. Steady-state kinetic studies with galactose as substrate have been interpreted in terms of an ordered, sequential bireactant or biproduct (Bi-Bi) turnover mechanism that requires the aldehyde product of alcohol oxidation to remain bound until O_2 reacts [32,36,44], although this interpretation has been challenged [32]. The behavior of galactose oxidase under normal assay conditions has been found to be quite complicated, with deviations from simple linear kinetics in phosphate buffer [38]. In a separate series of kinetics experiments where these complexities have been addressed, steady-state reaction rate data for dihydroxyacetone (an alternative substrate) has been interpreted as supporting a Bi-Bi ping-pong mechanism in which the first product forms before the second substrate reacts [38]. The stereochemistry of the reaction has also been examined, and galactose oxidase has been shown to exhibit prochiral specificity in the oxidation of alcohols. Elegant experiments on isotopically labeled galactose have demonstrated that the pro-S methylene hydrogen is stereospecifically abstracted on forming the product [4,45]

2.4. A Puzzle

Research on galactose oxidase over a period spanning two decades thus provides a wealth of information but leaves us with a puzzle: how does a mononuclear Cu active site catalyze two-electron redox chemistry? What is the active form of the enzyme?

3. RECENT DEVELOPMENTS

3.1. Biochemical Preliminaries: Pure Redox Forms of Galactose Oxidase

Our recent work on galactose oxidase developed from a systematic and quantitative extension of Hamilton's earlier work on redox activation of the purified enzyme. By controlling exposure to oxidants such as ferricyanide, carefully desalting from excess reagents, it has been possible to prepare pure, homogeneous, and stable redox forms of the enzyme without protein damage problems for the first time [14]. This stability has been important in developing an extensive characterization of all of the accessible redox forms by a combination of EPR, optical absorption, and CD spectroscopy. From this combination of spectroscopic approaches the basis for a preliminary description of a unique active site structure has emerged.

The optical absorption spectra for three forms of galactose oxidase are compared in Fig. 1. The absorption spectrum of a sample of Native

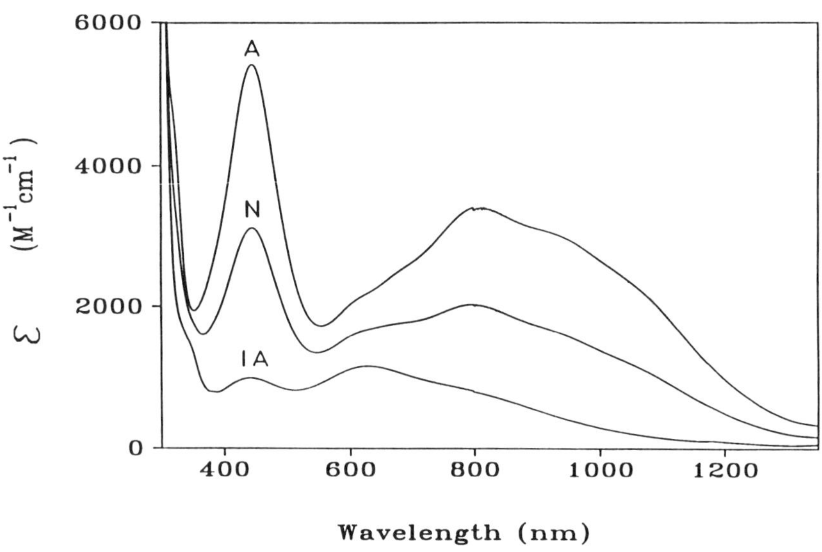

FIG. 1. Optical absorption spectra for Active (A), Native (N) and Inactive (IA) forms of galactose oxidase.

enzyme (N) as initially isolated from fungal culture filtrates is shown in the middle spectrum in Fig. 1 for a sample having approximately 50% maximal specific activity. The intensity of the visible absorption spectrum for the Native enzyme varies unpredictably between samples, with as much as fivefold variation from one batch to the next. Accurate measurements of the catalytic activity of enzyme samples have proven to be important in developing insight into this variability in the physical and catalytic properties of galactose oxidase. The original, indirect, coupled assay procedure was reported to be susceptible to artifacts through activation of galactose oxidase by peroxidase as well as inhibition by the chromogen product [25] and the oxygen electrode measurement in the presence of ferricyanide eliminates the differences between samples [38]. An alternative, direct assay procedure was developed by the Kosman group [25,27], in which the formation of aldehyde product is monitored spectrophotometrically. Using this direct assay procedure, we have found that a clear correlation can be established between the absorption intensity and the specific activity of enzyme samples [14]. Enzyme treated with ferricyanide and then desalted yields an intensely absorbing sample that gives the absorption spectrum at the top of Fig. 1. This spectrum is extremely unusual for either Cu metalloproteins or Cu small-molecule complexes. The spectrum never falls below 2000 M^{-1} cm^{-1} on a molar extinction basis over the entire visible region, and is actually growing in intensity into the infrared. Two major features dominate the spectrum: an intense feature is observed near 445 nm, and a second, low-energy absorption maximum occurs near 800 nm in the near-infrared (IR). Both of these features are similar to the spectra of galactose oxidase in the presence of oxidants reported in earlier studies. We find that conversion to the intensely absorbing green form of galactose oxidase yields enzyme with the maximum specific activity which is therefore identified as the Active form (A) of the enzyme. Treatment of native enzyme with a reductant, ferrocyanide, followed by desalting leads to the preparation of a relatively weakly absorbing blue form of galactose oxidase (Fig. 1, lower spectrum) which is devoid of catalytic activity that is described as the Inactive form (IA) of the enzyme. This inactivation is reversible, though, and blue and green forms of the enzyme can be readily interconverted by treatment with the redox agents. This behavior confirms Hamilton's earlier two-state model for the redox activation process.

These three forms can be related to galactose oxidase species previously studied by EPR spectroscopy [14], as shown in Fig. 2. The Native enzyme, as isolated, exhibits a well-resolved EPR spectrum characteristic of a tetragonal cupric complex and is identical to that previously reported for galactose oxidase. Spin quantitation of this EPR

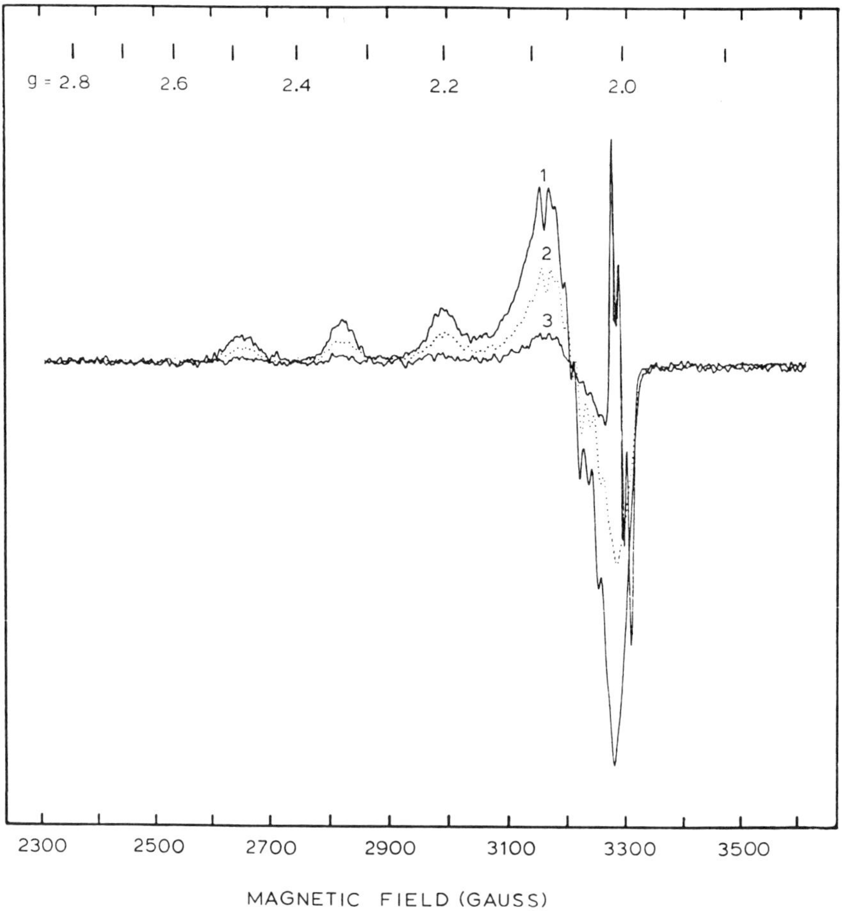

FIG. 2. EPR spectra for galactose oxidase species. 1, Reductively inactivated; 2, native; 3, oxidized galactose oxidase in 50 mM Na_2HPO_4 buffer pH 7.0. (From Ref. 14 by permission.)

signal by double integration of the spectrum and comparison with a standard indicates that only half of the enzyme-bound Cu contributes to this spectrum for a sample having 50% maximal specific activity. On oxidation, the EPR signal virtually disappears, leaving a 10% minority cupric EPR signal that appears to arise from a fraction of the active sites which are unable to undergo the redox activation step. In addition, a free radical EPR signal appears near $g = 2.005$, quantitating to only about 1 percent $S = \frac{1}{2}$ per active site, leaving its significance uncertain (but see below). On reduction, the full complement of enzyme-bound copper contributes to the intensity of the EPR signal. This result seems counterintuitive, since normally reduction of a copper complex produces an EPR-silent cuprous ion, while the oxidized complex is EPR-active. The fact that the cupric EPR signal appears on reduction of the enzyme needs to be accounted for.

It appears likely that the Native enzyme is a mixture of Active and Inactive forms but the lack of isobestic points in the optical absorption and EPR spectra leaves this ambiguous. In CD, however, clean isodichroic points are observed between these spectra [14]. This is illustrated in the series of spectra shown in Fig. 3, which follows the redox interconversion of Active and Inactive (green and blue) forms of the enzyme through equilibration with a series of ferri-/ferrocyanide redox buffers with the electrochemical potential of the solution being controlled by the ratio of the redox components. Isodichroic points at 337, 356, 390, and 587 nm are maintained throughout the interconversion, indicating that only two limiting forms exist in this series. Comparison of the CD spectra of the samples at high and low potential limits with the CD spectra of the pure Active and Inactive forms isolated from redox agents (Fig. 4) confirms the identity of the limiting species in the redox interconversion. The CD spectrum of the Native enzyme, which has never been exposed to any redox agents, intersects the spectra of the limiting forms at exactly the same points, requiring that it is simply a heterogeneous mixture of the Active and Inactive species and not a distinct form of the enzyme (Fig. 4). This implies that all of the variability of absorption and catalytic properties of galactose oxidase which were an obstacle in earlier studies on galactose oxidase were the result of a variable amount of inactivation occurring in the enzyme during purification.

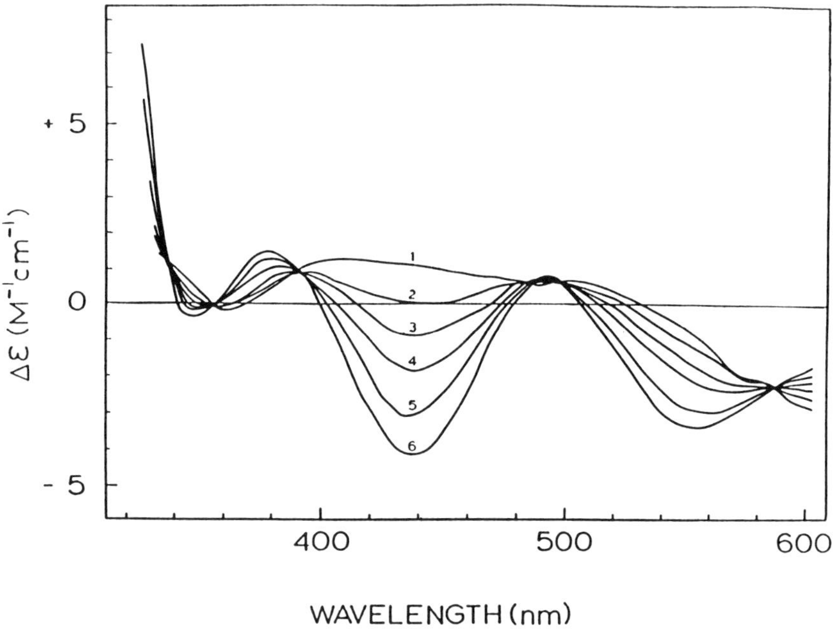

FIG. 3. Redox titration of galactose oxidase detected in CD spectra. Samples of galactose oxidase were incubated with ferri-/ferrocyanide redox buffers with log[Ox]/[Red] ranging from <-2 (spectrum 1) to >2 (spectrum 6). (From Ref. 14 by permission.)

3.2. Spectroscopic and Chemical Characterization of the Active Site

The spectral features in the absorption spectra of the Inactive, blue form of galactose oxidase are relatively easy to account for, since this form is known to contain a simple, isolated cupric ion. This leads to a plausible assignment [14] of the broad, low-energy absorption features near 600 nm as arising from ligand field (d→d) transitions for the cupric ion, an assignment which is supported by the appearance of an unusually strong CD in this region. Strong CD is often associated with d→d transitions, which have a relatively large magnetic dipole contribution to their transition moments, as illustrated in Fig. 5. The left side of this figure identifies the three d→d transitions for a cupric ion in C_{4v} site

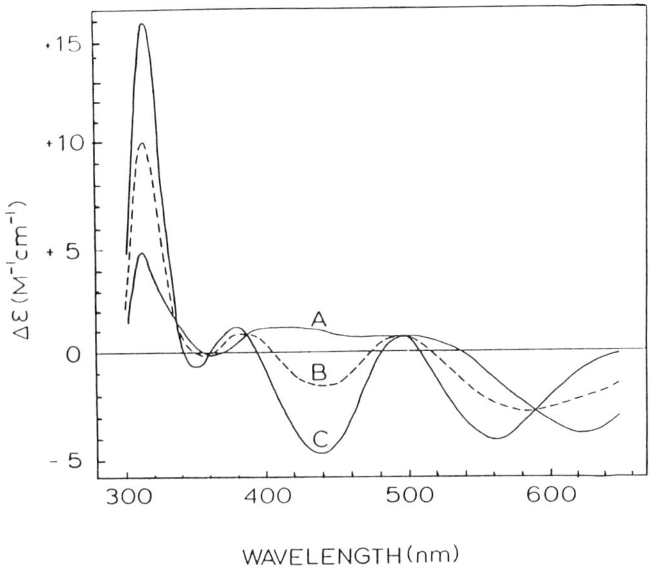

FIG. 4. CD spectra for galactose oxidase redox modifications. (A) Reductively inactivated. (B) Native. (C) Redox-activated galactose oxidase. (From Ref. 14 by permission.)

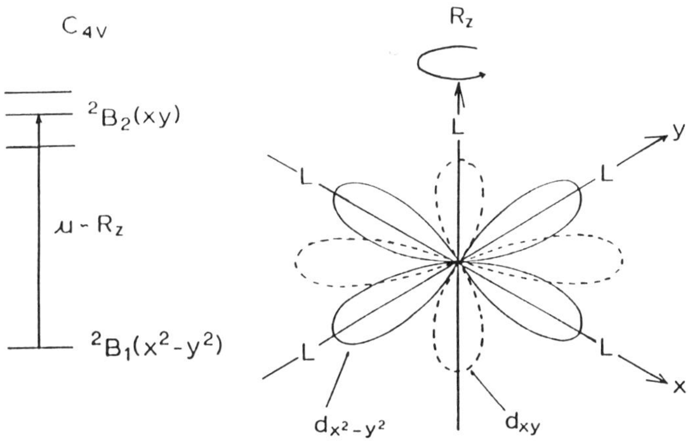

FIG. 5. Selection rules for ligand field transitions for a square pyramidal Cu^{2+} complex. *Left*: Magnetic dipole character in an electronic transition. *Right*: Pictorial presentation.

symmetry. Analysis of the group theoretical selection rules defines the $d_{x^2-y^2} \rightarrow d_{xy}$ ($^2B_{1g} \rightarrow {}^2B_{2g}$) transition as being magnetic dipole allowed, a result which is presented in more pictorial terms on the right side of Fig. 5, showing that the $d_{x^2-y^2} \rightarrow d_{xy}$ orbital transition involves a circulation of charge which will induce a magnetic dipole transition moment along the *z* axis. Group theoretical analysis of all three d→d transitions in this site symmetry confirm that the largest magnetic dipole character will be associated with the $^2B_{1g} \rightarrow {}^2B_{2g}$ transition. As a result, this transition will be expected to acquire a large CD anisotropy ($\Delta\epsilon/\epsilon$) under a chiral perturbation which removes the reflection symmetry in the complex. On the basis of this analysis, the strong CD band observed for blue IA galactose oxidase may be reasonably assigned as the $^2B_{1g} \rightarrow {}^2B_{2g}$ transition for the active site cupric ion. Transitions occurring at higher energy (near 435 and 330 nm in the near-UV) likely arise from ligand to metal charge transfer transitions involving the coordinating ligands, thus providing information on the ligand environment of the metal ion. The higher energy band near 330 nm probably arises from imidazole to Cu^{2+} π-charge transfer based on model studies [46], consistent with the evidence for nitrogen ligation from previous EPR studies [29]. On the other hand, the 435 nm absorption band is too low in energy to be associated with imidazole interactions and must be associated with a previously unidentified copper ligand.

The optical spectrum of the green, Active enzyme is more difficult to interpret. The absorption bands are unlike anything previously reported for any other copper metalloprotein and are associated with a unique EPR-silent oxidized copper complex in the enzyme. The appearance of a strong CD band in the ligand field region of the spectrum of the Active enzyme, again in a region of relatively weak absorption (actually a minimum of the absorption spectrum), suggests a d→d assignment and therefore a cupric oxidation state for the metal ion. Based on a tentative assignment of this relatively strong CD band as the $^2B_{1g} \rightarrow {}^2B_{2g}$ ligand field transition, the shift to a higher average energy of the d→d bands would be consistent with a change in geometry of the active site metal ion on redox activation. In any case, the disappearance of the d→d band associated with the Cu^{2+} site in IA requires a change in geometry of the metal complex occurring on activation. However, this prediction of a distortion of the site is clearly of secondary interest compared to the oxidation state assignment, which must be supported by more than a poorly characterized CD band.

In developing a reliable basis for the oxidation state assignment, we have explored the interactions of Active and Inactive enzyme with azide anion, serving as an exogenous ligand probe of metal oxidation state. Azide binds strongly to both active and inactive forms of galactose oxidase, converting them to complexes characterized by an intense near-UV absorption band near 390 nm (Fig. 6). This absorption feature is convenient for monitoring the progress of the anion titration, as shown in Fig. 7 for both forms of the enzyme. The titration data shown here are fit well by single-step binding isotherms, the solid line in the plots, indicating that a single anion coordinates in each complex, but with significantly different affinities. The azide is bound at least seven-fold tighter to the active than the inactive form of galactose oxidase. The intense absorption feature near 390 nm has been assigned in a number of copper metalloprotein and small-molecule cupric complexes [47–50] as arising from azide to metal π-charge transfer for an equatorially coordinated anion. The fact that the charge transfer bands lie at the

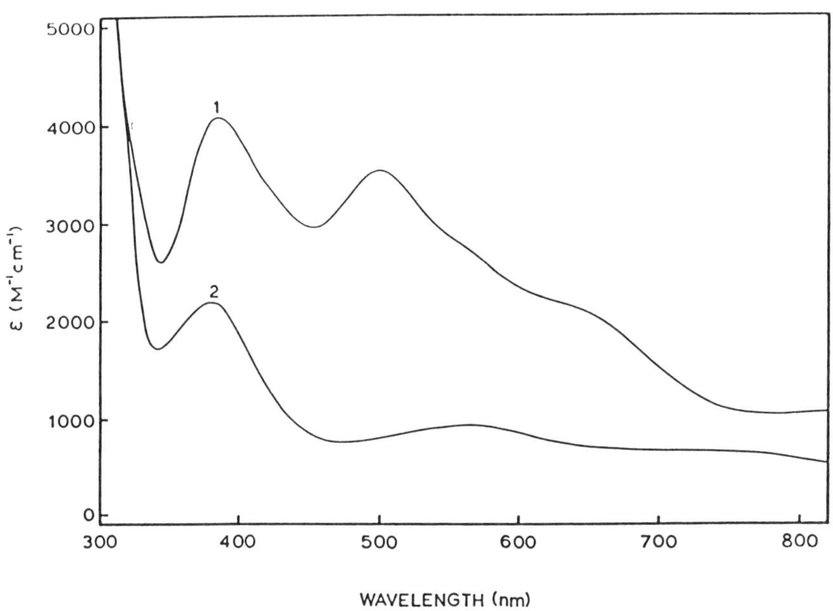

FIG. 6. Absorption spectra of azide complexes of galactose oxidase prepared by the addition of 10 mM KN_3 to oxidized (1) and reduced (2) enzyme in 50 mM Na_2HPO_4 buffer pH 7.0. (From Ref. 14 by permission.)

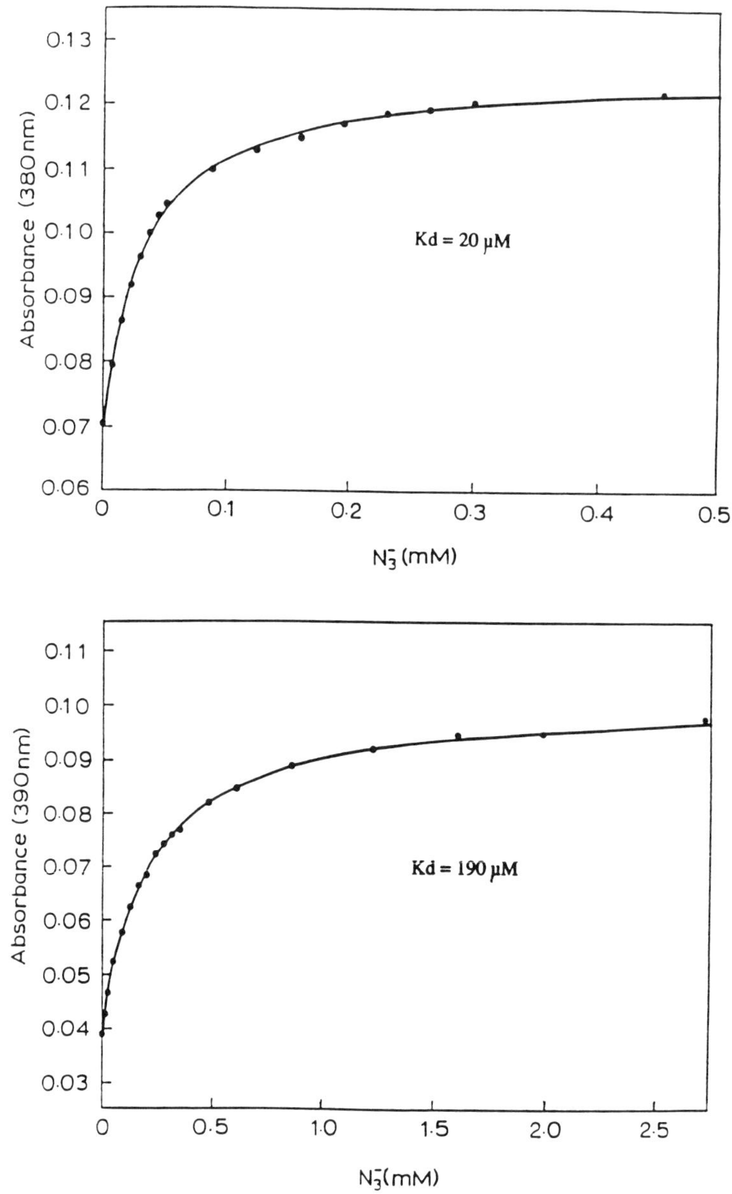

FIG. 7. Titration data for azide complex formation for active galactose oxidase (top) and inactive enzyme (bottom). (•) Experimental data; (—) theory.

same energy in both complexes requires that the oxidation state of the metal ion be the same, since an increase in metal oxidation state should result in a shift of ligand to metal charge transfer bands to lower energy. On the basis of quantitative EPR experiments, the metal is known to be cupric in the Inactive enzyme [14], so a cupric oxidation state assignment is required for the Active enzyme as well. The Active enzyme remains EPR-silent and retains the intense optical absorption features characteristic of this redox form on forming the azide complex, indicating that the azide is not simply reducing the active site to form the IA complex.

The oxidation state assignment is made incontrovertible by X-ray absorption spectral studies (XAS) on pure galactose oxidase species [41]. Previous XAS studies have been reported for this enzyme [51], but the earlier analysis of these near-edge structure (XANES) spectra is now complicated by the possibility of unrecognized redox heterogeneity. The lack of quantitation of redox components in those samples makes the significance uncertain although this information was previously adduced in arguing against a redox role for the active site metal ion in galactose oxidase. Repeating these experiments on samples in well-defined oxidation states confirms that the metal ion is not oxidized in the activation process [41]. X-ray absorption Cu K-edge spectra for galactose oxidase are shown in Fig. 8. The K-edge XANES for the Inactive enzyme, for which the metal is known to be cupric, occurs at 8989 eV. Redox-activated enzyme exhibits a Cu K edge at slightly lower energy, near 8988 eV, a shift in the wrong direction for a higher oxidation state, trivalent copper assignment. Corresponding spectra for the active enzyme in the presence of ferricyanide, added as a trap for photoelectrons preventing photoreduction of the enzyme, turn out to be the same. On the basis of these results, a cupric oxidation state is defined for the metal ion in the redox-activated enzyme, with the slight edge shift consistent with a change in geometry of the metal complex on oxidation. This poses a problem, since redox activation is known to involve oxidation of the enzyme by one electron, yet the metal ion does not reflect this oxidation. Some other site in the protein must be the site of oxidation, and since no other metal ions or cofactors [13] are present, this oxidation must involve the protein itself. One-electron oxidation of any component of the protein other than the metal ion will form something which will be correctly described as a free radical, allowing

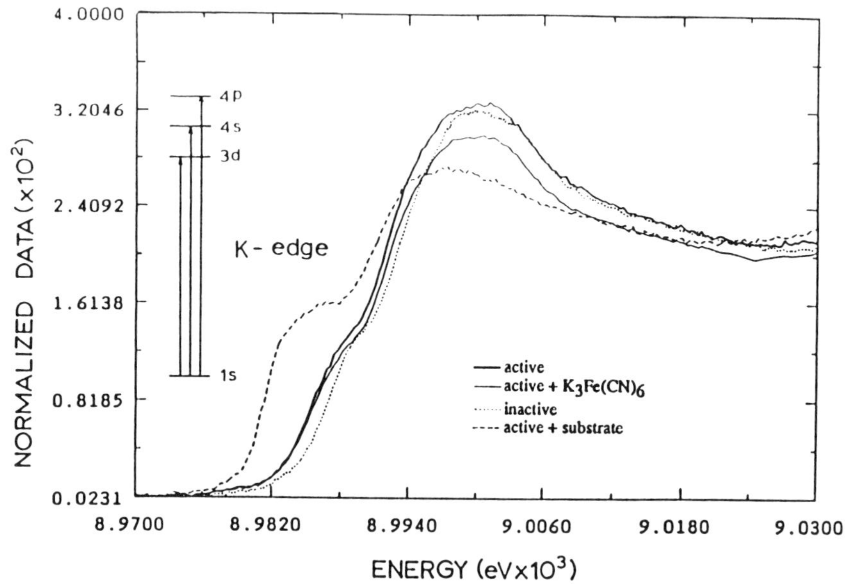

FIG. 8. X-ray absorption Cu K-edge spectra for galactose oxidase. The K-edge transition is indicated in the insert.

us to define both a cupric ion and a free radical in the active site of galactose oxidase.

The absence of a Cu^{2+} EPR spectrum, previously recognized as a striking characteristic of the redox-activated enzyme, can now be understood as a consequence of interactions between a paramagnetic free radical and the cupric ion. However, the absence of an EPR signal does not tell us anything about the strength of this interaction, since even quite weak dipolar coupling can lead to broadening of EPR spectra making them difficult to observe. Using magnetic susceptibility, it has been possible to show that the active enzyme has a diamagnetic ($S = 0$) ground state [52]. Superconducting quantum interference device (SQUID) susceptibility studies performed over a range of temperatures give evidence for thermal population of a paramagnetic excited state at higher temperatures, which can be interpreted as shown in Fig. 9 as the result of strong antiferromagnetic exchange coupling between the free radical ($X^{\bullet}$) and the cupric ion, leading to stabilization of an EPR silent singlet ground state, lying at least 200 cm^{-1} below the paramagnetic

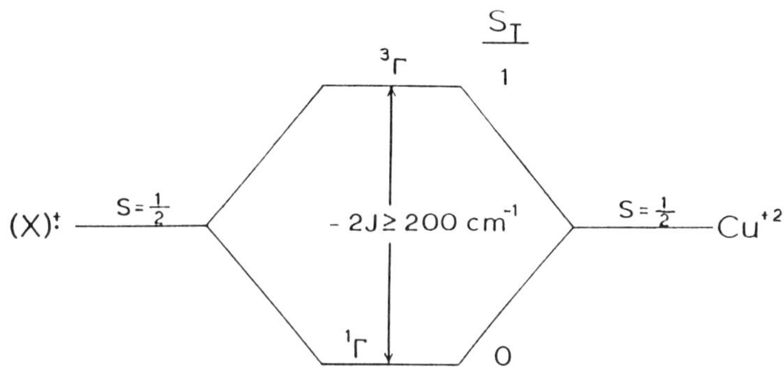

FIG. 9. Exchange splitting of a free radical–coupled cupric ground state.

component of the exchange-split ground state. The magnitude of the singlet-triplet splitting measured in these experiments is chemically important because it implies direct coordination of the free radical to the metal ion. Dipolar interactions of a noncoordinated radical would be expected to be much smaller, on the order of 1 cm^{-1} or less [53]. The magnitude of the exchange coupling also suggests an equatorial coordination site for the radical in the copper complex, since studies on metal-free radical interactions have shown that only very weak antiferromagnetic or even ferromagnetic coupling is observed when a radical such as a spin label is apically coordinated to a cupric ion [53]. This is easily understood as a consequence of the symmetry of the orbital overlaps which provide the mechanism for exchange in these complexes. For an apically coordinated radical, both σ and π overlaps of the ligand valence orbitals with the $d_{x^2-y^2}$ orbital of the metal ion vanish by symmetry, whereas these overlaps maximize for equatorial ligation. These results lead to the identification of a free radical–coupled copper complex in galactose oxidase, the first example of an entirely new type of metalloenzyme redox-active site.

The analysis of metal-free radical interactions provides an essential insight into the electronic interactions and a geometric feature of the active site metal complex, but does not give any clue to the identity of the free radical itself. A structural assignment for the free radical site in galactose oxidase requires a different type of spectroscopic probe, sensitive to the chemical structure of the radical. Since we have identified the intense absorption features in the active form of galactose oxidase as being associated with the presence of the free radical in the active site, resonance Raman spectroscopy provides a promising approach [54].

Resonance Raman spectra for the Active galactose oxidase and its azide adduct [55] are shown in Fig. 10. These spectra result from excitation of the sample at 659 nm in the red as indicated in the insert in Fig. 10, which shows the corresponding absorption spectra for the two forms. Both Raman spectra exhibit a feature near 980 cm^{-1} which arises from nonresonant scattering from 0.1 M sulfate added to the samples as an internal scattering intensity standard. For the Active enzyme, strong resonant enhancement is observed for a pattern of four vibrational features near 1170, 1247, 1484, and 1595 cm^{-1} Raman shift which form the characteristic signature of a ligated tyrosine residue under excitation within a ligand to metal charge transfer transition. These features have been observed in tyrosine–metal complexes in proteins, such as the iron-tyrosinate proteins and in their Cu analogs, as well as in small-molecule complexes with Cu-tyrosinate ligation [56,57]. The appearance of these features on excitation within the radical absorption bands indicates that a ligated tyrosine residue is somehow involved in the radical site. On closer inspection, further information is revealed: the tyrosine ring breathing modes (Wilson modes ν_{19a} and ν_{8a}) occur in this spectrum at unusually low frequency. These breathing modes are known to be particularly insensitive to environmental effects and vary only slightly, within a 10 cm^{-1} range; thus the shift of these features to 5 and 15 cm^{-1} lower than the previously reported limit in metal–tyrosine complexes is dramatic and must reflect a significant perturbation of the protein side chain. The types of perturbation consistent with the shift include π-stacking interactions of the ring which could remove electron density from the π-bonding system or add to the π-antibonding system, thus weakening the bonding in the ring and lowering vibrational frequencies, or a heavy-atom substitution of the ring system. The Raman spectra which are available for a tyrosine phenoxyl free radical have been reported to be distinctly different [58], with a single vibrational mode exhibiting strong resonance enhancement near 1483 cm^{-1}. This was in the past assigned as a phenoxyl $\nu_{\text{C-O,str}}$ C-O stretching and symmetric ring deformation mode, shifted to higher frequency by an increase in C-O double-bond character on oxidation of the π system. We

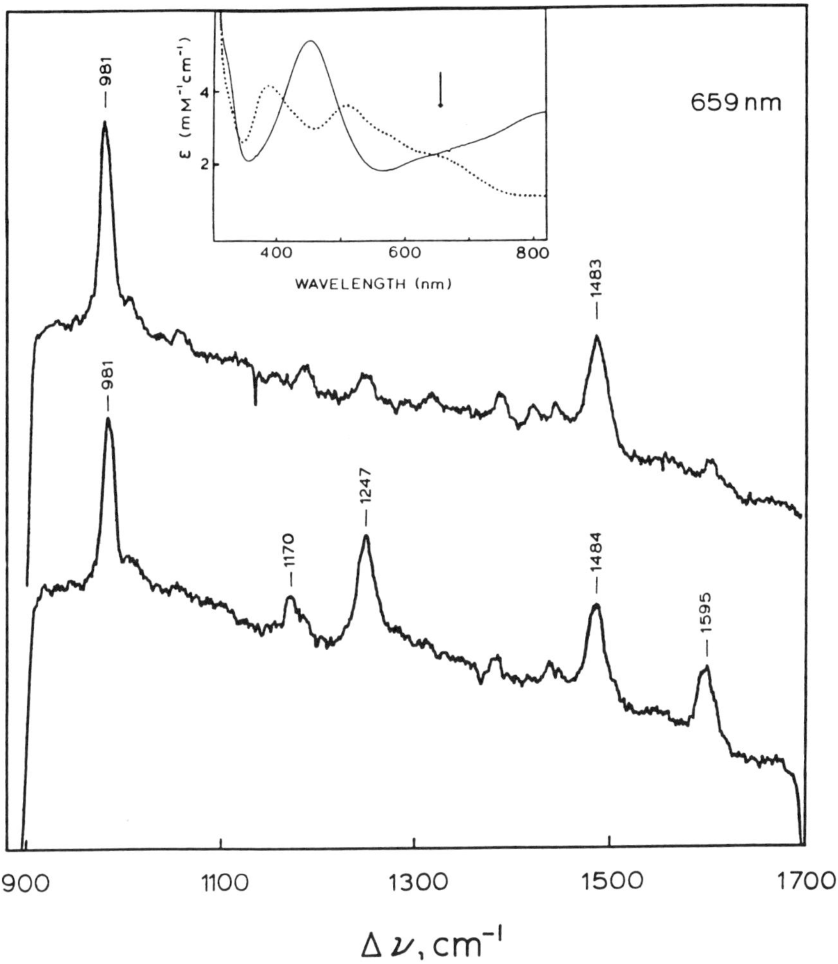

FIG. 10. Resonance Raman spectra for active galactose oxidase (bottom) and its azide adduct (top) with absorption spectra indicated in the insert. (From Ref. 55 by permission.)

observe this type of spectrum for the azide adduct of Active enzyme, as shown in the top spectrum in Fig. 10. Here, a single strongly enhanced Raman band is observed. However, lower intensity features also occur, which match exactly the frequencies of corresponding features observed in the spectrum of the unliganded protein. The two spectra therefore are actually nearly identical in their frequencies, differing primarily in the enhancements of the various features. This is understandable as a consequence of excitation in a distinct electronic transition, which would be expected to lead to a distinct pattern of enhancement for the Raman modes. It appears likely that both spectra reflect the vibrational features of a single group associated with the radical site, and the similarity of the spectra to those reported for tyrosinate complexes allows identification of a perturbed, ligated tyrosine residue in the active site. This new piece of structural information suggests an assignment of the lower energy charge transfer band near 440 nm in the visible spectrum of Inactive blue enzyme as arising from tyrosinate ligation to the cupric ion.

The spectra for model tyrosine [56,57] and phenoxyl radicals [58] provide our points of reference in interpretation of the spectra described above. Unfortunately, while reliable vibrational assignments are available for the tyrosinate complexes, the assignments of the vibrational spectra for the phenoxyl radical are more controversial. Calculated vibrational structures for a phenoxyl radical are in fact very similar to the parent tyrosine [59], in contrast to earlier suggestions that dramatic frequency shifts are associated with ring oxidation [58]. These considerations suggest the possibility that the resonance Raman spectra obtained for the Active galactose oxidase arise from a perturbed tyrosine radical and directly probe the vibrational structure of the radical. Additional resonance Raman data obtained very recently reveal that relatively weak features at frequencies typical of a normal tyrosine are also detectable in these spectra [60], providing evidence for two distinct tyrosine ligands in the copper binding site.

The combination of EPR, optical absorption, CD, XAS, magnetic susceptibility and resonance Raman analyses have provided consistent but indirect evidence for a free radical coupled Cu active site in galactose oxidase, without ever giving results that are immediately recognizable as arising from an organic free radical. In each case, the radical site is masked by interactions in the complex: coupling to the cupric ion masks the radical so that the typical EPR signature is not observed; in optical spectra, the radical is masked by perturbations that distinguish it from simple, well-characterized radical species such as observed in ribonucleotide reductase [16]. It has proven possible, however, to unmask the radical for direct spectroscopic analysis [61].

3.3. Direct Probes of the Radical Site

The unmasking of the radical site requires the removal of the active site metal ion, which can be readily accomplished by treating the enzyme with diethyldithiocarbamate [13]. The metal ion is chelated by this reagent and abstracted from the protein complex, leaving metal-free apoenzyme. To examine the protein for potential damage caused by this procedure the active site must be reconstituted and the catalytic activity of the renatured complex determined. The reconstitution procedure turns out to be less straightforward than might be imagined, and simple dialysis vs. cupric salts only partly reconstitutes the apoenzyme, suggesting a relatively low affinity for the cupric ion. This is surprising, since activation by dialysis against cupric solutions is one of the steps in the enzyme purification [25], and it has been assumed that this is a reconstitution process. Since cupric ion does not reconstitute the apoenzyme, this more likely reflects a redox activation of the partially inactive Native enzyme rather than reconstitution of Cu-depleted protein. The most effective reagent for reconstitution of apogalactose oxidase appears to be cuprous salts, which are efficiently and apparently selectively bound in the active site complex [61]. (This selectivity of the metal binding site for cuprous ion has interesting implications relating to the assembly of a functional active site; see below.) After anaerobic reconstitution, dialysis of the Cu^+-containing protein against aerated buffer results in quantitative reoxidation and formation of Inactive galactose oxidase having the maximum copper content, approximately one Cu per active site. However, not all of this copper is in productive complexes, although redox activation restores better than 80% of the activity of the starting enzyme. The optical spectrum of the active enzyme is also restored to a similar extent in this process, confirming that the extraction of Cu does not result in a damaged protein.

When the metal-free apogalactose oxidase is exposed to the same, mild oxidizing conditions which generate the Active enzyme in the copper-containing protein, a new signal is detected in the low-temperature EPR spectrum [61] (Fig. 11). This signal, with an average *g* value near 2.005, is sharp and occurs near the free electron *g* value, characteristic of an organic free radical, and is the same as a minority signal in the activated copper-containing protein [14] (Fig. 2) but at higher quantitation. This radical is very different from radicals previously generated in proteins by treatment with strong oxidants or ionizing radiation. Radicals formed under such conditions are typically unstable in aqueous solution under exposure to air, while the radical which is formed in apogalactose oxidase under relatively mild treatment with ferricyanide is stable to normal protein isolation procedures. It may be reduced by a variety of reagents known to be effective in quenching organic free

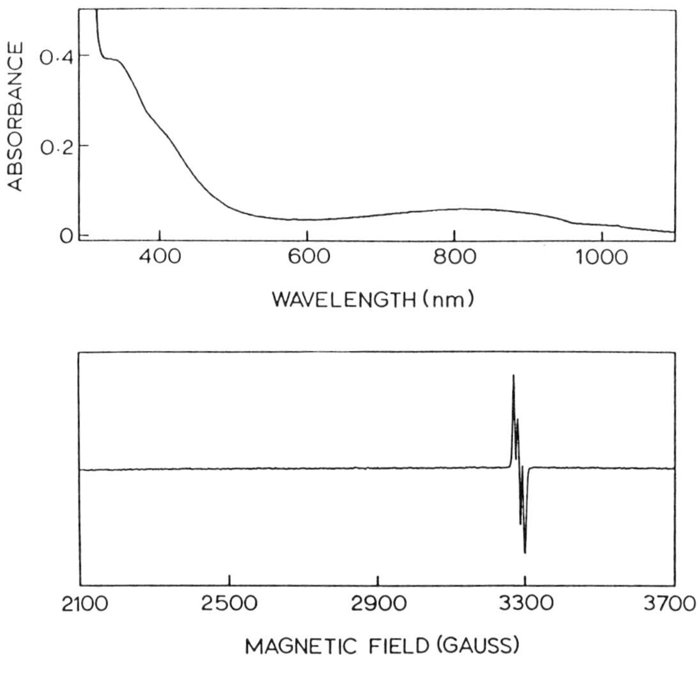

FIG. 11. Optical absorption (top) and EPR (bottom) spectra for oxidized metal-free apogalactose oxidase. (From Ref. 61 by permission.)

radicals, including ascorbate and hydroxyurea, but is unreactive toward substrate.

New optical features appear in the absorption spectrum of the radical-containing protein, including an intense absorption band in the near-UV near 340 nm and a broad absorption band in the red end of the spectrum of lower intensity. This spectrum is decidedly different from the spectrum of the Active green enzyme. Could it arise from the free radical? Anaerobic titration of the radical-containing apoprotein with ascorbate eliminates both optical absorption and EPR spectra with an equivalent amount of reductant at a stoichiometry consistent with the spin quantitation of the radical EPR signal, confirming that the optical spectrum arises from the free radical site in the protein. In order to assign the radical, other approaches are needed since both optical and EPR spectra are different from any radicals previously reported arising from amino acid side chain or cofactor radicals. Even the additional hyperfine structure resolved in the expanded EPR spectrum (Fig. 12, top) is not sufficient for an assignment. However, the average EPR g value measured from this spectrum, $g = 2.005$, is fairly characteristic of phenolic free radicals, motivating an isotopic labeling experiment to provide a structural assignment.

Normally, such selective labeling experiments are performed on amino acid auxotrophs, mutant organisms unable to produce a given amino acid themselves which efficiently utilize exogenously added amino acids in protein synthesis. This has not proven possible with *D. dendroides*, the source of galactose oxidase. The lack of a sporulation phase in the reproductive cycle of a mycelial fungus like *D. dendroides* [62] makes genetic manipulation difficult: mutants can be produced but not selected or isolated straightforwardly. Even more perverse, *D. dendroides*, when presented with media containing supplementary aromatic amino acids, prefers to convert them to black pigments rather than incorporate them into fungal protein. An alternative strategy in labeling experiments is to suppress the endogenous biosynthesis of aromatic amino acids, making the fungus effectively auxotrophic, using a specific inhibitor of the biosynthetic pathway. Glyphosate (N-phosphonomethylglycine), the active ingredient in the weedkiller Roundup™, inhibits 5-enolpyruvoylshikimate-3-phosphate synthase, a committed step in the biosynthesis of phenylalanine, tyrosine, and tryptophan as

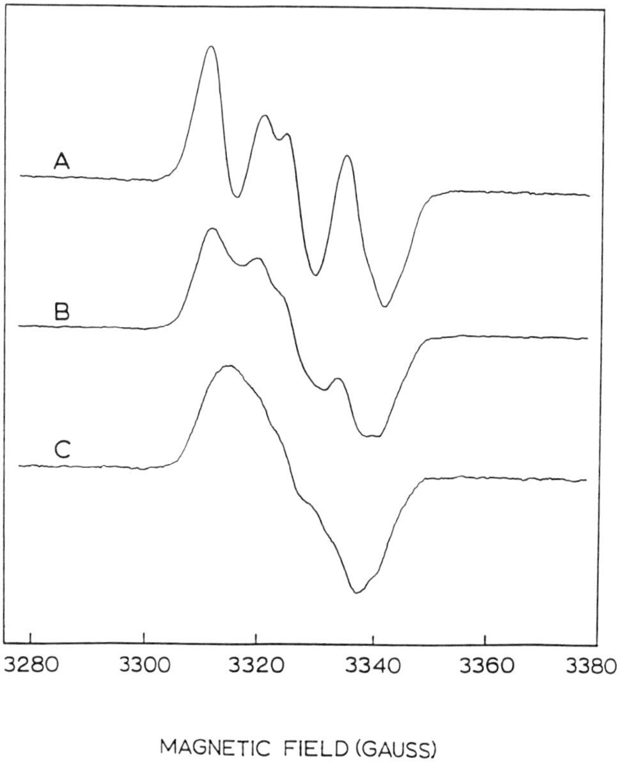

FIG. 12. EPR spectra for oxidized apogalactose oxidase (A); [β,β-^{2}H]tyrosine-labeled enzyme (B); and difference EPR spectrum (C). (From Ref. 61 by permission.)

the basis of its herbicidal action [63]. Cultivating *D. dendroides* in minimal media containing millimolar concentrations of glyphosate blocked fungal growth, which was restored by supplementing the culture media with all three aromatic amino acids. Radiotracer experiments indicated that in the presence of high levels of glyphosate, up to 70% of the tyrosine in the galactose oxidase produced under these conditions was incorporated from the media. The remaining fraction of unlabeled tyrosine likely represents a biosynthetic leak from the phenylalanine which must also be added to permit growth to occur. This glyphosate treatment was used to incorporate [β,β-^{2}H]methylene-deuterated tyrosine into galactose oxidase. On forming the apoprotein of the labeled enzyme and oxidizing, a radical was expressed for EPR analysis.

The free radical EPR signal from protein labeled with β-methylene-deuterated tyrosine appears distinctly perturbed relative to the unlabeled control (Fig. 12, middle). The signal is comparatively broad, and the hyperfine structure relatively poorly resolved. Recalling that the labeled protein contains approximately 30% of unlabeled tyrosine, it is appropriate to subtract this fraction of the spectrum of the unlabeled protein to obtain a difference spectrum arising from just the labeled sites. The result is a broad, relatively featureless EPR spectrum which is precisely what is expected for a tyrosine π radical if the hyperfine coupling to the β-methylene protons is removed by deuteration as a consequence of the relatively small coupling of the unpaired electron spin to the deuterium nucleus (Fig. 12, bottom). This result conclusively assigns the free radical that forms in apogalactose oxidase as deriving from a tyrosine residue in the protein in a complex that leaves the β-methylene group intact, and thus provides further support for the interpretation of the resonance Raman spectra. The presence of a pyrroloquinoline quinone structure in the active site, which was raised at one point [64], is ruled out by this result and is also inconsistent with the optical data for the protein.

Still, the assignment is incomplete. The existence of an unusual radical-forming site in the protein is convincingly demonstrated in these experiments, and the association with a tyrosine-related side chain is clear, yet both the EPR and optical spectra of the free radical in the apoenzyme, and the optical and resonance Raman spectra of the Active enzyme, are distinct from well-characterized tyrosine phenoxyl free radicals in other proteins and in models [65–67]. These differences emphasize the fact that the radical site is perturbed relative to a normal tyrosine: its redox potential is significantly lower (450 mV compared to >800 mV for tyrosine); the ring-breathing modes are dramatically lower in frequency than is observed in models; and the hyperfine coupling in the free radical EPR spectrum is distinct. Simulation of the EPR spectrum permits some qualitative insight to be gained into the nature of the differences. Figure 13 shows the experimental EPR spectrum for the apogalactose oxidase free radical and, below, a computed powder spectrum generated from a set of spin Hamiltonian parameters. The axial g anisotropy that is assumed is characteristic of the g values of a phenoxyl π radical, and the values chosen for the simulation are within the range typically observed. Coupling to a single β-methylene proton is consistent with results from a wide range of protein and model studies on tyrosine radicals, a consequence of a relatively stable conformation which directs one of the C-H bond vectors nearly normal to the ring, resulting in one dominant proton hyperfine coupling of approximately 15 G. Hyperfine coupling to the 3- and 5-ring α protons is relatively small with an average value of approximately 9 G. Using this parameter set, an excellent fit to the experimental spectrum may be

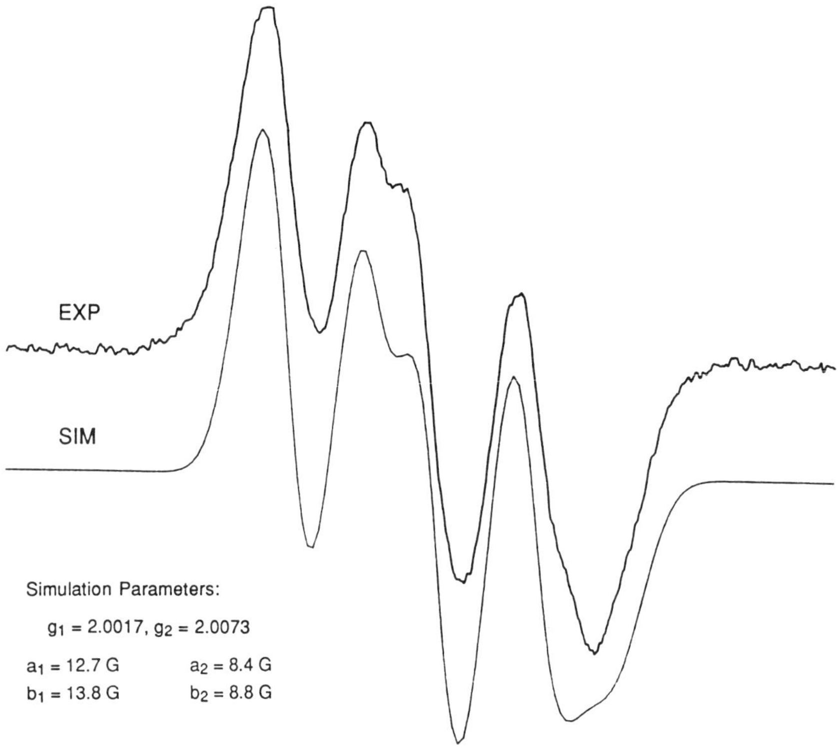

FIG. 13. Experimental and simulated EPR spectra for oxidized apogalactose oxidase.

obtained, as shown. This excellent agreement requires, however, that a *single ring proton* is magnetically interacting with the unpaired electron spin. This is very unusual, since normally the two ortho ring protons in phenoxyl radicals are coupled symmetrically to the electron spin. This suggests a modification of a tyrosine in the radical-forming site, but a test for the presence of dopa (3,4-dihydroxyphenylalanine), a covalent modification of tyrosine known to occur in certain proteins [68], gave negative results with galactose oxidase. It is important to point out that such a simulation of a free radical EPR spectrum is generally not unique and that it may in fact be easier to obtain a satisfactory fit with unphysical parameters than with reliable ones. The EPR simulation therefore cannot provide a firm basis for analysis, and other data (e.g., from electron nuclear double resonance, or ENDOR, spectroscopy; see below) is often required for an adequate interpretation.

To summarize, these spectroscopic studies have provided a basic assignment of the absorption spectra for both the Active and Inactive forms of galactose oxidase. For the Inactive blue enzyme, the broad ligand field absorption extends over the red end of the visible spectrum; for the Active green enzyme, intense absorption features near 445 and 800 nm signal the presence of the free radical in the active site. Using these assignments, we have been able to explore some of the essential mechanistic aspects of galactose oxidase chemistry.

The most interesting active site interactions are those with the substrates, galactose and O_2. Figure 14 shows the result of titrating an anaerobic solution of the green, Active enzyme with a little more than one stoichiometric equivalent of the substrate, β-OCH_3 galactose. The optical spectrum is bleached over the entire Vis–near-IR spectral range, to give a colorless enzyme solution [14]. This bleaching eliminates not only the intense absorption features of the radical but also the relatively weak absorption bands arising from the cupric center. Since each of these sites is a one-electron redox unit, we can count electrons in this experiment and define a two-electron reactivity for the active site of galactose oxidase, which matches the two-electron chemistry of alcohol oxidation to aldehyde. This result provides the first experimental evidence for this type of two-electron reactivity in this active site and for the participation of the metal in redox catalysis [14]. This analysis implies that the metal is cuprous in the substrate-reduced complex, and

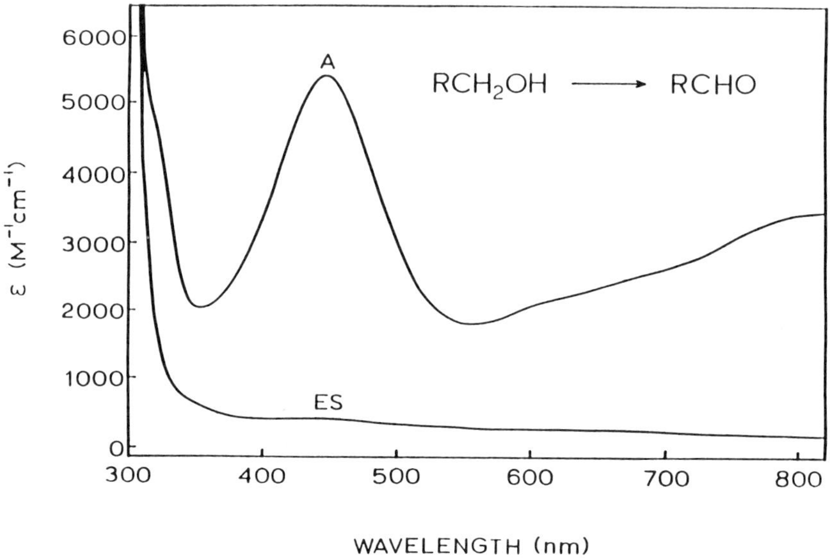

FIG. 14. Optical absorption spectra for active (A) and substrate-reduced (ES) complexes of galactose oxidase.

this is confirmed by Cu K-edge XANES experiments for which this complex exhibits the typical edge energy (8984 eV X-ray energy) and line shape of a classic cuprous coordination complex [41]. This cuprous complex is also competent for O_2 reduction, and exposing the preformed cuprous intermediate to air restores the intense absorption spectrum of the Active enzyme complex [14]. This suggests that it is possible to step the turnover reaction through two distinct half-reactions by omitting O_2 in the first step, making likely a mechanism in which aldehyde release precedes dioxygen reduction, in contrast to the sequential, ordered Bi-Bi mechanism proposed earlier based on steady-state kinetic analysis [32].

3.4. The Emerging Crystal Structure—Novel Covalent Modification of a Cu Ligand

These spectroscopic studies on the active site of galactose oxidase stimulated interest among crystallographers in solving the structure of the protein. Crystallization of galactose oxidase in small hexagonal plates was reported in the earliest work [22], and, interestingly, similar platey crystals have afforded the crystallographers a solution to the protein structure. A structure for galactose oxidase was recently reported at nearly atomic resolution (1.7 Å), the result of work by Nabatushi Ito with Simon Phillips and Peter Knowles [23] at Leeds University. The emerging crystal structure reveals the protein in delightful detail, exhibiting a number of interesting structural features. The overall protein fold is unusual, a seven-member β barrel formed from a sevenfold repeat of a slat motif of four-stranded β-sheet structure. This fold appears to be an especially stable form of a propeller assembly of β sheets [69] and was previously reported in another protein, methylamine dehydrogenase [70]. Into the interior of this barrel, a peptide strand winds to contribute a metal ligand to the Cu binding site. The Cu complex itself is formed from two histidine imidazoles, two tyrosines, and water (or acetate, depending on crystallization conditions), forming a five-coordinate, roughly square-pyramidal complex (Fig. 15). The crystallographers encountered an unusual feature in the active site structure: one of the tyrosine ligands makes unusually close contacts with the sulfur of a cysteine residue. This could potentially be an artifact of the structure determination, but appears consistently through refinement without any constraints being placed on this contact. The structure that emerges is that of a unique, cova-

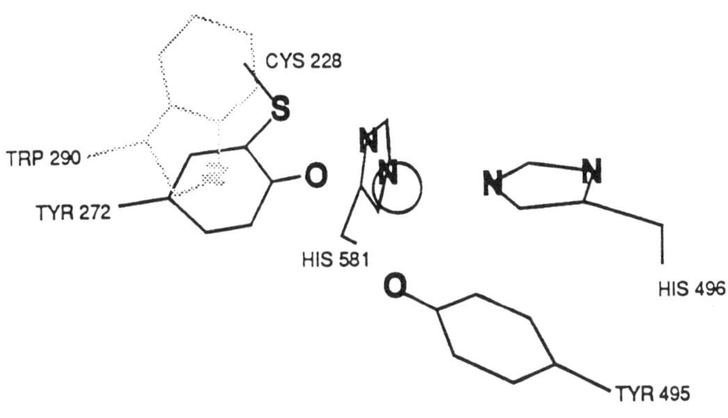

FIG. 15. Galactose oxidase active site structure. (Based on Ref. 23.)

lently modified tyrosine residue, crosslinked to a cysteine residue remote in the polypeptide chain by formation of a C-S bond at a position ortho to the hydroxyl group. The confirmation of this structure through isolation of a peptide containing this covalent crosslink has not yet been reported. This unique modification clearly provides a strong candidate for the radical site identified in galactose oxidase by spectroscopic studies. The thioether substitution ortho to the hydroxyl group could plausibly perturb the redox characteristics of the phenol, stabilizing the phenoxyl oxidation product by several hundred millivolts. The substitution pattern would also account for the observed hyperfine coupling to a single ring proton in the apoenzyme radical. In addition, the heavy-atom substitution of the tyrosine ring may provide the perturbation which is detected in the lowered ring-breathing mode frequencies in resonance Raman spectra of the Active enzyme radical.

This crystal structure may provide further insights into free radical sites in other proteins. In galactose oxidase, the putative radical site, the Tyr-Cys crosslink, forms a planar unit containing the ring, the sulfur, and the cysteine β-methylene in eclipsed conformation. The significance of this arrangement is yet to be defined, but possibly maximizes the sulfur–ring interactions [23]. The thioether sulfur itself appears not be coordinating to the metal ion. This entire unit is overlaid by a tryptophan indole side chain forming a π-stacking contact with both the tyrosine ring system and the thioether bridge. This interaction raises the possibility that the tryptophan participates in delocalization of the unpaired electron in the radical thus contributing to its stability [23]. Stacking interactions were proposed earlier as a possible charge transfer perturbation accounting for a low-energy absorption band in the optical spectrum of the Active enzyme [55].

The presence of an unusual amino acid crosslink in the active site of galactose oxidase raises interesting questions. Is enzymatic post-translational modification required for its formation, or does this structure form spontaneously in the active site? A final answer to these questions will await biochemical studies on the expression of galactose oxidase, but a spontaneous assembly of the active site structure seems quite possible. We have shown that the metal binding site is receptive specifically to cuprous ion [61], suggesting that cuprous ion bound to a proenzyme, lacking the amino acid crosslink, could activate O_2 by one-electron reduction to form metal-bound superoxide. Subsequent reaction with thiolate would generate a thiyl radical adjacent to the phenolic ring of the tyrosine poised to react to give the covalently cross-linked structure.

The crystal structure has not as yet been able to provide much information on the substrate interactions of the active site, since it has not yet proven possible to prepare a substrate complex of the crystalline enzyme [23]. In part, this may be due to the form of the enzyme stabilized in the crystal. Based on the sensitivity of galactose oxidase to alcohols and the method of preparation of the protein crystals, it seems very likely that the Inactive form is present in the crystals reported so far [23]. The absence of significant structural change on addition of the oxidant and activator hexachloroiridate(IV) to the crystals was previously considered evidence for the presence of the active enzyme in the crystal [23]. This needs to be reconsidered, since this oxidant will be rapidly reduced to hexachloroiridate(III) by polyethylene glycol stabilizers and it is likely this reduced complex which is observed as a heavy-atom derivative in these crystals. No protein crystal structure will be able to define the oxidation state of a redox cofactor: one electron more or less is beyond the resolution of the X-ray diffraction experiment. The overall structure of the Inactive protein may be very similar to that of the Active form, but spectroscopic studies on the active site, particularly the CD spectra of green and blue enzymes [14], implies that the copper site changes geometry on redox activation. Although the exact nature of this structural change is uncertain, the data appear to be consistent with a more tetragonal structure in the activated enzyme. This change in geometry of the active site is also reflected in the relative affinities of the active site metal for exogenous anion binding. As a consequence, extrapolation from the known crystal structure to the structure of the Active enzyme should be approached with caution. The appearance of a distorted tetrahedral coordination geometry for the crystallographically defined Cu site at neutral pH has suggested a possible analogy with the "blue" electron transfer proteins such as plastocyanin, azurin, and stellacyanin [23], but important differences exist in the chemistry performed by these redox-active sites. The exceptionally fast redox chemistry that makes the blue Cu sites efficient partners in outer sphere redox reactions is thought to relate to the small reorganization barrier for $Cu^{+/2+}$ interconversion within a distorted tetrahedral complex geometrically poised to stabilize both oxidation states of the copper complex. In galactose oxidase, however, the redox chemistry is likely to be inner sphere in character, consistent with the appearance of an open or labile coordination site on copper in the crystal structure. While geometric changes for the metal complex will certainly be important in this mechanism, in which both cupric and cuprous metal centers appear, the participation of a tetrahedral Cu site will probably not be essential.

3.5. ENDOR Analysis of the Radical Site

ENDOR experiments on the radical site in apogalactose oxidase and radicals formed from thioether-substituted phenols have contributed to a more detailed understanding of the nature of the free radical electronic and geometric structure [71]. Low-temperature proton ENDOR spectra for the enzyme are dominated by strong features, appearing in the matrix region of the spectrum, arising from weakly coupled protons giving rise to ENDOR transitions near the free proton Larmor precession frequency (ν_p = 14.7 MHz). These signals are altered in D_2O and identified as exchangeable protons in the vicinity of the radical and assigned on the basis of line shape to hydrogen-bonding protons interacting with the phenoxyl oxygen. Protonation of the apoenzyme radical at oxygen possibly replaces Cu^{2+} interactions. Features are also observed near 5 and 35 MHz, associated with a fairly strong coupling and near axially symmetric hyperfine tensor, characteristic of hyperconjugative coupling to a single β proton. Similar features are observed for a wide range of phenoxyl radicals including *p*-cresol radical, and the protein-associated radicals in Photosystem II of green plants, ribonucleotide reductase, among others. Based on analysis of ENDOR data for galactose oxidase, the conformation of the β-methylene group in the radical has been experimentally examined, and a dihedral angle of θ = 34° calculated, to be compared with a crystallographically defined dihedral angle estimated at 16° for the metal-containing protein. This suggests that a rearrangement of the group occurs, either on reducing the radical site or, more likely, on forming the apoenzyme.

Sulfur interactions in the radical appear surprisingly small. The observed *g* anisotropy in both the protein and the 2-SCH_3 *p*-cresol model radical is not significantly different from unsubstituted phenoxyl radicals, indicating a relatively small degree of participation of sulfur valence orbitals through covalent mixing in the free radical ground state. This unexpected result may be a consequence of a particular conformation of the thioether side chain, which will determine the spatial orientation of nonbonding orbitals on S and control the degree of covalent overlaps between sulfur and ring π-valence orbitals and therefore the electronic perturbation. These studies on the radical site and an effective model for the protein redox unit support an assignment of the radical to the crystallographically defined tyrosine-cysteine crosslink in the protein structure.

3.6. Ligand Interactions

The chemistry of active site ligand interactions can be explored through a detailed analysis of anion perturbations. The spectroscopic consequence of binding anions such as azide, cyanate, and cyanide is conversion of the Inactive blue enzyme to a red form and the Active green enzyme to a yellow complex. For the Inactive enzyme, this change reflects the loss of the 450 nm band, assigned to phenolate-to-Cu charge transfer transition, and a shift of the cupric ligand field absorption to higher energy. The absence of phenolate charge transfer spectra in the anion adducts suggests a displacement of the unmodified tyrosinate ligand on formation of the anion complexes.

The chemical consequences of anion interactions are less obvious. Weakening metal interactions will increase the basicity of the phenolate oxygen, generating a moderately strong base ($pK_a = 10$) in the active site. The appearance of a base with $pK_a > 9$ has recently been detected, demonstrating proton-coupled anion binding to the active site metal ion (Fig. 16). The experiments involve titration of the enzyme by anions in solutions buffered only by a pH indicator dye, providing sensitivity to protonic equilibria and permitting quantitative measurement of proton uptake by the enzyme during titration. Using thymol blue indicator at pH 8.5 and bromthymol blue indicator at pH 7, uptake of 0.8 proton per Cu ion is observed on saturation of galactose oxidase with azide. There is evidence that ionization of coordinated water occurs at low temperature as an analog of deprotonation of coordinated alcohols, suggesting important mechanistic implications for this chemistry.

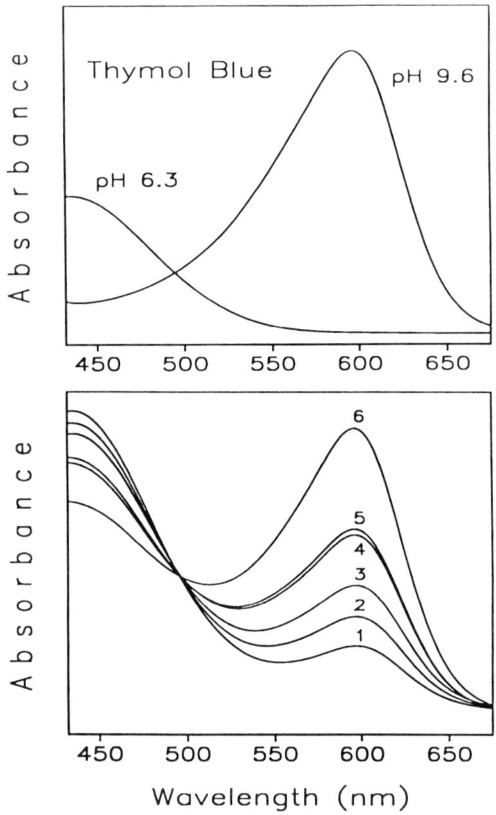

FIG. 16. (top) Titration of thymol blue indicator dye with base, showing characteristic spectral changes. (bottom) Proton-coupled anion binding in galactose oxidase. 1, Thymol blue–buffered galactose oxidase. 2,3, Addition of aliquots of base for calibration. 4,5, Two successive equal additions of saturating NaN_3 (10 times K_d). 6, Further addition of base.

3.7. Mechanistic Insights Arising from Structural Studies

Based on the studies [14,41,52,55,61] described above, it is possible to outline basic features of a mechanism for galactose oxidase (Fig. 17). The turnover chemistry involves only the redox-activated form of galactose oxidase. The form which contains an EPR-detectable cupric ion is a catalytically inactive form of the enzyme, in which the copper appears

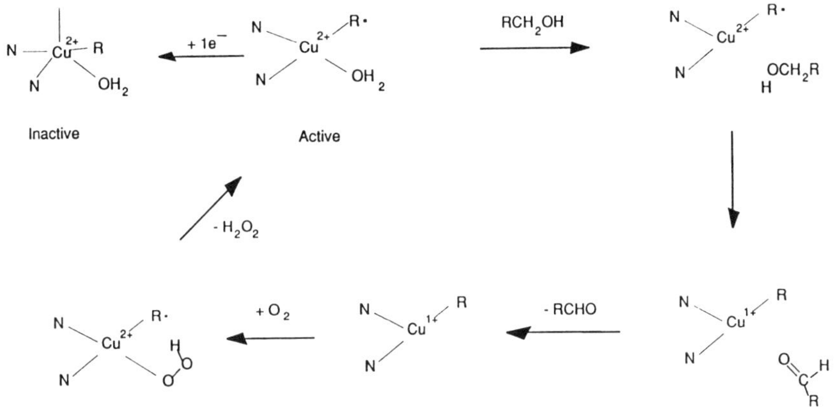

FIG. 17. Mechanistic scheme for catalytic cycle of galactose oxidase based on spectroscopic, magnetic, and chemical studies.

to be bound in roughly square-pyramidal geometry. Redox activation occurs by the removal of one electron per active site. The electron does not come from the metal, but from ligand valence orbitals to form a unique *free radical–coupled copper complex*, the catalytically competent unit in the enzyme. This EPR-silent complex interacts with hydroxylic substrates, and in some combination of electron and proton transfers is reduced by two electrons to give a cuprous intermediate that has been characterized by XAS [41]. The presence of a radical component in the active site makes it likely that a free radical mechanism occurs (see below). Dissociation of the aldehyde is expected to occur at this point, based on our observation that the oxidative and reductive half-reactions can be separated into two distinct steps. The cuprous complex probably has a lower coordination number than the starting cupric site, consistent with the typical coordination chemistry of the cuprous ion. The cuprous complex is competent for dioxygen reduction, most likely at the metal ion, and may form a peroxide intermediate for which the azide complex described earlier may serve as an effective model. Dissociation of peroxide completes the turnover cycle. The alternative reaction of galactose oxidase, the enzymatic oxidation of aldehydes to uronic acids, could occur similarly, from the hydrated acetal (*gem*-diol) form of the aldehyde substrate.

Galactose oxidases were recently isolated from two other fungal sources, *Gibberella fujikuroi* [72] and *Fusarium graminearum* [73]. The biological function of galactose oxidase is becoming clearer from metabolic studies on other wood rot fungi, particularly *Phanerochetes chrysosporium*. In the latter case, an enzyme analogous to galactose oxidase, glyoxal oxidase [74], is expressed and secreted as a component of an armamentarium of lignin degradation enzymes which also include the lignin peroxidases. These peroxidases, key to the cleavage of the highly crosslinked lignin polymer into soluble products, require hydrogen peroxide as a substrate. For *P. chrysosporium*, the extracellular production of hydrogen peroxide as a cosubstrate for the peroxidases appears to be the function of glyoxal oxidase. In *D. dendroides*, this function is likely performed by galactose oxidase, accounting for its unusually low specificity for organic substrate. Thus, the biological function of galactose oxidase is probably to serve as a factory for hydrogen peroxide, supporting lignin peroxidase reactions, rather than the biosynthesis of aldehydes. It is interesting to note that galactose oxidase may be redox activated in the presence of peroxidases [25,75], suggesting that a closed self-activating extracellular enzyme system can be assembled from these components. To investigate this question further, we examined the culture filtrates of *D. dendroides* grown under conditions which maximize galactose oxidase production but found no significant lignin peroxidase activity. This may reflect different optimal conditions for expression of the oxidase and peroxidase enzymes.

4. MODEL REACTIONS AND THE ROLE OF THE RADICAL IN CATALYSIS

An intriguing report by Semmelhack et al. appearing in 1984 describes the use of a catalytic cocktail prepared from 2,2,6,6-tetramethyl-1-piperidinyloxy (TEMPO) nitroxyl spin label and copper salts for the selective oxidation of primary hydroxyl groups in a variety of alcohols to aldehydes [76] [Eq. (2)]:

$$\mathrm{RCH_2OH + O_2 \xrightarrow{\text{TEMPO + CuCl}} RCHO + H_2O} \tag{2}$$

The reaction exhibits an unusual regioselectivity for primary alcohol oxidation even in the presence of secondary alcohol groups, although the catalytic species has never been isolated in this reaction and the original report proposed that the oxidizing agent is a nitronium ion ($>N=O^+$). However, the association of a free radical (TEMPO) with copper in an alcohol oxidation systems suggests the possibility of participation of a free radical–coupled Cu complex in the observed reaction, modeling the chemistry we defined for galactose oxidase. To explore this possibility further, we have prepared a structurally defined complex ([Cu(hfac)$_2$TEMPO], hfac = hexafluoroacetylacetonate) originally described by Lim and Drago [77,78]. However, no redox chemistry appears to occur within this complex either in the presence or in the absence of O_2. Thus the association of a free radical with a cupric ion is not itself sufficient to create an alcohol oxidation catalyst. Recently, a nitroxyl-based copper reagent for alcohol oxidation was reported [79]. The reagent, formed from the reaction of spin label free radical with $CuBr_2$, was structurally defined and shown to react stoichiometrically (non-catalytically) with alcohols to form aldehyde products. The reaction is selective for primary hydroxyl groups and supports the proposal presented above that a free radical–coupled copper complex is involved in the Semmelhack reaction.

In order to examine the chemistry and spectroscopy of the radical-forming tyrosine–cysteine crosslink, we prepared a thioether-substituted phenol model accurate to the first alkyl groups of the side chain. This compound has provided insight into the interpretation of EPR and ENDOR spectra for the protein radical, as described earlier, and is presently furnishing valuable information on the coordination chemistry of this novel ligand design. A cupric complex of methylthiocresol has been synthesized as a starting point of a structural model for active site interactions. The crystal structure of [Cu(PMDT)(2-SCH$_3$-*p*-cresol)](ClO$_4$) (PMDT = pentamethyl diethylenetriamine) shown in Fig. 18 reveals that direct coordination of S to the metal ion occurs in this complex, resulting in a twist of the alkyl side chain out of the plane of the phenolic ring. This demonstrates that metal interactions with both oxygen and sulfur atoms may occur, although the short Cu-O bond distance implies relatively strong phenolate interactions in the complex.

The evidence from a combination of physical and chemical characterization of the active site of galactose oxidase demonstrated that a free radical formed from a protein group is essential for catalysis. The

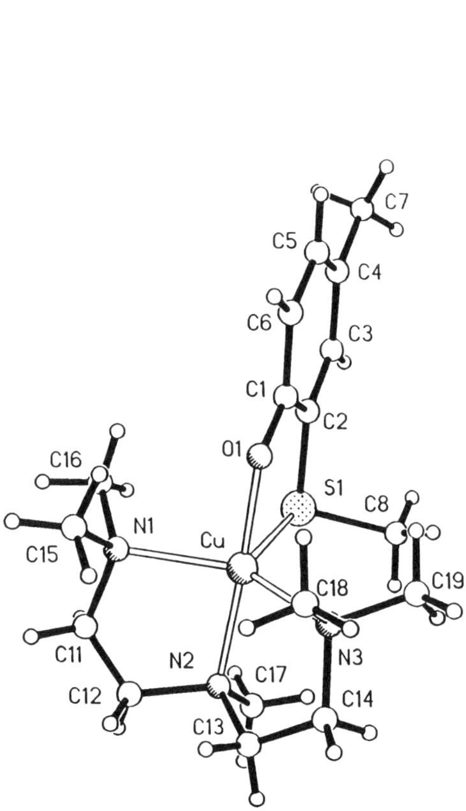

FIG. 18. Crystallographic structure for [Cu(PMDT)(2-SCH_3-*p*-cresol)](ClO_4) model.

radical forms part of a free radical–coupled Cu biological complex, a two-electron redox unit that is active in alcohol oxidation chemistry. Many questions remain relating to the role of the free radical in the catalytic mechanism. What advantage is provided by the free radical over other two-electron reactive groups, such as a binuclear metal complex, or the nicotinamide cofactor known to participate in alcohol oxidation chemistry in alcohol dehydrogenase [80]?

In a speculative vein, a detailed mechanism for the substrate oxidation steps indicated in the earlier mechanistic scheme is outlined in Fig. 19. Coordination of the substrate hydroxyl group to the metal ion will increase the acidity of the hydroxyl proton, and proton abstraction by a basic group in the protein may occur. As described above, the unmodified tyrosinate ligand appears to be involved in proton-coupled anion binding and likely serves as the endogenous base in this reaction. Within the alcoholate complex, hydrogen atom transfer from the meth-

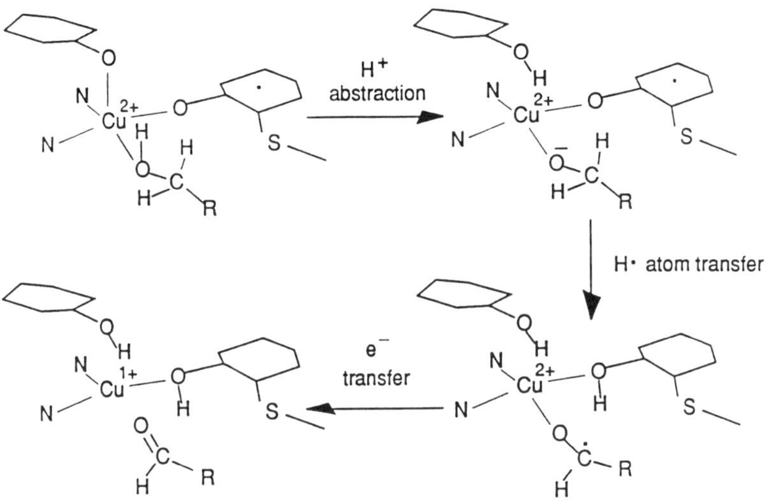

FIG. 19. Proposed free radical mechanism for alcohol oxidation by galactose oxidase.

ylene position in the substrate to the coordinated phenoxyl oxygen of the radical site, generating a ketyl radical as shown. The odd-alternant spin distribution experimentally determined for the radical site in galactose oxidase in ENDOR studies implies significant unpaired electron density on oxygen, facilitating this radical chemistry. A second electron transfer reducing the cupric center completes the formation of aldehyde product, which is subsequently released. The advantage of a radical structure over a binuclear metal two-electron oxidant may relate to relatively favorable energetics expected for a hydrogen atom transfer to an organic free radical in redox chemistry compared to transfer of $H^{\cdot}$ from alcohol to an inorganic acceptor. (Similar considerations may apply in the mechanism of amine oxidases for which free radical species also appear to be involved in catalysis [81]). Free radical chemistry is expected to provide access to essential new reaction pathways in catalysis beyond the relatively well-characterized closed shell reactions. Clearly these ideas require further experimental examination before their validity can be judged. It remains that in galactose oxidase free radical–metal interactions have evolved as a novel approach to efficient alcohol oxidation catalysis.

ACKNOWLEDGMENT

This research was supported by NSF DMB 8717058 and NIH GM 46749.

ABBREVIATIONS AND DEFINITIONS

Bi-Bi	bireactant or biproduct
CD	circular dichroism
EDTA	ethylenediamine-N,N,N′,N′-tetraacetate
ENDOR	electron nuclear double resonance
EPR	electron paramagnetic resonance
IA galactose oxidase	inactive galactose oxidase
IR	infrared
NHE	normal hydrogen electrode
PMDT	pentamethyldiethylenetriamine
SQUID	superconducting quantum interference device
TEMPO	2,2,6,6-tetramethyl-1-piperidinyloxy
XANES	X-ray absorption near-edge structure
XAS	X-ray absorption spectroscopy

REFERENCES

1. G. Avigad, C. Asensio, D. Amaral, and B. L. Horecker, *Biochem. Biophys. Res. Commun.*, *4*, 474–477 (1961).
2. G. Avigad, D. Amaral, C. Asensio, and B. L. Horecker, *J. Biol. Chem.*, *237*, 2736–2743 (1962).
3. F. M. Kelleher and V. P. Bhavanandan, *J. Biol. Chem.*, *261*, 11045–11048 (1986).
4. A. M. Klibanov, B. N. Alberti, and M. A. Marletta, *Biochem. Biphys. Res. Commun.*, *108*, 804–808 (1982).
5. T. O. Slykehouse and J. A. Fee, *J. Biol. Chem.*, *251*, 5472–5477 (1976).
6. B. B. Keele, J. M. McCord, and I. Fridovich, *J. Biol. Chem.*, *245*, 6176–6181 (1970).

7. J. S. Valentine and M. W. Pantoliano, in *Copper Proteins* (T. G. Spiro, ed.), Wiley-Interscience, New York, 1986, pp. 292–358.
8. L. C. Sieker, R. E. Stenkamp, and L. H. Jensen, in *The Biological Chemistry of Iron* (H. B. Dunford and D. Dolphin, ed.), D. Reidel, Boston, 1982, pp. 161–175.
9. A. Volbeda and W. Hol, *J. Mol. Biol*, *209*, 249–279 (1989).
10. Y. Kono and I. Fridovich, *J. Biol. Chem.*, *258*, 6015–6019 (1983).
11. M. Wikström, K. Krab, and M. Saraste, *Cytochrome Oxidase*, Academic Press, New York, 1981.
12. G. T. Babcock, in *New Comprehensive Biochemistry: Photosynthesis* (J. Amesz, ed.), Elsevier, New York, 1987.
13. D. Amaral, L. Bernstein, D. Morse, and B. L. Horecker, *J. Biol. Chem.*, *238*, 2281–2284 (1963).
14. M. M. Whittaker and J. W. Whittaker, *J. Biol. Chem.*, *263*, 6074–6080 (1988).
15. M. L. Randolf, in *Free Radicals in Biological Systems* (M. S. Blois, Jr., H. W. Brown, R. M. Lemmon, R. O. Linblom, and M. Weissbluth, eds.), Academic Press, New York, 1961, pp. 249–275.
16. B.-M. Sjöberg and A. Gräslund, *Adv. Inorg. Biochem.*, *5*, 87–110 (1983).
17. J. Stubbe, *Ann. Rev. Biochem.*, *58*, 257–285 (1989).
18. J. Stubbe, *Biochemistry*, *27*, 3893–3900 (1988).
19. J. A. D. Cooper, W. Smith, M. Bacila, and H. Medina, *J. Biol. Chem.*, *234*, 445–448 (1959).
20. D. Amaral, F. Kelly-Falcoz, and B. L. Horecker, *Meth. Enzymol.*, *9*, 87–92 (1966).
21. M. K. Nobles and C. Madhosingh, *Biochem. Biophys. Res. Commun.*, *12*, 146–147 (1963).
22. Sh. Bauer, G. Blauer, and G. Avigad, *Isr. J. Chem.*, *5*, 126 (1967).
23. N. Ito, S. E. V. Phillips, C. Stevens, Z. B. Ogel, M. J. McPherson, J. N. Keen, K. D. S. Yadev, and P. F. Knowles, *Nature*, *350*, 87–90 (1991).
24. A. R. Shatzman and D. J. Kosman, *J. Bacteriol.*, *130*, 455–463 (1977).
25. P. S. Tressel and D. J. Kosman, *Meth. Enzymol.*, *89*, 163–171 (1982).
26. D. J. Kosman, M. J. Ettinger, R. E. Weiner, and E. J. Massaro, *Arch. Biochem. Biophys.*, *165*, 456–467 (1974).

27. P. Tressel and D. J. Kosman, *Anal. Biochem.*, *105*, 150–153 (1980).
28. M. W. C. Hatton and E. Regoeczi, *Meth. Enzymol.*, *89*, 172–176 (1982).
29. R. D. Bereman and D. J. Kosman, *J. Am. Chem. Soc.*, *99*, 7322–7325 (1977).
30. R. D. Bereman, M. J. Ettinger, D. J. Kosman, and R. J. Kurland, in *Bioinorganic Chemistry*, Vol. 2 (K. Raymond, ed.), American Chemical Society, Washington, D.C., 1977, pp. 263–280.
31. D. J. Kosman, R. D. Bereman, M. J. Ettinger, and R. S. Giordiano, *Biochem. Biophys. Res. Commun.*, *54*, 856–861 (1973).
32. B. Marwedel, L. Kwiatkowski, D. Melnyk, P. Tressel, R. D. Bereman, R. J. Kurland, M. J. Ettinger, and D. J. Kosman, in *Oxidases and Related Redox Systems* (T. E. King, ed.), Pergamon Press, Ltd., Oxford, 1982, pp. 225–242.
33. W. E. Blumberg, B. L. Horecker, F. Kelly-Falcoz, and J. Peisach, *Biochim. Biophys. Res. Acta*, *96*, 336–338 (1965).
34. M. J. Ettinger, *Biochemistry*, *13*, 1242–1246 (1974).
35. M. J. Ettinger and D. J. Kosman, *Biochemistry*, *13*, 1247–1251 (1974).
36. D. J. Kosman, in *Copper Proteins and Copper Enzymes*, Vol. 2 (R. Lontie, ed.), CRC Press, Boca Raton, 1985, pp. 1–26.
37. M. J. Ettinger and D. J. Kosman, in *Copper Proteins* (T. G. Spiro, ed.), Wiley-Interscience, New York, 1982, pp. 219–261.
38. G. A. Hamilton, P. K. Adolf, J. de Jersey, G. C. DuBois, G. R. Dyrkacz, and R. D. Libby, *J. Am. Chem. Soc.*, *100*, 1899–1912 (1978).
39. G. R. Dyrkacz, R. D. Libby, and G. A. Hamilton, *J. Am. Chem. Soc.*, *98*, 626–628 (1976).
40. F. P. Bossu, K. L. Chellappa, and D. W. Margerum *J. Am. Chem. Soc.*, *99*, 2195–2203 (1977).
41. K. Clark, J. E. Penner-Hahn, M. M. Whittaker, and J. W. Whittaker, *J. Am. Chem. Soc.*, *112*, 6433–6434 (1990).
42. G. A. Hamilton, in *Copper Proteins* (T. G. Spiro, ed.), Wiley-Interscience, New York, 1982, pp. 193–218.
43. J. M. Johnson, H. B. Halsall, and W. R. Heineman, *Biochemistry*, *24*, 1579–1585 (1985).
44. L. D. Kwiatkowski, M. Adelman, R. Pennely, and D. J. Kosman, *J. Inorg. Biochem.*, *14*, 209–222 (1981).

45. A. Maradufu, G. M. Cree, and A. S. Perlin, *Can. J. Chem.*, *49*, 3429–3437 (1971).
46. J. L. Hughey, IV, T. G. Fawcett, S. M. Rudich, R. A. Lalancette, J. A. Potenza, and H. J. Schugar, *J. Am. Chem. Soc.*, *101*, 2617–2623 (1979).
47. L. Morpurgo, C. Giovagnoli, and G. Rotilio, *Biochim. Biophys. Acta*, *322*, 204–210 (1973).
48. Y. Agnus, R. Louis, and R. Weiss, *J. Am. Chem Soc.*, *101*, 3381–3384 (1979).
49. J. A. Fee and B. P. Gabor, *J. Biol. Chem.*, *247*, 60–65 (1972).
50. D. M. Dooley and K. C. Golnick, *J. Biol. Chem.*, *258*, 4245–4248 (1983).
51. W. E. Blumberg, J. Peisach, D. J. Kosman, and H. S. Mason, in *Oxidases and Related Redox Systems* (T. E. King, ed.), Pergamon Press, Oxford, 1982, pp. 207–224.
52. M. Sendova, J. Peterson, E. P. Day, M. M. Whittaker, and J. W. Whittaker, unpublished results.
53. S. S. Eaton and G. R. Eaton, *Coord. Chem. Rev.*, *26*, 207–262 (1978).
54. T. G. Spiro, in *Chemical and Biochemical Applications of Lasers*, Vol. 1 (C. B. Moore, ed.), Academic Press, New York, 1974, pp. 29–70.
55. M. M. Whittaker, V. L. DeVito, S. A. Asher, and J. W. Whittaker, *J. Biol. Chem.*, *264*, 7104–7106 (1989).
56. L. Que, Jr., R. H. Heistand, II, R. Mayer, and A. L. Roe, *Biochemistry*, *19*, 2588–2593 (1980).
57. Y. Tomimatsu, S. Kint, and J. R. Scherer, *Biochemistry*, *15*, 4918–4924 (1976).
58. G. N. R. Tripathi and R. H. Schuler, *J. Phys. Chem.*, *92*, 5129–5133 (1988).
59. S. L. Anderson, L. Goodman, K. Krough-Jespersen, A. G. Ozkabak, R. N. Zare, and C.-F. Zheng, *J. Chem. Phys.*, *82*, 5329–5339 (1985).
60. D. Eads, T. G. Spiro, M. M. Whittaker, and J. W. Whittaker, unpublished results.
61. M. M. Whittaker and J. W. Whittaker, *J. Biol. Chem.*, *265*, 9610–9613 (1990).
62. C. J. Diapp, D. Kubicki, and D. Amaral, *Braz. J. Med. Biol. Res.*, *16*, 297–303 (1983).

63. G. M. Kishore and D. M. Shah, *Ann. Rev. Biochem.*, *57*, 627–663 (1988).

64. R. A. van der Meer, J. A. Jongejan, and J. A. Duine, *J. Biol. Chem.*, *264*, 7792–7794 (1989).

65. B.-M. Sjöberg, P. Reichard, A. Gräslund, and A. Ehrenberg, *J. Biol. Chem.*, *253*, 6863–6865 (1978).

66. R. Karthein, R. Dietz, W. Nastainczyk, and H. H. Ruf, *Eur. J. Biochem.*, *171*, 313–320 (1988).

67. B. A. Barry and G. T. Babcock, *Proc. Natl. Acad. Sci. USA*, *84*, 7099–7103 (1987).

68. J. H. Waite and C. V. Benedict, *Meth. Enzymol.*, *107*, 397–413 (1984).

69. A. G. Murzin, *Proteins*, *14*, 191–201 (1992).

70. F. M. D. Vellieux, F. Huitema, H. Groendijk, K. H. Kalk, J. Frank, Jzn, J. A. Jongejan, J. A. Duine, K. Petratos, J. Drenth, and W. G. J. Hol, *EMBO J.*, *8*, 2171–2178 (1989).

71. G. T. Babcock, M. K. El-Deeb, P. O. Sandusky, M. M. Whittaker, J. W. Whittaker, *J. Am. Chem. Soc.*, *114*, 3727–3734 (1992).

72. K. Aisaka and O. Terada, *Agr. Biol. Chem.*, *46*, 1191–1197 (1982).

73. O. V. Koroleva, M. L. Rabinovich, T. T. Buglova, and A. I. Yaropolov, *Prik. Biokhim. Mikrobiol.*, *19*, 632–637 (1983).

74. P. J. Kersten and T. K. Kirk, *J. Bacteriol.*, *169*, 2195–2201 (1987).

75. L. Cleveland, R. E. Coffman, P. Coon, and L. Davis, *Biochemistry*, *14*, 1108–1115 (1975).

76. M. F. Semmelhack, C. R. Schmid, D. A. Cortés, and C. S.Chou, *J. Am. Chem. Soc.*, *106*, 3374–3376 (1984).

77. Y. Y. Lim and R. S. Drago, *Inorg. Chem.*, *11*, 1334–1338 (1972).

78. M. H. Dickman and R. J. Doedens, *Inorg. Chem.*, *20*, 2677–2681 (1981).

79. J. Laugier, J.-M. Latour, A. Caneschi, and P. Rey, *Inorg. Chem.*, *30*, 4474–4477 (1991).

80. K. Dalziel, in *The Enzymes*, Vol. 11, 3rd ed. (P. Boyer, ed.), Academic Press, New York, 1975, p. 1.

81. J. Z. Pedersen, S. El-Sherbini, A. Finazzi-Agro, and G. Rotilio, *Biochemistry*, *31*, 8–12 (1992).

12
Amine Oxidases

Peter F. Knowles[1] and David M. Dooley[2]

[1]Department of Biochemistry and Molecular Biology, University of Leeds, Leeds LS2 9JT, England

[2]Department of Chemistry, Amherst College, Amherst, Massachusetts 01002, USA

This review is dedicated to the memory of Hermann Blaschko, who greatly stimulated studies on the enzymology and biology of amine oxidases.

1. INTRODUCTION

1.1. Discovery and Classification of Amine Oxidases

Amine oxidases catalyze the oxidative deamination of mono-, di-, and polyamines according to the following general reactions which involve attack at either primary amine sites (1) or secondary amine sites (2):

Mono- and diamines

$$RCH_2NH_2 + O_2 + H_2O \rightarrow RCHO + H_2O_2 + NH_3 \quad (1)$$

Polyamines

$$RCH_2NHCH_2R' + O_2 + H_2O \rightarrow RCH_2NH_2 + R'CHO + H_2O_2 \quad (2)$$

There is a broad division of amine oxidases into two classes; one class contains flavin and has the Enzyme Commission classification EC 1.4.3.4 while the "nonflavin" class is coded EC 1.4.3.6. Both classes of amine oxidase are widely distributed in nature though the flavin class is associated particularly with peroxisomes and mitochondria [1,2]. Current evidence indicates that only the flavin class of amine oxidase can catalyze specific attack at both primary and secondary amine sites [see reactions (1) and (2) above]. There is some confusion in the literature on this point since nonflavin amine oxidases can utilize polyamines, such as spermidine $H_2N\text{-}(CH_2)_3NH(CH_2)_4NH_2$, as substrates though attack is always at the primary amine site, proceeding according to reaction (1).

This chapter is concerned solely with nonflavin amine oxidases.

Subclassification into mono- and diamine oxidases based on substrate preferences and inhibitor sensitivity was in common usage until recently [3]. However, they are all grouped by the Enzyme Commission as "amine oxidase (copper-containing) EC 1.4.3.6" with the systematic name amine:oxygen oxidoreductase (deaminating) (copper-containing).

Discovery of amine oxidases is credited to Zeller [4] and the period 1940–1970 saw important advances in understanding the biochemistry and pharmacology of these enzymes [5–7]. The quality of the biochemical work carried out during this period was excellent. Examples to illustrate this point are the purification and crystallization of pig plasma amine oxidase by Buffoni and Blaschko [8], the first demonstration that this class of enzyme contained copper [9], and the elegant use of physical techniques, particularly electron paramagnetic resonance (EPR) spectroscopy, to probe active site structures [10].

During the 1970s, attempts were made to establish whether there was a second cofactor in addition to copper and, if so, to identify it. Most research workers in the field were not satisfied with evidence presented which appeared to favor pyridoxal phosphate as the cofactor. The literature covering this confusing period has been reviewed critically [11]. A breakthrough, which has greatly stimulated the field of amine oxidases, came from the reports in 1984 that the cofactor was pyrroloquinoline quinone [12,13]. Later work, principally by Klinman and collaborators, established that the organic cofactor in many, if not all, amine oxidases is a quinone derived from trihydroxyphenylalanine [14,15]: this work will be discussed in detail later. Thus nonflavin amine oxidases are now more positively classified as "copper- and quinone-containing" amine oxidases.

1.2. Distribution of Amine Oxidases

Amine oxidases are widespread in mammals, plants, and both prokaryotic and eukaryotic microorganisms. They can be intracellular or extracellular in their location. An excellent recent review on amine oxidases [16] includes a comprehensive list of the locations recorded for amine oxidases. A further addition to this list is that copper-containing amine oxidases have now been found in gram-negative bacteria [17], which indicates that the distribution of these enzymes is ubiquitous in nature.

There are two general comments which emerge from consideration of the distribution of amine oxidases. First, differentiated tissue may contain more than one "type" of amine oxidase, and second, each type of amine oxidase may consist of a number of isozymes. Examples to illustrate these points will now be given.

1.2.1. Diversity of Amine Oxidases

(i) Intestine. A microsomal fraction from human small intestine has both flavin-containing and copper-containing amine oxidases [18]. The copper-containing form is membrane-associated and requires 1% Triton X-100 detergent for its extraction. This indicates that it has some kind of membrane anchor [19].

(ii) Aorta. The lysyl oxidase in bovine aorta is a monomer, molecular weight 32 kD, containing one copper [20], and is distinct in structure and substrate specificity from the "semicarbazide-sensitive" amino oxidase found in this tissue [21], which is a dimer, molecular weight 180 kD, and, like the intestinal enzyme [18], is membrane-associated. The elegant work of Palcic and coworkers [21,22] further establishes clear stereochemical patterns in the oxidative deamination reaction catalyzed by different amine oxidases, which distinguishes the "semicarbazide-sensitive" amine oxidase from other copper-containing amine oxidases. These stereochemical patterns are a strong indicator that amine oxidases have specific cellular roles.

(iii) Placenta. Lysyl oxidase, a monomer of molecular weight 32 kD, has been isolated from human placenta [23] and is quite distinct from the amine oxidase, a monomer with molecular weight 75 kD, found in this tissue [24].

(iv) Eukaryotic microorganisms. Penicillin species produce an extracellular flavin-containing polyamine oxidase [25] which is a dimeric enzyme, molecular weight 128 kD, an intracellular flavin-containing polyamine oxidase [26], which is a dimeric enzyme, molecular weight 160 kD, and an intracellular copper-containing amine oxidase [27] which is also a dimeric enzyme, molecular weight 160 kD. The factors which control intracellular or extracellular location, or cofactor type in penicillin species, are intriguing.

Similar observations have been made for yeasts. Growth of *Candida boidinii* on the polyamine spermidine leads to production of a peroxisomal amine oxidase whose cofactor has not been fully identified

[28], whereas growth on primary amines leads to two distinct amine oxidases, both of which appear to contain copper [29]. Two amine oxidases have also been isolated from *Pichia pastoris* [30,31], one of which has been shown [31] to have lysyl oxidase activity. Note, however, that the *P. pastoris* "lysyl oxidase", molecular weight 106 kD, is quite different from mammalian lysyl oxidase, molecular weight 32 kD [20].

(v) Plants. This topic has been reviewed by Smith [32]. Amine oxidases, which have a substrate preference for diamines such as putrescine, cadaverine, and spermidine, but also oxidize aliphatic and aromatic monoamines, have been reported for many plants but are particularly active in Leguminosae. These amine oxidases have been studied extensively and definitely contain copper and quinone cofactors [33]. The broad substrate specificity, which includes polyamines, is characteristic of this type of amine oxidase and is distinct from the specificity of the amine oxidases found in Gramineae (cereals), which act specifically on polyamines such as spermidine and spermine. It is not fully established whether these latter "polyamine" oxidases contain copper/quinone or flavin cofactors [34,35]. A useful review comparing the properties of polyamine oxidases from bacteria, fungi, plants, and mammals is available [36].

1.2.2. Existence of Amine Oxidase Isozymes

It has been reported [37,38] that lysyl oxidase from bovine aorta occurs as isozymes. The amine oxidase from pig plasma has also been partially resolved into discrete molecular forms by chromatography and by isoelectric focusing: these forms differ from each other only in their carbohydrate content [39]. No effects of this heterogeneity on catalytic properties have been detected. The functional significance of amine oxidase isozymes is obscure.

2. STRUCTURAL PROPERTIES OF AMINE OXIDASES

2.1. Gene Sequence and Protein Structure

Extensive characterization of several amine oxidases indicated that these enzymes were rather large proteins with M_r 170–190 kD, generally composed of dimers of two indistinguishable and presumably identical subunits. Current evidence suggests that the amine oxidase from *E. coli* K 12 may have a single subunit (R. A. Cooper, personal communication). Eukaryotic amine oxidases were also shown to be glycoproteins with carbohydrate contents estimated at 3–10%. These and many additional aspects of the protein chemistry and structure of amine oxidases have been thoroughly reviewed [11,16,40,41]. Our understanding of amine oxidase protein structure has been greatly increased by the recent success in cloning and sequencing amine oxidase genes from the yeast *Hansenula polymorpha* [42], the bacterium *Arthrobacter* P1 [43], lentil seedlings [44], and lysyl oxidase from rat aorta [45]. An overall alignment of the first three sequences is shown in Fig. 1A; the alignment for regions that may be involved in the formation of the active site is presented in Fig. 1B. It should be noted that the alignment in Fig. 1 differs from that in Ref. 44. A comparison of the three sequences in Fig. 1 with the lysyl oxidase sequence indicates that there is very little homology between lysyl oxidase and the other amine oxidases. Surprisingly, lysyl oxidase displays significant homology to the rrg protein, which may have a role in inhibiting tumorigenesis [46,47]. As is evident from Fig. 1, the other three amine oxidases are highly homologous. Each amine oxidase gene codes for a polypeptide that is less than half of the M_r of the isolated protein. Unless the subunits are modified differently following their synthesis, the unambiguous conclusion is that these amine oxidases are indeed dimers of two identical subunits. One surprise was that the M_r values inferred from the translated gene sequence are considerably smaller than the values previously estimated from sodium dodecyl sulfate (SDS) gel electrophoresis [11,16,40] or analytical ultracentrifugation [11], suggesting that amine oxidases are nonspherical or that the carbohydrate content has been underestimated. Pea seedling amine oxidase was recently crystallized [48] and its M_r determined by density measurements to be 131 $\pm$ 13 kD; this is consistent with the M_r values inferred from the sequences as well as estimates of the carbohydrate content. The sequence data also revealed that there are significant differences in the size of amine oxidase polypeptides (Fig. 1).

The alignment in Fig. 1 indicates that there are four conserved histidine residues, of which three are probably copper ligands (see below). The fourth histidine might be the required active site base [44–51], but there are also several conserved aspartates and glutamates that could serve as the base [50,51]. Notably, there are two conserved cysteine residues among the three amine oxidases in Fig. 1. Earlier work has pointed to two possible roles for specific cysteines: (1) conformational stabilization [52–54], perhaps via an intersubunit disulfide bridge, and (2) an active site base [55,56]. Abeles and coworkers demonstrated that 1 Cys per subunit became accessible to sulfhydryl reagents following reduction of bovine plasma amine oxidase by substrate and that this Cys is essential for activity [55]; similar results have been reported for other amine oxidases [5,56]. Additional work is necessary to conclusively identify the critical active site base and to establish the structural and functional significance of the unique cysteine, if it is not an active site residue.

Several hydrophobic and aromatic residues are conserved. One of the these, Tyr412 in Fig. 1, occurs in the active site consensus sequence Asn-Tyr-Asp/Glu that defines the trihydroxyphenylalanine (topa) residue derived from tyrosine by posttranslational modification (see Sec. 2.2.2) [57,58]. The tryptophan located several residues away also appears to be modified in several amine oxidases [58], but it is not yet known if this is a property of the native proteins or an artifact of the isolation procedure for the derivatized active site peptides. Additional sequence information and comparisons are presented in Sec. 2.2.2. It is anticipated that a great deal of new information will be forthcoming as additional amine oxidases are sequenced. Successfully solving the crystal structure of pea seedling amine oxidase would provide answers to many important questions concerning the structure of these enzymes [48].

2.2. Active Site Structure

Two longstanding questions in the study of amine oxidases are whether there are variations in active site features or catalytic mechanisms among these enzymes and whether any real differences correlate with known variations in function or specificity. At times the apparent differences among amine oxidases have caused considerable debate and uncertainty. For example, at one time or another, it has not been clear that the identity of the organic cofactor or the cofactor–protein stoichiometry were the same for all copper-containing amine oxidases [11,16, 40]. The role of copper and the structure of the copper sites in

```
                                                                                                   80
APAO  ..........  .......LVG  VSHPLDPLSR  VEIARAVAIL  KEG.PAAAES  FRFISVELRE  PSKDDL....  .RAGVAVARE
 YAO  MERLRQIASQ  ATAASAAPAR  PAHPLDPLST  AEIKAATNTV  K.S.YFAGKK  ISFNTVTLRE  PARKAYIQWK  EQGGPLPPRL
LSAO  ..........  .....FTPLH  TQHPLDPITK  EEFLAVQTIV  QNKYPISNNK  LAFHYIGVDD  PEKDLVLKYE  TSPTLISIPR
                          •    H★★★★••   ★•  •  •••  •      ••••  • ★  ••• •  ★••• •        ••    •
                                                                                                  160
APAO  ADAVLVDRAQ  ARSFEAVVDL  EAGTVDSWKL  LAENIQPPFM  LDEFAECEDA  CRKDPEVIAA  LAKRGLTNLD  L..VCFEPWS
 YAO  AYYVILEAGK  PGVKEGLVDL  ASLSVIETRA  L.ETVQPILT  VEDLCSTEEV  IRNDPAVIEQ  CVLSGIPANE  MHKVYCDPWT
LSAO  KIFV.VAIIN  SQTHEILIDL  TIKSIVSDNI  HNGYGFPVLS  AAEQFLAIDL  PLKYPPFIAS  VNKRGL...N  ISEIVCSSFT
         ★ •   •  •   ★ ••★★  •  •• • •      •   ★ •   • •     •• • •  ★  ★•   •  ★•   • •  •  ••••
                                                                                                  240
APAO  VGYFGEDNEG  RRLMRALVFV  RDEADDSPYA  HPIENFIVFY  DLNAGKVVRL  EDDQAIPVPS  ..ARGNYLPK  Y....VGEAR
 YAO  IGYDERWGTG  KRLQQALVYY  RSDEDDSQYS  HPLD.FCPIV  DTEEKKVIFI  DIPNRRRKVS  KHKHANFYPK  HMIEKVGAMR
LSAO  MGWFGEEKNS  RTVR..VDCF  MKESTVNIYV  RPITGITIVA  DLDLMKIVEY  HDRDTEAVPT  AENTEYQVSK  ....QSPPFG
      •★• •  •••  • •    •    •••• • ★• •★•• •       ★ •  ★••        •      •     ••   •★       •••
                                                                                                  320
APAO  TDLKPLNITQ  PEGASFTVTG  NHVTWADWSF  RVGFTPREGL  VLHQLKFKDQ  G..VDRPVIN  RASLSEMVVP  YGDTAPVQAK
 YAO  PEAPPINVTQ  PEGVSFKMTG  NVMEWSNFKF  HIGFNYREGI  VLSDVSYNDH  G..NVRPIFH  RISLSEMIVP  YGSPEFPHQR
LSAO  PKQHSLTSHQ  PQGPGFQING  TSVSWANWKF  HIGFDVRAGI  VISLASIYDL  EKHKSRRVLY  KGYISELFVP  YQDPTEEFYF
      ••   •••   ★ ★•★ •★ ••★  • ••★••••★  ••★★• ★•★•  ★•    •   ★  •     ★•••    •  •★★• ★★ ★ •••
                                                                                                  400
APAO  KNAFDSGEYN  IGNMANSLTL  GCDCLGEIKY  FDGHSVDSHG  NPWTIENAIC  MHEEDDSILW  KHFDFRE..G  TAETRRSR.K
 YAO  KHALDIGEYG  AGYMTNPLSL  GCDCKGVIHY  LDAHFSDRAG  DPITVKNAVC  IHEEDDGLLF  KHSDFRDNFA  TSLVTRAT.K
LSAO  KTFFDSGEFG  FGLSTVSLIP  NRDCPPHAQF  IDTYIHSADG  TPIFLENAIC  VFEQYGNIMW  RHTETGIPNE  SIEESRTEVD
      ★  •★ ★★••  ★  • •★   • ★C    ••  •★••  •  ★  •★  ••★★•C  • ★• •••••  •H •    • •     ★•  •
                                                                                                  480
APAO  LVISFIATVA  NYEYAFYWHL  FLDGSIEFLV  KATGIL...S  TAGQLPGEK.  NPYGQSLNND  GLYAPIHQHM  FNVRMDFELD
 YAO  LVVSQIFTAA  NYEYVLYWVF  MQDGAIRLDI  RLTGIL...N  TYILGDDEEA  GPWGTRVYPN  .VNAHNHQHL  FSLRIDPRID
LSAO  LAIRTVVTVG  NYDNVLDWEF  KTSGWMKPSI  ALSGILEIKG  TNIKHKDEIK  EEIHGKLVSA  NSIGIYHDHF  YIYYLDFDID
      ★••• • ★••  ★Y★ •• ★ •    •★ •   •    •★★★  • ★    •★     •    •• •     • H•H• •   •★  •★
                                                                                                  560
APAO  GVKNAVYEVD  .......MEY  PEHNPTGTAF  MAVDRLLETE  QKAIRKTNEA  KHRFWKIANH  ESKN.LVNEP  VAYRLIPTNG
 YAO  GDGNSAAACD  AKSSPYPLGS  PE.NMYGNAF  YSEKTTFKTV  KDSLTNYESA  TGRSWDIFNP  NKVNPYSGKP  PSYKLVSTQC
LSAO  GTQNSFEKTS  LKTVRIVDEV  QEK....SYW  TTETQTAKTE  SDAKITIGLA  PAE.LVVVNP  NIKTAVGNE.  VGYRLIPAIP
      ★ ★•   •             •  •★    • •  •  •  •★  ••  • • ★      • • ★  •  • •••   •★•★•••
                                                                                                  640
APAO  IQLAARDDAY  VSKRAQFARN  NLWVTAYDRT  ERFAAGEYPN  Q.ATGADDGL  HIWTQKDR.N  IVDTDLVVWY  TFGMHHVVRL
 YAO  PPLLAKEGSL  VAKRAPWASH  SVNVVPYKDN  RLYPSGDHVP  QWSGDGVRGM  REWIGDGSEN  IDNTDILFFH  TFGITHFPAP
LSAO  AHPLLTEDDY  PQIRGAFTNY  NVWVTQ....  ........II  ELKNGLVDFM  LI........  ..........  ..........
       •   ••••    ★•••••  •• ★•                      •  ••    •
                                                                                    705
APAO  EDWPVMPRQN  IGFMLEPHGF  FNQNPTLNLP  TSTSTTQTGE  ADTCCHNGK.  ..........  .....
 YAO  EDFPLMPAEP  ITLMLRPRHF  FTENPGLDIQ  PSYAMT.TSE  AKRAVHKETK  DKTSRLAFEG  SCCGK
LSAO  ..........  ..........  ..........  ..........  ..........  ..........  .....
```

(A)

```
       1                                                   50
APAO   .......... .......LVG VSHPLDPLSR VEIARAVAIL KEG.PAAAES
 YAO   MERLRQIASQ ATAASAAPAR PAHPLDPLST AEIKAATNTV K.S.YFAGKK
LSAO   .......... .....FTPLH TQHPLDPITK EEFLAVQTIV QNKYPISNNK
                          •    H★★★★•• ★•  • ••• •      ••••

(alignment residues 51-320 not shown)

       321                                                370
APAO   KNAFDSGEYN IGNMANSLTL GCDCLGEIKY FDGHSVDSHG NPWTIENAIC
 YAO   KHALDIGEYG AGYMTNPLSL GCDCKGVIHY LDAHFSDRAG DPITVKNAVC
LSAO   KTFFDSGEFG FGLSTVSLIP NRDCPPHAQF IDTYIHSADG TPIFLENAIC
       ★  •★ ★★•• ★  • •★   • ★C    •• •★••  •  ★ •★  ••★★•C

       371                                                420
APAO   MHEEDDSILW KHFDFRE..G TAETRRSR.K LVISFIATVA NYEYAFYWHL
 YAO   IHEEDDGLLF KHSDFRDNFA TSLVTRAT.K LVVSQIFTAA NYEYVLYWVF
LSAO   VFEQYGNIMW RHTETGIPNE SIEESRTEVD LAIRTVVTVG NYDNVLDWEF
       • ★• ••••• •H •       • •     ★•  • ★••• • ★•• ★Y★ •• ★ •

       421                                                470
APAO   FLDGSIEFLV KATGIL...S TAGQLPGEK. NPYGQSLNND GLYAPIHQHM
 YAO   MQDGAIRLDI RLTGIL...N TYILGDDEEA GPWGTRVYPN .VNAHNHQHL
LSAO   KTSGWMKPSI ALSGILEIKG TNIKHKDEIK EEIHGKLVSA NSIGIYHDHF
         •★ •   •  •★★★   • ★      •★  •      •• •      • H•H•

(alignment residues 461-705 not shown)

KEY

★ perfect match     H conserved His          APAO, 639 residues
• well conserved    C conserved Cys           YAO, 692 residues
.. gaps             Y TOPA (modified Tyr)    LSAO, 569 residues
```

(B)

FIG. 1. Amino acid sequence alignment of amine oxidases from *Arthrobacter* P1 (APAO), *Hansenula polymorpha* (YAO), and lentil seedling (LSAO). (A) Complete alignment; (B) selected portions of alignment to show conserved Cys, His, and active site residues.

amine oxidases have also been persistent sources of debate and uncertainty. Particularly disparate results have been reported for mammalian amine oxidases purified from bovine or porcine plasma, despite the understandable expectation that these enzymes should be similar. Recent results have rationalized or corrected many of the conflicting claims in the literature regarding the active site structure and reactivity of amine oxidases and these results will be discussed here.

2.2.1. Copper Site Structure

Numerous spectroscopic studies led to the model for the Cu(II) sites in amine oxidases shown in Fig. 2. Although only limited data are available, the Cu(II) site in lysyl oxidase appears likely to have a similar structure [47,59]. Initial extended X-ray absorption fine structure (EXAFS) experiments established that the first coordination shell of Cu(II) in bovine plasma amine oxidase (BPAO) was composed of N,O scatterers [60]; the overall EXAFS was very similar to that of the model $[Cu(imid)_4]^{2+}$. Only a single first-shell scatterer (ligand) distance was required to fit the first-shell data of oxidized BPAO. A subsequent study of porcine plasma amine oxidase (PPAO) was largely consistent with this picture but indicated that the distances in the first coordination shell were split [61]. A heavy scatterer, originally attributed [61] to an axial sulfur, is now thought to be Cl^- derived from the buffer (P. F. Knowles, unpublished result). Thus the EXAFS data are generally consistent with 3–4

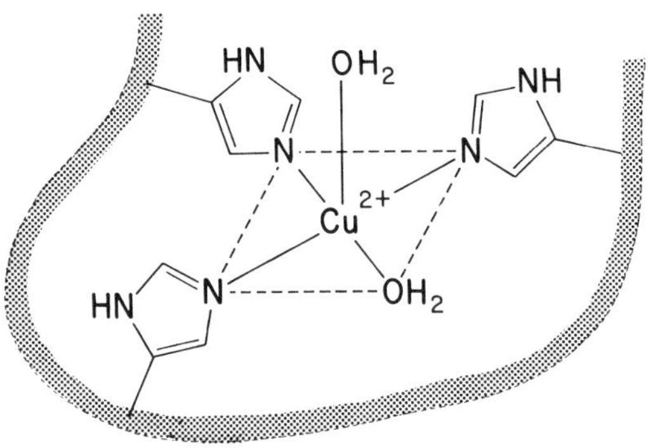

FIG. 2. Model for the Cu(II) site in amine oxidases.

(N,O) ligands and further suggest that the Cu(II) site structures are similar, except for shorter Cu(II)–(N,O) bonds in PPAO. The model in Fig. 2 has received strong support from EPR and electron nuclear double resonance (ENDOR) experiments on porcine plasma amine oxidase [62,63], and pulsed EPR studies (utilizing the electron spin-echo envelope modulation, ESEEM, technique) of the bovine plasma and porcine kidney amine oxidases [64]. First, two populations of magnetically distinct histidyl imidazoles were identified as Cu(II) ligands for both native proteins from the ESEEM experiments. Second, water was confirmed as a Cu(II) ligand by both ENDOR and ESEEM. N_3^- and CN^- were shown to substitute for an equatorial D_2O (H_2O) ligand, displacing it from the first coordination shell of amine oxidase Cu(II). However, another D_2O (H_2O) molecule, with ESEEM parameters that are very similar to those of axially coordinated D_2O (H_2O) in Cu(II) model complexes, probably remains bound to the Cu(II) in the anion complex. Anion binding abolishes the distinction in magnetic coupling between the two populations of imidazoles in both amine oxidases. In the ESEEM context, an axial water ligand is defined as one for which the interactions of the protons with the unpaired electron on the Cu(II) are weak, owing to the longer distance and smaller contact contribution to the hyperfine coupling, i.e., off the plane of the electronic wavefunction. Thus the ESEEM results indicate that amine oxidase Cu(II) has both equatorial and axial water ligands in addition to the coordinated histidine imidazoles. The ESEEM results are fully consistent with extensive spectroscopic and ligand substitution experiments conducted earlier and reviewed elsewhere [41].

In order to further define copper binding and the coordination structures of the Cu(II) sites, the paramagnetic effects of enzyme-bound Cu(II) on solvent water relaxation were investigated. The approach was to measure the nuclear magnetic relaxation dispersion (NMRD) profiles of amine oxidases as a function of temperature, substrate and inhibitor binding, and copper content [65]. Complete NMRD profiles were obtained for the resting (oxidized) forms of amine oxidases from bovine plasma, porcine plasma, porcine kidney, and *Arthrobacter* P1. These measurements provided new insights into the structure and dynamics of the Cu(II) sites in amine oxidases and indicated that amine oxidases examined may be divided into two classes on the basis of three factors: the paramagnetic contribution to the relaxivity $(T_{1p})^{-1}$; the temperature dependence of $(T_{1p})^{-1}$; and the effects of azide, a Cu(II) ligand, on $(T_{1p})^{-1}$. The key results are displayed in Fig. 3 and 4 and Table 1. Note that the relaxivities for all the amine oxidases are high but that the enzymes may be divided into two classes based on temperature dependence. Specifically, the magnitude and dispersion of $(T_{1p})^{-1}$ are consistent with the presence of at least one liganded water molecule in rapid exchange with bulk solvent in amine oxidases from porcine plasma, porcine kidney (PKAO), bovine plasma, and *Arthrobacter* P1 (APAO). As is evident in Fig. 4, the $(T_{1p})^{-1}$ values for amine oxidases display similar dispersions as that for Cu,Zn-superoxide dismutase (SOD) but have higher magnitudes. Since $(T_{1p})^{-1}$ of SOD arises from rapid exchange of an axial water, the NMRD profiles of amine oxidases suggest that the labile water molecule is also an axial ligand. The data in Figs. 3 and 4 indicated that an additional exchangeable water ligand is present in PPAO and APAO. This water exchanges rapidly at 25°C but relatively slowly at 5°C, and is probably displaced upon binding anionic inhibitors, e.g., azide (Table 1). Exogenous anionic inhibitors, such as azide,

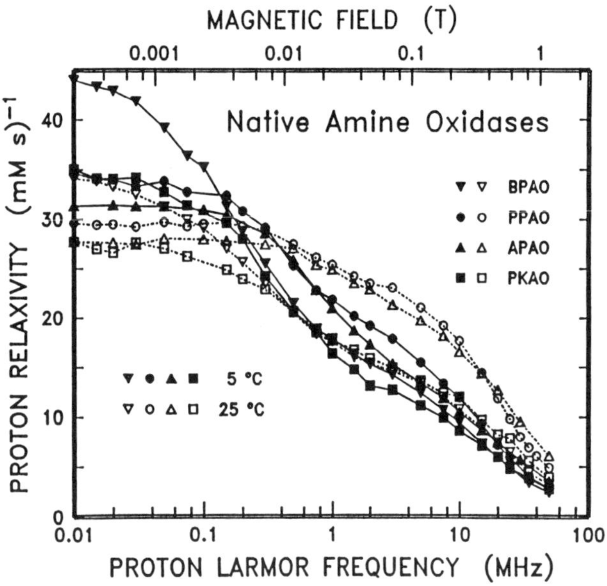

FIG. 3. NMRD profiles (total relaxivities, T_1^{-1}) of native amine oxidases at pH 7.0, plotted per mM protein.

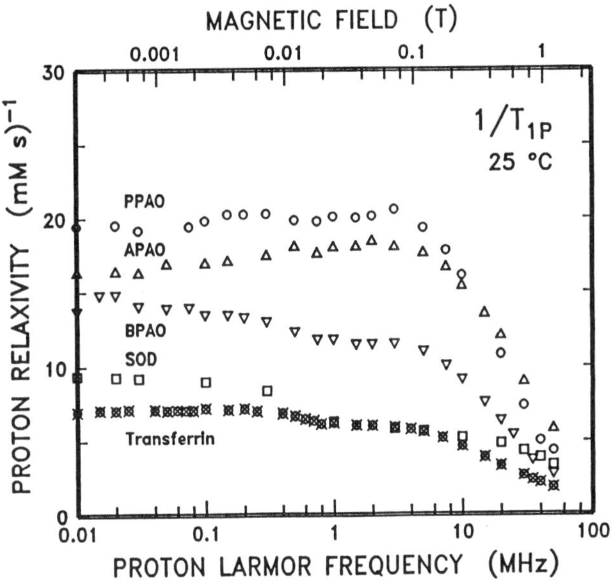

FIG. 4. Comparison of the paramagnetic contributions $(T_{1p})^{-1}$ of PPAO, APAO, and BPAO to that of Cu_2Zn_2SOD and Cu transferring at 25°C per mM per protein.

are known to bind equatorially to tetragonal Cu(II) ions in amine oxidases [9,25]; nonetheless, azide binding decreases only slightly $(T_{1p})^{-1}$ for BPAO and PKAO but significantly reduces $(T_{1p})^{-1}$ for PPAO. Collectively these results suggest that the equatorial water molecule is sufficiently labile to contribute to the NMRD profiles of the porcine plasma and *Arthrobacter* amine oxidases at 25°C. Therefore the data accord fully with the proposed structural model for the Cu(II) sites in amine oxidases (Fig. 2). Both equatorial and axial water molecules are ligands, but the dynamics of equatorial ligand substitution or exchange may vary. It should be noted that an axial histidine or other N,O ligand [to give a six-coordinate Cu(II) structure] would not be inconsistent with the published ESEEM, EXAFS, or NMRD results.

For both porcine plasma and bovine plasma amine oxidase it was found that the paramagnetic contribution to the relaxivity, $(T_{1p})^{-1}$, the specific activity, and the intensity of the principal absorption band are linear functions of the copper content; data for PPAO are shown in Fig.

TABLE 1

Effect of Azide on NMRD of Amine Oxidases

Enzyme	[Azide] (mM)	Relaxivity[a]
BPAO	0	13.3
	10	13.2
	50	13.3
	100	12.9
	200	12.7
	400	12.1
PKAO	0	11.6
	3	10.9
	10	10.8
	160	10.4
PPAO	0	16.2
	10	12.9
	20	12.6
	40	12.3
	80	11.9
	200	11.3

[a]Measured at 5 MHz, 5°C, where the paramagnetic contribution to the total relaxivity dominates.

5. Simulations of the Q-band EPR spectrum of PPAO consistently indicate that the two Cu(II) ions in the resting enzyme are inequivalent, in contrast to other amine oxidases examined [66]. One site is axial at 35 GHz but the other appears rhombic. When copper is added to the metal-depleted protein the Cu(II) ions are equally distributed between the two sites at all copper–protein stoichiometries. However, the two copper sites cannot be distinguished by NMRD measurements. The data may be reconciled in two ways: either copper binding to the apoprotein is highly cooperative (i.e., the samples are mixtures of apo- and fully reconstituted proteins), or the copper sites have equal affinity for Cu(II), which means that the populations of the two sites are always equal, in

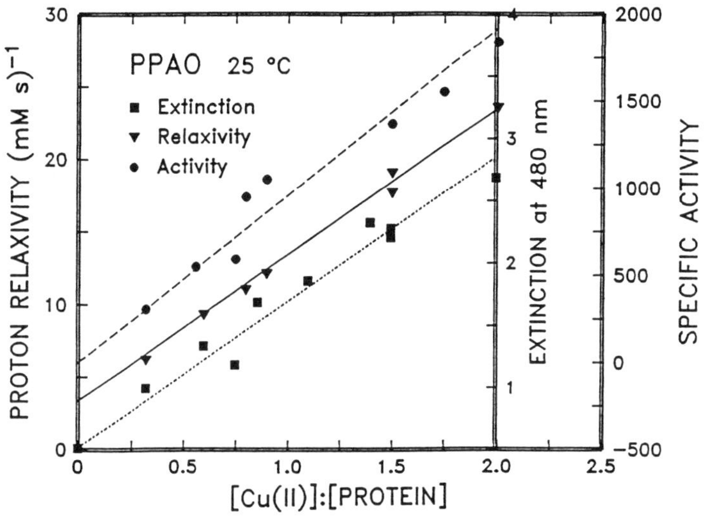

FIG. 5. Visible absorbance at 480 nm (■), total relaxivity at 2 MHz (▼), and specific activity (●) of PPAO as a function of Cu(II): protein ration. Note that the absorbance spectra and the relaxivities were measured at 25°C but the activities were measured at 37°C. Extinction at 480 nm is reported (mM^{-1} cm^{-1}).

both bovine and porcine plasma amine oxidase. These results are inconsistent with previous (and conflicting) claims that a one-Cu derivative of the bovine enzyme could be prepared that had either no, or full, activity [67–69]. Finally, we must note that there is little information on the coordination structure of the Cu(I) center, a situation that warrants attention given the potential importance of the Cu(I)-semiquinone state in turnover (see Sec. 3.3).

2.2.2. *Organic Cofactors*

One of the most difficult obstacles to developing a sound understanding of the structure and reactivity of amine oxidases has been the uncertainty surrounding the identity of the active site carbonyl group. Amine oxidases were originally identified as pyridoxal enzymes, despite considerable difficulties in reconciling some of the chemical, spectroscopic, and mechanistic data with the properties of pyridoxal. A previous review [11] may be consulted for a critical assessment of the situation through the early 1980s. The suggestion in 1984 that amine oxidases contained "covalently bound" pyrroloquinoline quinone (PQQ) provided a superior rationalization for all the available data and greatly stimulated additional research [12,13,15]. Ultimately, Klinman and coworkers succeeded in isolating a labeled active site peptide from BPAO in high yield and structurally characterized it via nuclear magnetic resonance (NMR) and mass spectroscopy [14]. Their results established that the cofactor in the resting (oxidized) enzyme was the quinone of trihydroxyphenylalanine (topa, or 6-hydroxydopa). The carbonyl reagents phenylhydrazine and *p*-nitrophenylhydrazine were used to label the quinone. Resonance Raman spectroscopy using these carbonyl reagents as a probe of the quinone cofactor in amine oxidases [70] has been developed. Following the demonstration by Klinman and coworkers that topaquinone was the organic cofactor in BPAO, two additional, and critical, points were subsequently established [71]. First, comparisons of the resonance Raman spectra of the intact, derivatized proteins and the isolated, labeled peptide established that the topa structure was not formed during the isolation and purification of the active site peptide. Second, the resonance Raman spectra of the phenylhydrazine and *p*-nitrophenylhydrazine derivatives of the BPAO peptide were identical to the spectra of the corresponding derivatives of topaquinone, and very different from the spectra of PQQ or pyridoxal (Fig. 6). Hence, resonance Raman spectroscopy provided very strong, independent support for the presence of topaquinone in amine oxidases. Examination of the amine oxidases from the yeast *Hansenula polymorpha* [42] and from *Arthrobacter* [43] establish that both of these enzymes contain topa and that a tyrosine codon corresponds to topaquinone at the active site. Therefore topa is produced in the active site via posttranslational modification. The resonance Raman method has now been extended to peptides isolated from several amine oxidases; the results shown in Fig. 7 establish that topaquinone is the cofactor in all the enzymes examined [57,58,70,72]. A rapid, simple, and clean spectrophotometric assay for topa (in intact enzymes) has also been developed that correlates with the resonance Raman spectroscopy approach [58]. The *p*-nitrophenylhydrazone derivatives of amine oxidases at pH 7.2 absorb at 457–463 nm and at 575–587 nm in 1–2 M KOH. Derivatives of active site peptides and the *p*-nitrophenylhydrazone of topaquinone hydantoin display practically identical shifts, as shown in Fig. 8. Thus the pH

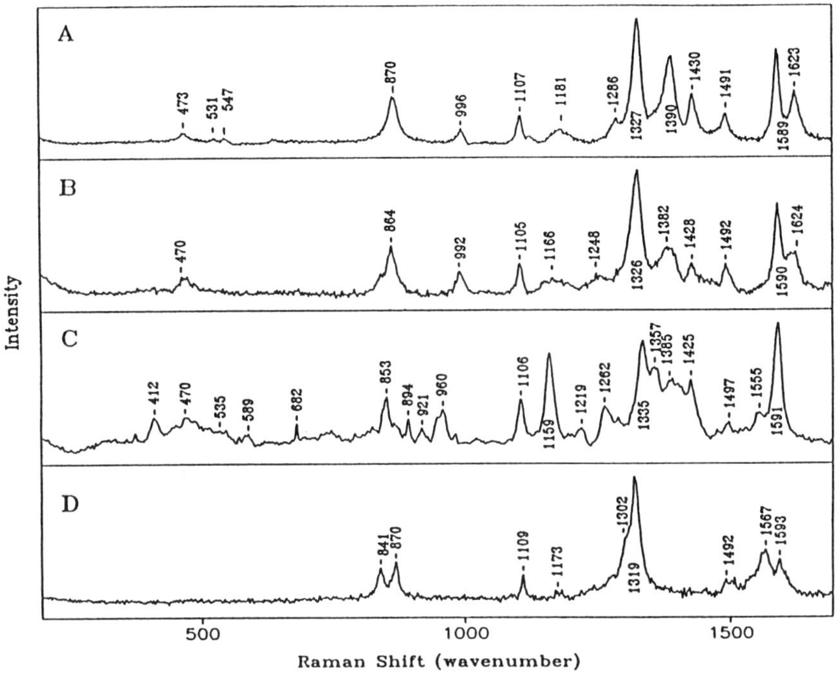

FIG. 6. Comparison of the resonance Raman spectra of the *p*-nitrophenylhydrazones of topahydantoin (A), the active site peptide of BPAO (B), PQQ(C), and pyridoxal phosphate (D).

dependence of the topaquinone *p*-nitrophenylhydrazone absorbance appears to be diagnostic. Sequences of the active site peptides from several amine oxidases are shown in Table 2; the high degree of homology evident in Table 2 and Fig. 1 provides further support for the conclusion that all of these amine oxidases contain topa. The consensus sequence for topa deduced from these comparisons is Asn-Tyr-Asp/Glu [58], which falls in a region of high homology among sequenced amine oxidases (Fig. 1). It should again be noted that mammalian lysyl oxidase is not homologous to this group of amine oxidases [45,47]. Although resonance Raman data are consistent with a quinone cofactor similar to topa, the lysyl oxidase spectrum is not identical to the spectra of the other amine oxidases [73]; the structural basis for this difference is not understood.

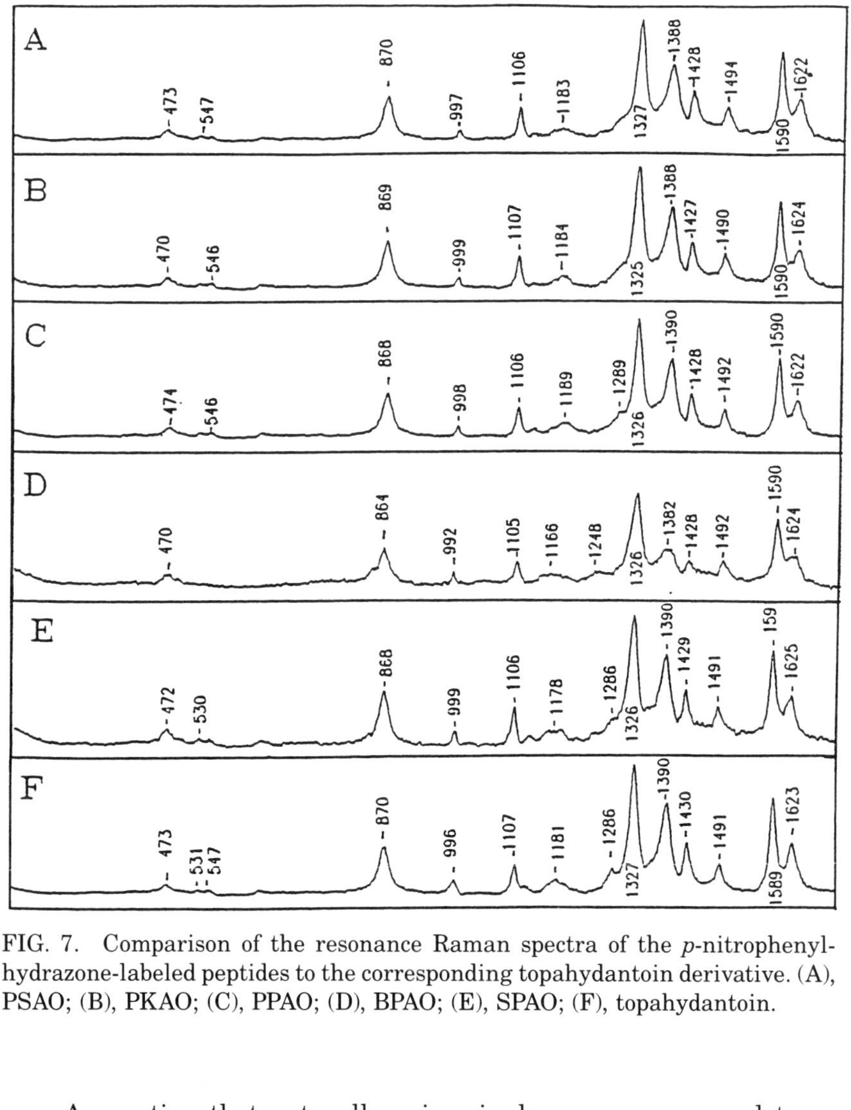

FIG. 7. Comparison of the resonance Raman spectra of the *p*-nitrophenylhydrazone-labeled peptides to the corresponding topahydantoin derivative. (A), PSAO; (B), PKAO; (C), PPAO; (D), BPAO; (E), SPAO; (F), topahydantoin.

A question that naturally arises is, how are copper and topaquinone arranged in amine oxidase active sites, and is topa a copper ligand? Numerous experiments indicate that copper and topa interact [11,16,40,41,65,74–76], but it is not known if the observed interactions are direct or protein-mediated. Three experiments have addressed this question directly [77–79]. The conclusion from this work is that the

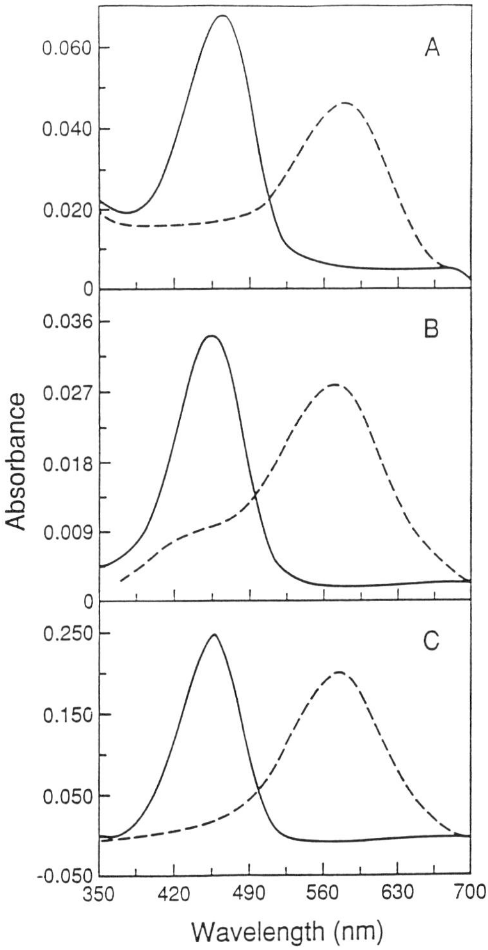

FIG. 8. Visible absorption spectra of the *p*-nitrophenylhydrazones of native PSAO (A), the active site peptide of PSAO (B), and topahydantoin (C) in 0.1 M potassium phosphate buffer pH 7.2 (—) and either 2 M (A) or 1 M (B, C) KOH (---).

TABLE 2

Sequences for Active Site Thermolytic Peptides from Copper Amine Oxidases

BPAO	PPAO	PKAO	PSAO	YAO
		Val	Val	Val
Leu	Leu	Tyr	Gly	Ala
Asn	Asn	Asn	Asn	Asn
X	X	X	X	X
Asp	Asp	Asp	Asp	Glu
Tyr	Tyr	Tyr	Asn	Tyr
			Val	Val

reactive carbonyl and Cu(II) are separated by 10 Å or more. It is possible that the C-2 carbonyl is weakly bonded to Cu(II), perhaps as an axial ligand [72], but this remains to be established.

3. CATALYTIC MECHANISM OF AMINE OXIDASES

3.1. Overview

When amine oxidases were reviewed in the mid-1980s [11,80] it could not be decided with complete certainty whether an imine or an aminotransferase mechanism was operative (see Fig. 9). Both alternatives are consistent with the "ping-pong" steady-state kinetic data [81–83], which require that interaction of enzyme with amine substrate [step (i)] leads to formation of a reduced enzyme intermediate that is subsequently reoxidized by oxygen in step (ii). In the case of the pig plasma amine oxidase, stopped-flow kinetic studies had shown that steps (i) and (ii) occur at readily distinguishable rates [84,85]. Thus discrimination between the imine mechanism and the aminotransferase mechanism could be achieved by testing whether the transient kinetics of ammonia release conforms to the rate of the reduction process [step (i)] or the rate of the reoxidation process [step (ii)]. The results of quenched flow kinetic

Amino transferase mechanism

$$\text{(i)}\quad \text{E-CHO} + \text{RCH}_2\text{NH}_2 \rightarrow \text{E-CH}_2\text{NH}_2 + \text{RCHO}$$

$$\text{(ii)}\quad \text{E-CH}_2\text{NH}_2 + \text{O}_2 + \text{H}_2\text{O} \rightarrow \text{E-CHO} + \text{H}_2\text{O}_2 + \text{NH}_3$$

Imine mechanism

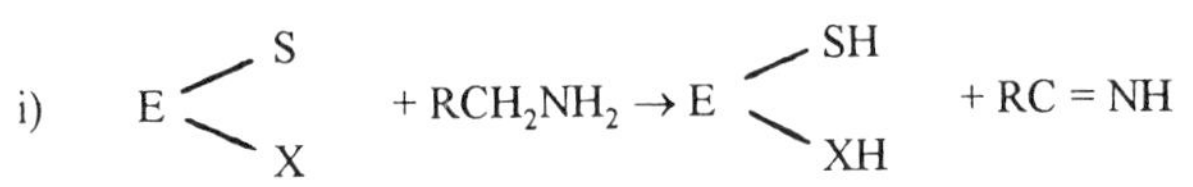

$$\text{RCH} = \text{NH} = \text{H}_2\text{O} \rightarrow \text{RCHO} + \text{NH}_3$$

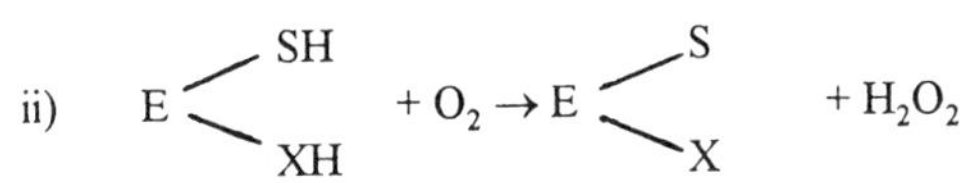

FIG. 9. Proposed mechanisms for copper-containing amine oxidases.

studies [86,87] established unequivocally in favor of an aminotransferase mechanism.

It may be noted that these experiments to distinguish between the two mechanisms were particularly favorable for the pig plasma amine oxidase where k_{red} in step (i) is $> k_{ox}$ in step (ii); the experiments would have been less clear-cut with the bovine plasma oxidase where $k_{red} \approx k_{ox}$ or the pea seedling enzyme where $k_{red} < k_{ox}$. It has, however, been shown convincingly for the bovine enzyme that ammonia is not formed in the reductive half cycle, again a result consistent with the aminotransferase mechanism [88].

During the 1970s, systematic kinetic studies on the mechanism of pig plasma amine oxidase were carried out by Pettersson and coworkers [80]. Although these studies were based on the false premise that pyridoxal phosphate is the cofactor, the studies have provided an excellent foundation for understanding the catalytic mechanisms. Since 1983, elegant studies on the mechanism of bovine plasma amine oxidase have been made by Klinman and coworkers; these have led to a

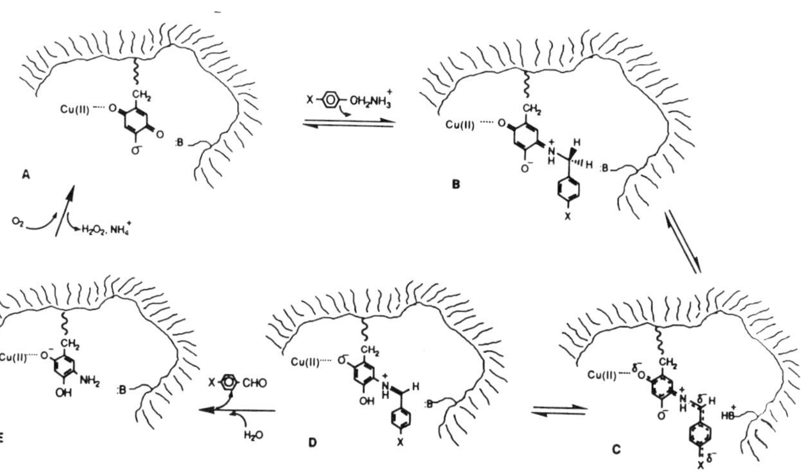

FIG. 10. Detail reaction mechanism for plasma amine oxidase. (A) The active site cofactor, topaquinone shown in proximity to an active site base and copper ion. (B) The covalent imine complex between substrate and cofactor, generated subsequent to simple E/S complex formation. (C) The carbanion intermediate, generated upon proton transfer from substrate to active site base. (D) Product imine complex, in which the substrate-derived proton has been transferred to the oxy anion at C-4 of the cofactor. (E) The aminoquinol of the cofactor, formed by hydrolysis of aldehyde from the product imine complex. (Reproduced from Ref. 89 with permission.)

proposed catalytic mechanism [89] consistent with all available evidence including the observations by Neuman *et al.* [90] that phenylacetaldehyde is formed from β-chlorophenethylamine by a proton abstraction process. The overall catalytic mechanism proposed by Hartmann and Klinman [89] is shown in Fig. 10. It is clear from this figure that our understanding in molecular detail of the steps leading to reduced enzyme is much greater than our understanding of the reoxidation steps. Mechanistic studies on amine oxidases have been comprehensively and critically reviewed recently by McIntire and Hartmann [16].

3.2. Reductive Phase of the Catalytic Cycle

A general mechanism showing the intermediate involved in substrate binding and formation of the reduced enzyme is given in Scheme 1.

$$E_{ox} + RCH_2NH_3 \Leftrightarrow E_{ox}RCH_2NH_3 \Leftrightarrow E_{red}{-}N{=}CHR$$

$$E_{red}{-}N{=}CHR \rightarrow E_{red}{-}NH_2 + RCHO$$

Scheme 1. Steps involved in the reductive half cycle of amine oxidases according to an aminotransferase mechanism.

It can be seen that the process involves two steps: first, binding of substrate to the topaquinone cofactor, followed by proton abstraction with concomitant enzyme reduction and, second, hydrolysis of the product imine intermediate to liberate the aldehyde product.

The mechanism shown in Scheme 1 is certainly an oversimplification as is revealed by steady-state kinetic studies. For example, inhibition at high substrate concentrations is clearly evident for the pig plasma amine oxidase and has been interpreted in terms of a species E·S·S involved in catalysis [16,91]. Again, the initial rate data for methylamine oxidase from *Candida boidinii* indicate an ordered rather than a ping-pong mechanism [29]. This has been discussed in terms of the low K_m for oxygen in the case of methylamine oxidase (10 μM) compared to 500 μM for pig plasma amine oxidase [81], and is perhaps related to the distinct physiological roles played by these two enzymes. More detailed studies on the catalytic mechanisms of amine oxidases from microorganisms are needed.

Spectroscopic and kinetic evidence indicates that the substrate binds to the amine oxidase through its amine group. There are discrepancies in the literature on whether the amine binds in its unprotonated ($-NH_2$) or protonated ($-\overset{+}{N}H_3$) state [50,84]. In the case of bovine plasma amine oxidase, the evidence is convincingly in favor of monoamines binding in their protonated states [16,89]; the protonation state of monoamines binding to other amine oxidases requires further study. It would be interesting to examine the behavior of the bovine plasma amine oxidase with diamine substrates in view of the report by Bardsley and coworkers [92] that for the pig kidney enzyme the second amine group in its protonated state binds to a negatively charged site on the enzyme and that a hydrophobic patch on the enzyme contributes to substrate binding.

Kinetic isotope effects have been found to be large (k_H/k_D = 13–15) for reduction of bovine plasma amine oxidase by primary amines [93]. This indicates the accumulation of the E/S complex prior to the rate-limiting bond cleavage step. The Schiff base complex in bovine plasma amine oxidase has been "trapped" by reduction with [^{3}H]NaCNBH$_3$ in the presence of ^{14}C-labeled benzylamine [94]. There was stoichiometric incorporation of ^{14}C label but not ^{3}H label into the protein. As is made clear by Fig. 11, these results establish a substrate–imine complex as an intermediate in the catalytic mechanism. Careful studies by Janes and Klinman [88] showed that incubation of highly purified bovine plasma amine oxidase with amine substrate under anaerobic conditions results in production of 2 mol of aldehyde per mol of enzyme but no release of ammonia; thus the nitrogen from the substrate must have been transferred to the quinone cofactor. These studies provide strong support for two active sites for enzyme dimer in the case of bovine plasma amine

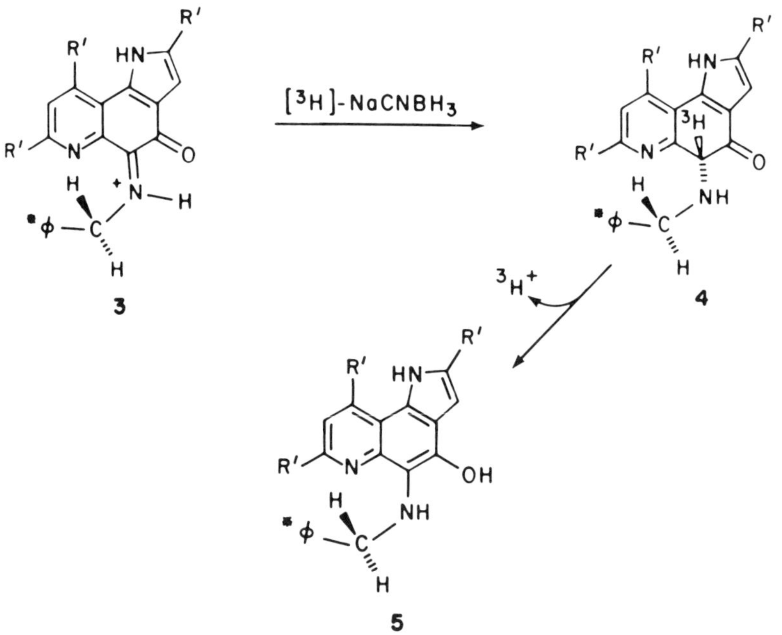

FIG. 11. Reductive trapping of substrate to bovine plasma amine oxidase. ^{14}C-labeled substrate (*) forms on Schiff's base with the proposed cofactor which undergoes reduction by [^{3}H]NaCNBH$_3$. Although the enzyme should be labeled initially, subsequent enolization will generate the more stable form of the cofactor (5) which does not contain ^{3}H. Note that the cofactor is given as PQQ but is now known to be topaquinone: the mechanistic explanation for the labeling pattern still holds true. (Reproduced from Ref. 94 with permission.)

oxidase and it is argued [62] that this is true also for other amine oxidases. Evidence has been presented on the basis of active site titrations that pig plasma amine oxidase has one active site per protein dimer [66,95,96]. 35-GHz EPR spectroscopic studies [66] indicate that the two coppers in pig plasma amine oxidase, unlike other amine oxidases, are chemically distinguishable, which might reflect a difference in reactivity of the two Cu(II)/topaquinone sites. These results might be interpreted in terms of half-site reactivity. However, continued efforts must be made to improve the purification procedures for the pig plasma enzyme to the standards set for the bovine enzyme [88] before half-site reactivity can be shown to be fact or myth. In this context, the possibility that ammonia may remain at least partially bound at the end of each catalytic cycle should be considered [16] since the presence of topaquinone and topaiminoquinone species could affect both spectroscopic and reactivity characteristics.

The aminotransferase mechanism shown in Scheme 1 is, by analogy with mechanisms involving pyridoxal phosphate, expected to be base-catalyzed and to lead to transient formation of a resonance-stabilized carbanion intermediate. Convincing evidence has been presented for a base-catalyzed process at the active site of bovine plasma amine oxidase [51]. This is consistent with conclusions drawn from studies on mechanism-based inactivators of bovine plasma amine oxidase by Abeles and coworkers [90,97,98]. The mechanism of C-H bond cleavage during substrate oxidation by bovine plasma amine oxidase has been investigated through structure–function correlations involving a series of ring-substituted benzylamine derivatives [89] which establish proton removal as an elementary step. Similar studies with pig plasma amine oxidase did not allow a distinction to be made between homolytic and heterolytic C-H bond cleavage [99]; this is a reflection of the much lower kinetic isotope effect ($k_H/k_D = 2.8$) in the reductive half cycle of this enzyme. The observation of the large primary isotope effects in oxidation of benzylamine by bovine plasma amine oxidase [93] suggests the possibility of hydrogen tunneling. This has been studied in more detail [100] and the conclusion drawn that significant tunneling indeed occurs. This is the first example of quantum mechanical tunneling in an enzyme-catalyzed proton abstraction reaction [101].

Palcic and coworkers determined the stereochemical trends among copper-containing amine oxidases in abstraction of the proton at C-1 and also studied the solvent exchange characteristics at C-2 for substrates such as dopamine and tyramine [21,22,102,103]. A summary of their findings is given in Table 3. It can be seen that, unlike most enzymes, the amine oxidases do not exhibit uniform stereospecificity between species in substrate oxidation. Of particular interest are the range of stereospecificities and solvent exchange characteristics for amine oxidases from different tissue in the same organism (e.g., pig) and the observation [16] that, whereas the plant amine oxidases studied to date show uniform stereospecificity, the enzymes from mammalian sources do not.

Brunori and coworkers [104,105] carried out careful transient kinetic studies of the catalytic cycle for the amine oxidase from lentil

TABLE 3

Stereochemistry at C-1 of Tyramine and Solvent Exchange Characteristics at C-2 Catalyzed by Semicarbazide-Sensitive and Copper Amine Oxidases

Enzyme source	C-1 proton abstraction	C-2 solvent exchange	Ref.
Semicarbazide-Sensitive Amine Oxidases			
Porcine aorta	pro-S	Yes	21
Bovine aorta	pro-S	Yes	21
Copper Amine Oxidases (EC 1.4.3.6)			
Pea seedling	pro-S	No	102
Porcine kidney	pro-S	No	22
Soybean seedling	pro-S	No	22
Chick pea seedling	pro-S	No	22
Porcine plasma	pro-R	Yes	102
Bovine plasma	nonstereospecific	Yes	102
Sheep plasma	nonstereospecific	Yes	22
Rabbit plasma	nonstereospecific	Yes	22
Lysyl oxidase			
Bovine aorta	pro-S	Yes	103

seedlings. For plant amine oxidases, the reductive phase of the catalytic cycle is rate limiting and this has allowed spectral and kinetic characterization of the intermediates. The steps involved are consistent with Scheme 1 but additionally indicate an internal redox step following release of aldehyde product which leads to a yellow intermediate. Unlike the earlier steps in the reductive phase where copper appears not to be required [106], formation of the yellow intermediate absolutely requires the presence of copper. EPR studies [107] have established that the yellow intermediate is a free radical, probably originating from the semiquinone state of the topa cofactor and where the Cu(II) cofactor has been reduced to Cu(I).

3.3. Reoxidation of the Substrate-Reduced Enzyme

It has been clear for some time that substrate amines form covalent adducts with topaquinone. Subsequently an active site base abstracts a proton from the imine leading to the two-electron reduction of the quinone to an aminophenol, as discussed above. NH_3 and H_2O_2 are released during the reoxidation of the substrate-reduced enzyme by O_2 [87,108]. An intermediate has been detected in the reoxidation reaction that differs from both the substrate-reduced and the resting enzyme [85]. The role of copper in amine oxidases has been uncertain since early EPR measurements failed to detect any change in the copper oxidation state following substrate addition [11,16,40]. Rapid freeze-quench EPR experiments appeared to rule out the possibility of a copper valence change at any stage of the catalytic cycle [109]. In 1991, it was shown that substrate amines do reduce copper under anaerobic conditions, generating a Cu(I)-semiquinone radical state [107]. Figure 12 illustrates the key results. If the EPR spectra are obtained at room temperature, a sharp signal at $g \sim 2$ is observed for all five amine oxidases, although the yield of this species is variable. The Cu(II) signals of PSAO and APAO are significantly decreased, and a minor decrease is apparent in the PKAO spectrum. Quantitatively, the Cu(II) signals of the pea seedling and *Arthrobacter* amine oxidases decrease by 40%, with the semiquinone radical signal intensity corresponding to about half of the copper change. The difference was attributed to disproportionation of the semiquinone and this has been confirmed by independent experi-

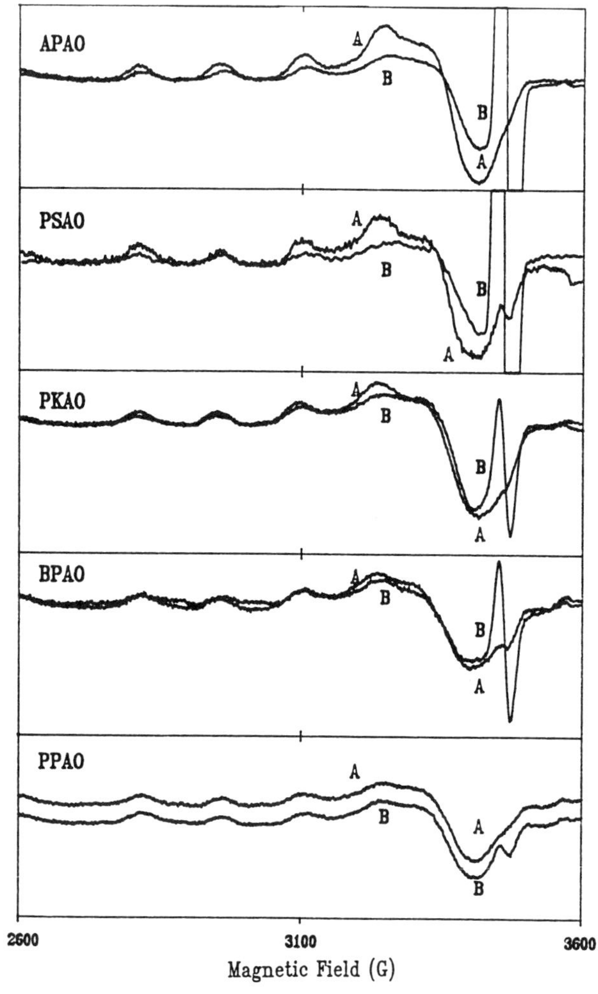

FIG. 12. Room temperature X-band EPR spectra of amine oxidases before (A) and after (B) anaerobic addition of substrate. The PPAO spectra are offset for clarity.

ments [110]. Circular dichroism (CD) spectra of APAO (Fig. 13) and PSAO (not shown) are quantitatively consistent with copper reduction by substrates as inferred from the EPR data. Collectively the data point to a temperature-dependent equilibrium:

$$\text{Cu(II)-reduced quinone} \rightleftharpoons \text{Cu(I)-semiquinone}$$

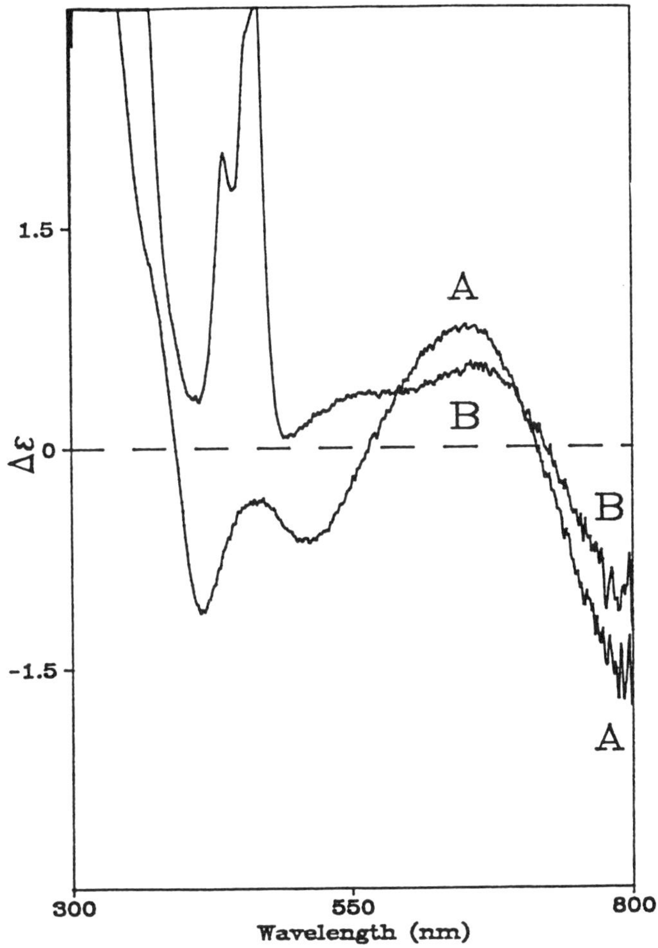

FIG. 13. Room temperature CD spectra ($\Delta\epsilon = \epsilon_L - \epsilon_R$) of native APAO (A) and of APAO after anaerobic addition of a fourfold excess of benzylamine substrate (B), in 0.1 M potassium phosphate buffer, pH 7.2.

The Cu(II) state is favored at low temperatures, and thus the Cu(I)-semiquinone state was not observed in previous EPR experiments, which were invariably conducted at low temperature. Based on these and other results, and a wealth of chemical precedent, the catalytic cycle in Fig. 14 has been proposed [107]. Note that in this figure, the Cu(I)-semiquinone is the form of the enzyme that reacts directly with

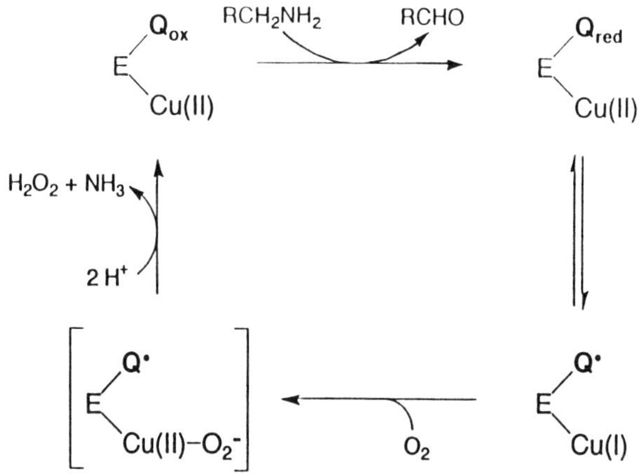

FIG. 14. Possible catalytic cycle of copper-containing amine oxidases. The species in brackets is a hypothetical intermediate, shown here to emphasize the possibility of sequential one-electron steps in the reduction of O_2. O_{ox} = topaquinone, $Q^{\bullet}$ = topasemiquinone, and Q_{red} = topa.

oxygen. Indirect evidence for sequential one-electron reduction of oxygen via a bound superoxide intermediate has been previously described [111–113]. Even if the Cu(I)-semiquinone is present only as a relatively small percentage of the total enzyme, it may be so reactive that product formation is channeled exclusively through this state. This possibility will require a great deal of additional work to evaluate and thus establish the role of copper in amine oxidases. Nevertheless the facile formation of the Cu(I)-semiquinone via substrate reduction does suggest that copper has a redox function in the reoxidation of substrate-reduced amine oxidases.

Recent work has corroborated important aspects of Fig. 14. Absorbance and EPR features, which were very similar to those of the putative amine oxidase intermediates, could be generated using 6-hydroxydopamine as a model for the substrate-reduced cofactor [114]. These results strongly support the assignment of the $g = 2$ EPR signal to the semiquinone. Kinetic evidence has also been presented that implicates a semiquinone intermediate in the catalytic cycle of lentil seedling amine oxidase [104,105].

The results discussed above and the mechanistic studies summarized in Fig. 10 indicate that the semiquinone form in Fig. 14 is actually an iminosemiquinone, i.e., that the substrate-derived nitrogen is covalently attached to the topa moiety. Continuous wave (CW) and pulsed EPR spectroscopy provides a critical test of this idea [115]. In these experiments, cyanide was used to stabilize the semiquinone state so that it could be examined at 4.2 K, as required by the ESEEM method; it has been shown that the EPR signals of the radicals generated in the presence or absence of cyanide are identical [107]. The pulsed EPR results confirm that the radicals in *Arthrobacter* and porcine kidney amine oxidases are identical. The ESEEM spectra provide strong evidence that at least one nitrogen nucleus is magnetically coupled to the paramagnet. Isotopic substitution experiments using [^{14}N]- and [^{15}N]methylamine with *Arthrobacter* methylamine oxidase establish that this coupled nitrogen nucleus is derived from substrate [74,115]. Stimulated echo ESEEM data collected at X band frequency for [^{14}N]methylamine reveal a broad peak in the Fourier transform at 3.3 MHz, which was not observed when [^{15}N]methylamine was used. Similar results were obtained for PKAO. Magnetic coupling and hyperfine parameters obtained by computer simulation are consistent with previous formulations of this radical as an iminosemiquinone. The hyperfine coupling tensor elements are in line with what one would expect for a radical where covalent attachment of the substrate amine nitrogen to an aromatic group has occurred. These results effectively rule out a tyrosine radical (or a radical derived from any other amino acid) as the source of the $g \sim 2$ signal in substrate-reduced amine oxidases. Both the ESEEM and other spectroscopic results support the aminotransferase mechanism. In conjunction with kinetic data the ESEEM data also suggest that the initial step in the oxidation of the reduced topa (and perhaps the first step in oxygen reduction) occurs prior to cleavage of the nitrogen–topa bond(s).

3.4. Outstanding Mechanistic Issues

As detailed above, the course of substrate oxidation and the concomitant two-electron reduction of the topa cofactor are understood in much greater detail than the subsequent reoxidation. Nevertheless some intriguing and important details of the substrate oxidation/enzyme reduction phase of amine oxidase catalysis remain to be elucidated. The identity of the active site base is unknown as are the active site structural features, which must involve the base, that determine the stereochemistry of proton abstraction. A variety of spectral intermediates have been detected by rapid-scanning stopped flow during the oxidation of substituted benzylamines [16,105], which are yet to be definitively identified. Although strong evidence for significant tunneling in H/D abstraction by BPAO has been presented [100], the origins and mechanistic significance of these phenomena are not completely understood.

With regard to the reoxidation phase more remains to be learned than has been discovered to date. For example, the site of oxygen reduction, whether Cu(I) or the topasemiquinone, is not known. The factors that control the yield of the semiquinone state and the rates of intramolecular electron transfer are not understood. Although proton transfer steps must be involved in oxygen reduction and product (peroxide) release, practically no information is available on the role and timing of such steps in the reoxidation reaction.

Half-of-the-sites reactivity phenomena in amine oxidases remain something of an enigma. Frankly, it is not clear that the amount of effort devoted to this issue is justified given the lack of evidence that negative cooperativity is physiologically significant. However, the possibility that negative cooperativity may be an *in vitro* manifestation of regulatory behavior involving subunit interactions is intriguing and worth pursuing. The regulation of amine oxidases is a subject that certainly deserves closer scrutiny.

4 BIOLOGICAL ROLES OF AMINE OXIDASES

4.1. Microorganisms

Amine oxidases are induced in prokaryotic organisms such as the bacteria *Arthrobacter* P1 [116,117] and *Escherichia coli* K-12 [118] when grown on monoamines as the sole carbon and energy source. The nutritional role of the amine oxidase in these organisms appears unambiguous. The role of amine oxidases induced in eukaryotic microorganisms by growth on different substrates is more complex [119]. Thus growth of *Penicillium chrysogenum* on glucose plus spermine leads to production of a flavin-containing polyamine oxidase [26] whereas growth on spermine as the sole carbon and nitrogen source [27] induces an amine oxidase of broad specificity. Further studies on amine oxidases from fungi and yeasts to elucidate their biological roles are needed; the relative simplicity of these organisms compared to plant and mammalian systems is an attractive feature of these studies.

4.2. Plants

It is convenient to consider separately the responses of the broadly specific amine oxidases (termed "diamine oxidases" in Ref. 3) and the more specific "polyamine oxidases". The diamine oxidases occur at very high concentrations in the cotyledons of legumes where they account for 3% of the total protein in pea seedlings [120]. A different diamine oxidase, existing in several isozymic forms, is found in embryos of pea seedlings. The diamine oxidase in the cotyledons is extracellular and associated with the cell wall. There is a positive correlation between the diamine and diamine oxidase levels along the entire length of the seedling and also a correlation between diamine oxidase and peroxidase activities which suggests a role for diamine oxidase in cell wall fabrication and lignin formation where hydrogen peroxide, generated through the action of diamine oxidase, is a required substrate. These events may form part of the healing response in plants. Diamine oxidases may also play a direct role in plant cell signaling either by controlling the levels of biogenic amines, such as spermidine, or through the production of specific effectors, such as pyrroline derivatives or alkaloids [32,119].

Polyamine oxidases, which have been studied most extensively in cereal plants, are extracellular enzymes located in the cell wall. The specificity of their substrate requirements indicates that they have defined roles in plant cells concerned, as with diamine oxidases, with cell signaling [32,119]. Further study, using modern molecular biology methods, to characterize the roles of biogenic amines and amine oxidases in plants is urgently needed.

4 3. Mammals

The diversity of amine oxidases in mammalian cells and their stereochemical precision [21,22] have been considered briefly in Sec. 1.2. Roles for amine oxidases, other than in scavenging biogenic amines, is indicated. Our knowledge of this subject is fragmentary and found largely in the medical and pharmacological literature. A recent review by McIntire and Hartmann [16] is excellent in documenting and summarizing these diverse observations.

Important advances have been made recently in defining the biological role or roles of lysyl oxidases. Much of this work has been carried out by Kagan and collaborators; an excellent recent review is available [47]. The gene for lysyl oxidase, an extracellular enzyme, has been cloned and sequenced [45]. The role of signal sequences in the gene for the synthesis, processing, and secretion of lysyl oxidase are reviewed in Ref. 47. Lysyl oxidase clearly has an important role in crosslinking of structural proteins such as elastin and collagen; pathogenic changes in the production or regulation of lysyl oxidase at the genetic level are implicated in fibrosis [45,47]. At the cellular level, it has been observed that lysyl oxidase activity is markedly reduced in malignantly transformed human cell lines [121]. Another indicator of a relationship between lysyl oxidase and tumorigenesis has emerged recently. The expression of the rrg transcript in host mouse cells is markedly decreased following transformation by the Ha-ras sarcoma virus. High levels of rrg expression are restored when this transformed tumorigenic cell line reverts to a nontumorigenic cell line. The specific importance of rrg to the reversion is emphasized by the fact that normal cells develop the transformed cell phenotype when transfected with rrg antisense c-DNA [122]. It was recently shown that rrg is 98% homologous at the protein level with lysyl oxidase [46]. These exciting results indicate strongly that a lysyl oxidase–like protein plays a critical role in maintaining the nontumorigenic state of these cell lines. The studies further indicate that similar molecular biological approaches to elucidate the roles of other mammalian (and plant) amine oxidases will be revealing.

Much less is known about the biosynthesis, secretion, and function of another class of mammalian, extracellular amine oxidases—the plasma amine oxidases [11,16,80]. Even the site of biosynthesis of the protein is uncertain though this is likely to be the liver [16,123,124]. The concentration of amine oxidases in serum is high (125 mg/L) [81]. There is a dramatic increase in the amine oxidase activity of human material blood and placenta during pregnancy [16,125,126] which indicates cell-signaling events. The roles in mammalian cells of amine oxidases which act on primary amine groups ("monoamine" and "diamine" oxidases) and at secondary amine sites ("polyamine" oxidases) are intriguing. The conclusion drawn by Morgan [36] is that "polyamines and derivatives formed from them by enzyme-catalysed oxidation are parts of an integrated biochemical system for the regulation of cell proliferation and growth". In attempts to understand these regulation processes, the following questions need to be addressed:

1. What is the interplay between flavin-containing and copper-containing amine oxidases in cellular function?
2. What is the significance of the different stereochemical patterns observed for different amine oxidases in terms of cellular function?
3. What are the roles of amine oxidases which act at primary and secondary amine sites in cellular processes?
4. How are amine oxidases biosynthesized, processed, and secreted in mammalian cells?
5. What is the biological significance of membrane attachment of amine oxidases, such as the "semicarbazide-sensitive" amine oxidase of aorta and intestine?

These and other questions can be answered using the range of molecular biological methods now available.

5. CONCLUSIONS AND PROSPECTS

Copper-containing amine oxidases are now recognized to be even more widespread than previously believed. Moreover, our knowledge of amine oxidase structure and function has advanced considerably in recent years. The unequivocal identification of the cofactor [14,57,58,71] and the evidence for a copper redox role [107] provide a new, well-grounded framework for additional mechanistic investigations. Important aspects of the active site structure and mechanism of these enzymes remain uncertain, however, though several productive approaches to the remaining questions have been identified and will generate considerable new information in the near future. Cloning and sequencing additional amine oxidase genes will be helpful, as will the construction and analysis of site-directed mutants. The crystallographic analysis of pea seedling amine oxidase [48] and other amine oxidases will open an entirely new and richly rewarding phase of investigation of these important enzymes. Finally, the biological role or roles of these enzymes needs to be studied; the availability of the cloned genes and the use of modern molecular biological approaches will facilitate these studies.

ACKNOWLEDGMENTS

The authors thank C. Hartmann, H. M. Kagan, W. S. McIntire, and M. M. Palcic for information and manuscripts prior to publication.

ABBREVIATIONS

Ala	alanine
APAO	*Arthrobacter* P1 amine oxidase
Asn	asparagine
Asp	aspartic acid
BPAO/BSAO	bovine plasma (or serum) amine oxidase
CD	circular dichroism
CW-EPR	continuous wave electron paramagnetic resonance
Cys	cysteine
ENDOR	electron nuclear double resonance
EPR	electron paramagnetic resonance
ESEEM	electron spin-echo envelope modulation
EXAFS	extended X-ray absorption fine structure
Glu	glutamic acid
Gly	glycine
imid	imidazole
Leu	leucine
NMR	nuclear magnetic resonance
NMRD	nuclear magnetic relaxation dispersion
PKAO	pig kidney amine oxidase
PPAO	pig plasma amine oxidase
PQQ	pyrroloquinoline quinone

PSAO	pea seedling amine oxidase
rrg	*ras* recision gene
SDS	sodium dodecyl sulfate
SOD	superoxide dismutase
SPAO	sheep plasma amine oxidase
topa	trihydroxyphenylalanine
Tyr	tyrosine
Val	valine

REFERENCES

1. K. T. Yasunobu, H. Ishazaki, and N. Minamiura, *Mol. Cell. Biochem.*, *13*, 3 (1976).
2. B. Mondovi and A. Finazzi-Agrò, *Adv. Exp. Med. Biol.*, *148*, 141 (1982).
3. B. G. Malmström, L. E. Andréasson and B. Reinhammar, in *The Enzymes*, Vol. 12b, 3rd ed. (P. D. Boyer, ed.), Academic Press, New York, 1975, p. 507.
4. E. A. Zeller, *Helv. Chim. Acta*, *21*, 880 (1938).
5. E. A. Zeller, in *The Enzymes*, Vol.8, 2nd ed. (P. D. Boyer, H. A. Lardy, and K. Myrback, eds.), Academic Press, New York, 1963, p. 313.
6. H. Blaschko, Ref. 5, p. 337.
7. H. Tabor, *Pharmacol. Rev.*, *6*, 299 (1954).
8. F. Buffoni and H. Blaschko, *Proc. Roy. Soc. B.*, *161*, 153 (1964).
9. H. Yamada, K. Yasunobu, T. Yamano, and H. S. Mason, *Nature*, *198*, 1092 (1963).
10. B. Mondovi, G. Rotilio, M. T. Costa, A. Finazzi-Agrò, E. Chiancone, R. E. Hansen, and H. Beinert *J. Biol. Chem.*, *242*, 1160 (1967).
11. P. F. Knowles and K. D. S. Yadav, in *Copper Proteins and Copper Enzymes*, Vol. 2 (R. Lontie, ed.), CRC Press, Boca Raton, 1984, p. 103.
12. C. L. Lobenstein-Verbeck, J. A. Jongejan, J. Frank, and J. A. Duine, *FEBS Lett.*, *170*, 305 (1984).

13. M. Ameyama, M. Hayashi, K. Matsushita, E. Shinagawa, and O. Adachi, *Agr. Biol. Chem.*, *48*, 561 (1984).
14. S. M. Jones, D. Mu, D. Wemmer, A. J. Smith, S. Kaur, D. Maltby, A. L. Burlingame, and J. P. Klinman, *Science*, *248*, 981 (1990).
15. J. P. Klinman, D. M. Dooley, J. A. Duine, P. F. Knowles, B. Mondovi, and J. J. Villafranca, *FEBS Lett.*, *282*, 1 (1991).
16. W. S. McIntire and C. Hartmann, in *Principles and Applications of Quinoproteins* (V. L. Davidson, ed.), Marcel Dekker, New York, 1992, p. 97.
17. R. A. Cooper, P. F. Knowles, D. E. Brown, M. A. McGuirl, and D. M. Dooley, *Biochem. J.*, *288*, 337 (1992).
18. O. Suzuki and T. Matsumoto, *Biogen. Amines*, *4*, 45 (1987).
19. M. A. J. Ferguson, *Curr. Opinion Struct. Biol.*, *1*, 522 (1991).
20. S. N. Gacheru, P. C. Trackman, M. A. Shah, C. Y. O'Gara, P. Spacciapoli, F. T. Greenaway, and H. M. Kagan, *J. Biol. Chem.*, *265*, 19022 (1990).
21. C. H. Scaman and M. M. Palcic, *Biochemistry*, *31*, 6829 (1992).
22. A. A. Coleman, C. H. Scaman, Y. J. Kang, and M. M. Palcic, *J. Biol. Chem.*, *266*, 6795 (1991).
23. H. Kuivaniemi, E. R. Savolainen, and K. J. Kivirikko, *J. Biol. Chem.*, *259*, 6996 (1984).
24. M. J. C. Crabbe, R. D. Waight, W. G. Bardsley, R. W. Barker, I. D. Kelly, and P. F. Knowles, *Biochem. J.*, *155*, 679 (1976).
25. Y. Kobayashi and K. Horikoshi, *Biochim. Biophys. Acta*, *705*, 133 (1982).
26. K. Isobe, Y. Tani, and H. Yamada, *Agr. Biol. Chem.*, *44*, 2651 (1980).
27. K. Isobe, Y. Tani, and H. Yamada, *Agr. Biol. Chem.*, *46*, 1353 (1982).
28. G. W. Haywood and P. J. Large, *J. Gen. Microbiol.*, *130*, 1123 (1984).
29. G. W. Haywood and P. J. Large, *Biochem. J.*, *199*, 187 (1981).
30. J. Green, G. W. Haywood, and P. J. Large, *Biochem. J.*, *211*, 481 (1983).
31. S. S. Tur and K. Lerch, *FEBS Lett.*, *238*, 74 (1988).
32. T. A. Smith, *Biochem. Soc. Trans.*, *13*, 319 (1985).

33. D. E. Brown, M. A. McGuirl, D. M. Dooley, S. M. Jones, D. Mu, and J. P. Klinman, *J. Biol. Chem.*, *266*, 4049 (1991).
34. Y. Suzuki and H. Yanagisawa, *Plant Cell Physiol.*, *21*, 1085 (1980).
35. T. A. Smith, *Meth. Enzymol.*, *94*, 311 (1983).
36. D. M. L. Morgan, *Biochem. Soc. Trans.*, *13*, 322 (1985).
37. G. P. Vidal, J. J. Sheih, and K. T. Yasunobu, *Biochem. Biophys. Res. Commun.*, *64*, 989 (1975).
38. H. M. Kagan, K. A. Sullivan, T. A. Olsson, and A. L. Cronlund, *Biochem. J.*, *177*, 203 (1979).
39. M. C. Falk, A. J. Staton, and T. J. Williams, *Biochemistry*, *22*, 3746 (1983).
40. B. Mondovi (ed.), *Structure and Function of Amine Oxidases*, CRC Press, Boca Raton, 1985.
41. D. M. Dooley, *Life Chem. Rep.*, *5*, 91 (1987).
42. P. G. Bruienberg, M. Evers, H. R. Waterham, J. Kuipers, A. C. Arnberg, and A. B. Geert, *Biochim. Biophys. Acta*, *1008*, 157–167 (1989).
43. W. S. McIntire, unpublished results.
44. A. Rossi, R. Petruzzelli, and A. Finazzi-Agrò, *FEBS Lett.*, *301*, 253 (1992).
45. P. C. Trackman, A. M. Pratt, A. Wolanski, S.-S. Tang, G. D. Offner, R. F. Troxler, and H. M. Kagan, *Biochemistry*, *29*, 4863 (1990). Erratum: *Biochemistry*, *30*, 8282 (1990).
46. K. Kenyon, S. Contente, P. C. Trackman, J. Tang, H. M. Kagan, and R. F. Friedman, *Science*, *253*, 802 (1991).
47. H. M. Kagan and P. C. Trackman, in *Principles and Applications of Quinoproteins* (V. L. Davidson, ed.), Marcel Dekker, New York, 1992, p. 173.
48. V. Vignevich, D. M. Dooley, J. M. Guss, I. Harvey, M. A. McGuirl, and H. C. Freeman, *J. Mol. Biol.*, *229*, 243 (1993).
49. M. C. Summers, R. Markovic, and J. P. Klinman, *Biochemistry*, *18*, 1969 (1979).
50. M. Farnum, M. Palcic, and J. P. Klinman, *Biochemistry*, *25*, 1898 (1986).
51. M. F. Farnum and J. P. Klinman, *Biochemistry*, *25*, 6028 (1986).

52. G. Floris, A. Giartosio, and A. Rinaldi, *Arch. Biochem. Biophys.*, *220*, 623 (1983).
53. M. A. Shah and R. Ali, *Biochem. Int.*, *18*, 59 (1989).
54. T. M. Wang, F. M. Achee, and K. T. Yasunobu, *Arch. Biochem. Biophys.*, *128*, 106 (1968).
55. R. H. Suva and R. H. Abeles, *Biochemistry*, *17*, 3583 (1978).
56. H. Kumagi, H. Uchida, and H. Yamada, *J. Biol. Chem.*, *254*, 10913 (1979).
57. D. Mu, S. M. Janes, A. J. Smith, D. E. Brown, D. M. Dooley, and J. P. Klinman, *J. Biol. Chem.*, *267*, 7979 (1992).
58. S. M. Janes, M. M. Palcic, C. H. Scaman, A. J. Smith, D. E. Brown, D. M. Dooley, M. Mure, and J. P. Klinman, *Biochemistry*, *31*, 12147 (1992).
59. S. N. Gacheru, P. C. Trackman, M. A. Shah, C. Y. O'Gara, P. Spacciapoli, F. T. Greenaway, and H. M. Kagan, *J. Biol. Chem.*, *265*, 19022 (1990).
60. R. A. Scott and D. M. Dooley, *J. Am. Chem. Soc.*, *107*, 4348 (1985).
61. P. F. Knowles, R. W. Strange, N. J. Blackburn, and S. S. Hasnain, *J. Am. Chem. Soc.*, *111*, 102 (1989).
62. R. Barker, N. Boden, G. Cayley, S. C. Charlton, R. Henson, M. C. Holmes, I. D. Kelly, and P. F. Knowles, *Biochem. J.*, *177*, 289 (1979).
63. G. J. Barker, P. F. Knowles, K. B. Pandeya, and J. B. Rayner, *Biochem. J.*, *237*, 609 (1986).
64. J. McCracken, J. Peisach, and D. M. Dooley, *J. Am. Chem. Soc.*, *109*, 4064 (1987).
65. D. M. Dooley, M. A. McGuirl, C. E. Cote, P. F. Knowles, I. Singh, M. Spiller, R. D. Brown III, and S. H. Koenig, *J. Am. Chem. Soc.*, *113*, 754 (1991).
66. D. Collison, P. F. Knowles, F. E. Mabbs, F. X. Rius, I. Singh, D. M. Dooley, C. E. Cote, and M. McGuirl, *Biochem. J.*, *264*, 663 (1989).
67. B. Mondovi, A. Finazzi-Agrò, G. Rotilio, and S. Sabatini, in *Frontiers in Bioinorganic Chemistry*, (A. V. Xavier, ed.), VCH, Weinheim, 1986, p. 604.
68. S. Suzuki, T. Sakurai, and A. Nakahara, *Biochemistry*, *25*, 338 (1986).
69. L. Morpugo, E. Agostinelli, O. Befani, and B. Mondovi, *Biochem. J.*, *248*, 865 (1987).
70. D. M. Dooley and D. E. Brown, in *Principles and Applications of*

Quinoproteins (V. L. Davidson, ed.), Marcel Dekker, New York, 1992, p. 173.

71. D. E. Brown, M. A. McGuirl, D. M. Dooley, S. M. Janes, D. Mu, and J. P. Klinman, *J. Biol. Chem.*, *266*, 4049 (1991).

72. D. M. Dooley, D. E. Brown, A. W. Clague, J. N. Kemsley, C. D. McCahon, M. A. McGuirl, P. N. Turowski, W. S. McIntire, J. A. Farrar, and A. J. Thomson, in *Bioinorganic Chemistry of Copper* (K. D. Karlin and Z. Tyeklar, eds.), Chapman and Hall (in press).

73. P. R. Williamson, R. S. Moog, D. M. Dooley, and H. M. Kagan, *J. Biol. Chem.*, *261*, 16302 (1986).

74. D. M. Dooley, W. S. McIntire, M. A. McGuirl, C. E. Cote, and J. L. Bates, *J. Am. Chem. Soc.*, *112*, 2782 (1990).

75. D. M. Dooley, C. E. Cote, M. A. McGuirl, J. L. Bates, J. E. Perkins, R. S. Moog, I. Singh, and P. F. Knowles, in *PQQ and Quinoproteins* (J. A. Jongejan and J. A. Duine, eds.), Kluwer, Dordrecht, 1989, p. 307.

76. D. M. Dooley, M. A. McGuirl, J. Peisach, and J. McCracken, *FEBS Lett.*, *214*, 274 (1987).

77. M. S. Lamkin, T. J. Williams, and M. C. Falk *Arch. Biochem. Biophys.*, *261*, 72 (1988).

78. T. J. Williams and M. C. Falk, *J. Biol. Chem.*, *261*, 15949 (1986).

79. F. T. Greenaway, C. Y. O'Gara, J. M. Marchena, J. W. Poku, J. G. Urtiaga, and Y. Zou, *Arch. Biochem. Biophys.*, *285*, 291 (1991).

80. G. Pettersson, in *Structure and Function of Amine Oxidases* (B. Mondovi, ed.), CRC Press, Boca Raton, 1985, p. 105.

81. C. E. Taylor, R. S. Taylor, C. Rasmussen, and P. F. Knowles, *Biochem. J.*, *130*, 713 (1972).

82. A. Lindström, B. Olsson, and G. Pettersson, *Eur. J. Biochem.*, *35*, 70 (1973).

83. S. Oi, M. Imamasu, and K. T. Yasunobu, *Biochemistry*, *9*, 3378 (1970).

84. A. Lindström, B. Olsson, J. Olsson, and G. Pettersson, *Eur. J. Biochem.*, *64*, 321 (1976).

85. B. Olsson, J. Olsson, and G. Pettersson, *Eur. J. Biochem.*, *74*, 329 (1977).

86. P. F. Knowles, D. J. Lowe, J. Peters, R. N. F. Thorneley, and K. D. S. Yadav, in *The Co-ordination Chemistry of Metallo-enzymes* (I.

Bertini, R. S. Drago, and C. Luchinat, eds.), Reidel, Dordrecht, 1983, p. 159.

87. F. X. Rius, P. F. Knowles, and G. Pettersson, *Biochem. J.*, *220*, 767 (1984).
88. S. M. Janes and J. P. Klinman, *Biochemistry*, *30*, 4599 (1991).
89. C. Hartmann and J. P. Klinman, *Biochemistry*, *30*, 4605 (1991).
90. R. Neumann, R. Hevey, and R. H. Abeles, *J. Biol. Chem.*, *250*, 6362 (1975).
91. I. D. Kelly, P. F. Knowles, K. D. S. Yadav, W. G. Bardsley, P. Leff, and R. D. Waight, *Eur. J. Biochem.*, *144*, 133 (1981).
92. W. G. Bardsley, C. M. Hill, and R. W. Lobley, *Biochem. J.*, *117*, 169 (1970).
93. M. M. Palcic and J. P. Klinman, *Biochemistry*, *22*, 5957 (1983).
94. C. Hartmann and J. P. Klinman, *J. Biol. Chem.*, *262*, 962 (1987).
95. M. C. Falk, *Biochemistry*, *22*, 3740 (1983).
96. A. Lindstrom and G. Pettersson, *Eur. J. Biochem.*, *83*, 131 (1978).
97. R. C. Hevey, J. Babson, A. L. Maycock, and R. H. Abeles, *J. Am. Chem. Soc.*, *95*, 6125 (1973).
98. A. L. Maycock, R. H. Suva, and R. H. Abeles, *J. Am. Chem. Soc.*, *97*, 5613 (1975).
99. B. Olsson, J. Olsson, and G. Pettersson, *Eur. J. Biochem.*, *64*, 327 (1976).
100. K. L. Grant and J. P. Klinman, *Biochemistry*, *28*, 6597 (1989).
101. Y. Cha, C. J. Murray, and J. P. Klinman, *Science*, *243*, 1325 (1989).
102. A. A. Coleman, O. Hindsgaul, and M. M. Palcic, *J. Biol. Chem.*, *264*, 19500 (1989).
103. M. A. Shah, C. M. Scaman, M. M. Palcic, and H. M. Kagan, *J. Biol. Chem.*, *268*, 11573 (1993).
104. A. Bellelli, M. Brunori, A. Finazzi-Agrò, G. Floris, A. Giartosio, and A. Rinaldi, *Biochem. J.*, *232*, 923 (1985).
105. A. Bellelli, A. Finazzi-Agrò, G. Floris, and M. Brunori, *J. Biol. Chem.*, *266*, 20654 (1991).
106. A. Rinaldi, A. Giartosio, G. Floris, R. Medda, and A. Finazzi-Agrò, *Biochem. Biophys. Res. Commun.*, *120*, 242 (1989).
107. D. M. Dooley, M. A. McGuirl, D. E. Brown, P. N. Turowski, W. S. McIntire, and P. F. Knowles, *Nature*, *349*, 262 (1991).

108. A. Lindstrom and G. Pettersson, *Eur. J. Biochem.*, *84*, 479 (1978).

109. J. Grant, I. Kelly, P. Knowles, J. Olsson, and G. Pettersson, *Biochem. Biophys. Res. Commun.*, *83*, 1216 (1978).

110. D. M. Dooley, unpublished.

111. D. M. Dooley, C. E. Cote, and K. C. Golnik, *J. Mol. Catal.*, *23*, 243 (1989).

112. M. Younes and U. Weser, *Biochim. Biophys. Acta*, *526*, 644 (1978).

113. G. Rotilio, L. Calabrese, A. Finazzi-Agrò, and B. Mondovi, *Biochim. Biophys. Acta*, *198*, 618 (1970).

114. J. Z. Pedersen, S. El-Sherbini, A. Finazzi-Agrò, and G. Rotilio, *Biochemistry*, *31*, 8 (1992).

115. J. McCracken, J. Peisach, C. E. Cote, M. A. McGuirl, and D. M. Dooley, *J. Am. Chem. Soc.*, *114*, 3715 (1992).

116. J. Van Iersel, R. A. Van de Meer, and J. A. Duine, *Eur. J. Biochem.*, *161*, 415 (1986).

117. W. S. McIntire, *Meth. Enzymol.*, *188*, 227 (1990).

118. S. Parrott, S. Jones, and R. A. Cooper, *J. Gen. Microbiol.*, *133*, 347 (1987).

119. T. A. Smith and J. H. A. Barker, in *Progress in Polyamine Research* (V. Zappia and A. E. Pegg, eds.), Plenum Press, New York, 1988, p. 573.

120. H. Yanagisawa, E. Hirasawa, and Y. Suzuki, *Phytochemistry*, *20*; 2105 (1981).

121. H. Kuivaniemi, R. M. Korhonen, A. Vaheri, and K. I. Kivirriko, *FEBS Lett*, *195*, 261 (1986).

122. S. Contente, K. Kenyon, D. Rimoldi, and R. F. Friedman, *Science*, *249*, 796 (1990).

123. R. Lewinsohn, *J. Pharm. Pharmacol.*, *33*, 569 (1981).

124. L. A. Alper, in *Haematology* (W. J. Williams, E. Buetler, A. J. Erslew, and M. A. Lichtman, eds.), McGraw-Hill, New York, 1990, p. 1616.

125. U. Bachrach, in *Structure and Function of Amine Oxidases* (B. Mondovi, ed.), CRC Press, Boca Raton, 1985, p. 5.

126. C. Maslinka, T. Bieganski, W. A. Fogel, and M. E. Keller, in *Structure and Function of Amine Oxidases* (B. Mondovi, ed.), CRC Press, Boca Raton, 1985, p. 153.

13

Bacterial Transport of and Resistance to Copper

Nigel L. Brown,[1] Barry T. O. Lee,[2] and Simon Silver[3]

[1]School of Biological Sciences, University of Birmingham, Edgbaston, Birmingham B15 2TT, England

[2]Department of Genetics, University of Melbourne, Parkville, Victoria 3052, Australia

[3]Department of Microbiology and Immunology, University of Illinois College of Medicine, MC/90, Box 6998, Chicago, Illinois 60680, USA

1. INTRODUCTION

Because of their wide diversity of habitats, some bacteria have always been exposed to high concentrations of metal ores. It is not surprising, therefore, that specific resistance mechanisms have evolved to cope with toxic heavy metals. In addition to exposure from naturally occurring ores and during biogeochemical cycling of metals, anthropogenic release of metals has recently increased bacterial exposure to high levels of metals in some environments. Pre-1900 copper smelting gave rise to high localized concentrations of copper near smelting plants [1] and for the last 100 years copper sulfate (as Bordeaux mixture [2]) has been used as a fungicide in horticulture. More recently, copper sulfate has been used in animal feed [3] to increase the rate of weight gain by piglets. Such uses of copper provide suitable conditions in which copper resistance determinants can be selected and spread.

Resistances to mercury, cadmium, and arsenite (with arsenate and antimonite) have been studied in detail and reviewed extensively [4–7]. The resistance mechanisms usually involve detoxification (reduction of mercuric ions to elemental mercury, for example) or the energy-dependent export of the metal cation (e.g., for cadmium) or oxyanion (e.g., for arsenite). The resistance determinants are frequently found on extrachromosomal elements (plasmids) and the encoded proteins often function independently of other cellular processes, apart from those required to maintain the membrane potential or to provide ATP or NADPH. In contrast to the above metals, which have no known beneficial biological function in bacteria, copper is an essential micronutrient which is toxic in excess. Bacterial cells are likely to have systems for the uptake and utilization of copper during normal growth, and any mechanism of resistance to copper must accommodate the requirement for this micronutrient. This means that the mechanisms of resistance to copper must interact with the cellular mechanisms for uptake and utilization of copper, so that the bacterial cell has sufficient copper available for normal metabolism and growth and is protected from the toxic effects of high copper concentrations.

2. COPPER IN BACTERIAL NUTRITION

Copper is a metal that is essential to all living organisms that live in oxic (aerobic) environments. It is essential in a number of enzymes involved in electron transfer and in transport and catalysis of dioxygen, the oxygenases, and hydroxylases. In bacteria it is also required for anaerobic "respiration" in those organisms which reduce nitrate to dinitrogen with an anaerobic electron transport chain. Copper may also be required for reactions not primarily concerned with redox chemistry; for example, one of the isozymes of 3-deoxy-D-arabinoheptulosonate-7-phosphate synthase may be a copper metalloenzyme [8]. Calculations based on growth limitation experiments in bacteria suggest that about 1 mg of copper is required to give 100 g dry weight of biomass [9].

2.1. Role of Copper in Metalloproteins

One of the best understood roles of copper is in electron transport by the blue copper proteins, in which the copper atom is bound in an unusual coordination and has a strong charge-transfer complex absorption at 600 nm. Azurin from *Pseudomonas aeruginosa* is a member of this group of proteins, and it is the electron donor to cytochrome cd_1. The three-dimensional structures of this protein, its mutants, and metal-substituted derivatives have been studied at high resolution by both nuclear magnetic resonance spectroscopy (NMR) and X-ray diffraction (for example, see Refs. 10 and 11). There is one copper atom bound per molecule. Azurin is an excellent model system for studying electron transfer *in vivo* and *in vitro*. The plastocyanins of the cyanobacteria and rusticyanin from *Thiobacillus ferrooxidans* are also bacterial members of this widely distributed class of blue copper proteins.

Specialized enzymes in a large number of bacteria require copper metalloproteins. Copper is required for the electron transfer pathway in bacterial nitrate reduction. The nitrite reductases of *Achromobacter cycloclastes* and *Alcaligenes faecalis* contain copper [12,13] and N_2O reductase contains bound copper in *Pseudomonas* and *Alcaligenes* species [14,15]. Many oxidases and hydroxylases of bacteria are also copper enzymes. The tyrosinase of *Streptomyces* species is a copper-containing oxidase [16]. The majority of superoxide dismutases (SOD) in bacteria are iron-containing enzymes, but a Cu-Zn SOD was originally reported in *Photobacterium leiognathi* [17], and Cu-Zn SODs have subsequently been found in the periplasm of several bacterial species (see Ref. 18). In *Caulobacter crescentus* the enzyme is thought to protect periplasmic or membrane functions requiring Mg or Ca against oxidative damage [18,19]. In pathogenic bacteria Cu-Zn SOD may protect the bacterial cell against superoxide produced by phagocytes.

2.2. Toxicity of Excess Copper

Da Silva and Williams [20] conjectured that copper became an important biological element only with the advent of dioxygen in the earth's atmosphere. The very properties of copper which allow protection against the toxic effects of dioxygen cause excess concentrations of copper to be highly toxic in the presence of dioxygen. Excess copper is toxic to all cells. The concentrations of copper at which toxicity is detected vary between bacterial strains and depend on the form (i.e., the speciation) of the copper in the medium. *Escherichia coli* growing in a rich medium is resistant to about 4 mM $CuSO_4$, whereas *Klebsiella aerogenes* has been reported to be sensitive to concentrations below 1 μM in an unspecified medium [21]. Part of the toxicity is proposed to be due to the binding of free copper [as Cu(I) or Cu(II)] to cysteinyl groups in proteins, thus inactivating essential enzymes. Such copper–thiolate bonds may be oxidized and irreversibly inactivate proteins [22]. Cu(II) is reduced to Cu(I) during the oxidation of sulfhydryl groups to sulfides. Copper can also undergo a Fenton-like reaction [Eq. (1)] *in vivo* with superoxide radicals and hydrogen peroxide [23]. The hydroxyl radicals thereby produced can cause membrane damage by lipid peroxidation as well as enzyme inactivation [24].

$$Cu_2^{2+} + 2H_2O_2 \rightleftharpoons 2Cu^{2+} + 2OH^- + 2OH^\cdot \qquad (1)$$

The bactericidal effect of copper is particularly acute under anaerobic conditions as cuprous ions are formed [25] which are more toxic than cupric ions, probably due to their participation in the reaction shown in Eq. (1). The cytoplasm of bacterial cells is a reducing environment due to the high concentrations of glutathione present. By analogy with studies on eukaryotic cells [26], intracellular glutathione may complex excess copper in the cytoplasm as a copper thiolate. Glutathione concentrations of 5–10 mM are present in the *E. coli* cytoplasm, virtually all of which is in the reduced form [27], and this excess thiol concentration reduces Cu(II) to Cu(I) [Eq. (2)].

$$2Cu^{2+} + 2GSH \rightleftharpoons Cu_2^{2+} + GSSG + 2H^+ \qquad (2)$$

Cu(II)/glutathione mixtures can catalyze DNA cleavage, and there is evidence for the participation of hydroxyl radicals in this cleavage [28]; these radicals may be generated by reactions (2) and (1) above.

2.3. Relationships to Other Metals

Cu(II) is the strongest Lewis acid among the essential metals, and would preferentially bind ligand sites were it not for the remarkable selectivity of biological systems. Multiple coordination of metals by a single macromolecule is often responsible for this metal ion selectivity [20,29,30]. Interactions between copper and zinc in nutrition of animals and man is well established and high zinc intake may produce copper deficiency. Studies *in vivo* and *in vitro* of metal-dependent processes have also shown an interaction between Cu(II) and Zn(II). *E. coli* mutants which show altered susceptibility to Cu(II) sometimes also show altered Zn(II) sensitivity [31,32]. Copper affords protection of *E. coli* against the toxicity of silver [33], presumably by competition of Cu(I) and Ag(I) for the target sites affected by silver toxicity. (These may be sites required for normal copper metabolism by the bacterium.)

3. COPPER TRANSPORT AND UPTAKE

3.1. Uptake and Transport in *E. coli*

Kinetic analysis of uptake of a highly reactive metal cation in trace amounts is difficult in organisms that have complex polysaccharide surfaces and which cannot be depleted of the cation. Most studies of inorganic transport in bacteria [34] have concentrated on nutrients required in moderate amounts or on purely toxic metals of which the organism can be depleted. Studies of the uptake of copper by bacterial cells have mainly been limited to overall accumulation or adsorption of copper (see, for example, Refs. 35–37) or growth limitation (see, for example, Refs. 30 and 38). However, recent information on copper uptake in *E. coli* has come from genetic analysis of copper-sensitive (and copper-sensitive/dependent) mutants. Characterization of these mutants showed that they were altered in the uptake and accumulation of copper (measured using ^{64}Cu). This analysis led to the proposal of a general model for copper uptake and transport in bacteria [31,32]. The proposed model involves uptake, intracellular storage and transport, and export of copper (Fig. 1).

Seven genes on the *E. coli* chromosome have been identified by mutations which alter the sensitivity of the cells to $CuSO_4$. These genes

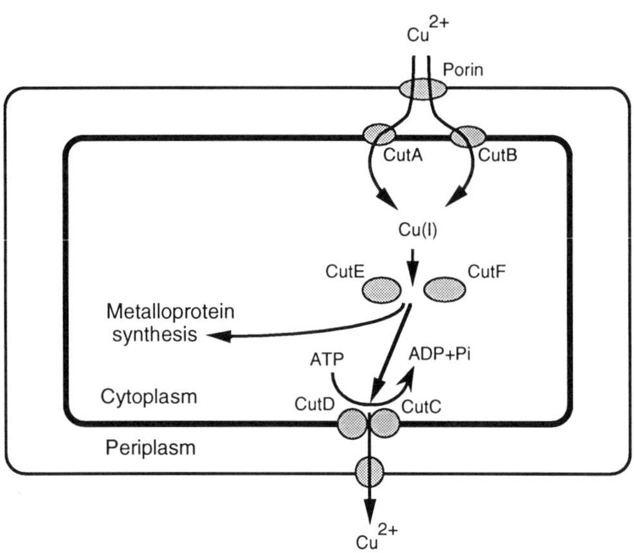

FIG. 1. Model for the uptake, intracellular storage, and homeostasis of copper in *E. coli*. The roles of the proteins are described in the text. (Modified with permission from Ref. 52.)

have been designated *cutA-cutF* and *cutR* (for *c*opper *u*ptake and *t*ransport). Mutations which totally abolish the function of some of these genes might be lethal, and therefore conditional (temperature-sensitive) mutations were isolated so that the physiological effects were only seen at the nonpermissive temperature (42°C) whereas the normal phenotype was exhibited at 37°C. The genetic and physiological properties of the mutants are given in Table 1.

The CutA and CutB proteins appear to be involved in the uptake of copper by the bacterial cell. Mutations in the corresponding genes alter the accumulation of copper by the cells and alter the sensitivity of the cells to other metal cations. The copper-sensitive *cutA* mutant has an increased apparent V_{max} for copper uptake and shows increased copper accumulation in the presence of valinomycin (which inhibits copper export). The *cutA* mutation also confers increased tolerance to zinc, and CutA may be part of the normal zinc uptake system. The *cutB* mutation confers reduced Cu^{2+} uptake (a lower apparent V_{max}) and the *cutB* strain requires elevated levels of copper for growth. It has a higher Ni^{2+} tolerance. Attempts to determine whether copper is taken up as Cu(II) or Cu(I) in *E. coli* by incubating cells with bathocuprein sulfonate in the presence of copper have failed to detect Cu(I) outside the cell (D. A. Rouch and N. L. Brown, unpublished data).

Once within the cytoplasm, free Cu(II) ions would be reduced to Cu(I) by glutathione and could cause oxidative damage to the cell by the generation of hydroxyl radicals as shown in Eq. (1). The products of the *cutE* and *cutF* gene products are thought to minimize this by acting as intracellular storage and transport proteins, sequestering copper and delivering it to the sites of synthesis of copper metalloproteins. The temperature-sensitive *cutE* mutant strain has normal uptake and export characteristics, but is copper-sensitive and copper-dependent. It is proposed that a deficiency in intracellular transport of copper confers a copper-dependent phenotype, as the mutant requires elevated concentrations of external copper to maintain intracellular concentrations within necessary limits for growth; the copper sensitivity is thought to be due to a deficiency in intracellular storage allowing toxic levels of free copper to accumulate. The *cutE* gene has been cloned and its sequence and position in the *E. coli* genome have been determined [39]. The CutE protein is predicted to be 512 amino acids long and to contain a single copper-binding motif His-Phe-Gln-Met-Ala-Arg-Met, similar

TABLE 1

Transport Properties of Copper-Sensitive Mutants of *E. coli*

Mutated gene	Cu tolerance	^{64}Cu accumulation[a]	Uptake kinetics		Inferred activity	
			Apparent K_m	Apparent V_{max}	Efflux	Uptake
cutA	Cu^S	+	N	+	N	+
cutB	$Cu^{S,DEP}$	−	N	+	N	−
cutC	Cu^S	+	N	N	−	N
cutD	Cu^S	+	N	+	−	+
cutE	$Cu^{S,DEP}$	N/+	N	N	N	N
cutF	$Cu^{S,DEP}$	+	N	N	−	N
(none)	Wild-type	N	N	N	N	N

[a]In the presence of valinomycin, copper accumulation increased in *cutB*, *cutA*, and wild-type strains; it remained the same in *cutC* mutant. The remaining strains were not tested. These data are taken with permission from [32].

N, normal; −, decreased; +, increased.

to the His-x-x-Met-x-x-Met motifs of the copper resistance proteins (Fig. 3; *vide infra*). To date this is the only *E. coli* gene sequenced that is known to be involved in copper handling in the absence of a specific copper resistance determinant.

The temperature-sensitive *cutF* mutant is copper-dependent and copper-sensitive. It shows increased accumulation of copper and reduced export. Because the mutation confers copper dependence in spite of the increased accumulation of copper, it was proposed that the CutF protein is responsible for intracellular transport of copper to the export system.

The genes *cutC* and *cutD* may encode export proteins by which excess copper is removed from the cytoplasm. The homeostasis of copper in *E. coli* is therefore maintained by balancing the activities of two uptake systems and one or more export systems. The uptake and export systems may be linked, as the copper-sensitive *cutD* mutant shows increased uptake and decreased export of copper at the nonpermissive temperature. The *cutC* mutant shows increased copper accumulation but no alteration in uptake kinetics compared with the wild type. It has a reduced response to the export inhibitor valinomycin, suggesting that the defect is in an export system. The *cutC* mutant shows increased sensitivity to both copper and zinc, which suggests that this system may export both metals.

The remaining chromosomal gene, *cutR*, was identified by its effect on the expression of a plasmid-borne copper resistance determinant (*vide infra*) and possibly encodes a regulatory protein. This protein may control the expression of the uptake and/or storage and/or export systems in order to maintain the intracellular concentration of copper within acceptable limits. The genetic regulation of copper homeostasis in *E. coli* is considered further in Sec. 4.3.

The role of glutathione (γ-glutamylcysteinylglycine) in normal copper uptake and homeostasis in bacterial cells is likely to be complex. Glutathione is the major small thiol in bacterial cells, occurring at concentrations of 5–10 mM [27]. As well as reducing Cu(II) to the more toxic Cu(I), it also sequesters copper, and may thus protect the cellular constituents against copper-induced damage. Indeed, preliminary experiments (S. Silver and B. T. O. Lee, unpublished data) show that mutants defective in glutathione synthesis [40–42] are more sensitive than wild-type cells to elevated concentrations of copper, although glutathione appears not to protect *E. coli* from oxidative stress by H_2O_2 [43].

3.2. Uptake and Transport in Other Bacteria

Although the *E. coli* system is the most thoroughly described in respect of the number of proteins involved in copper uptake and homeostasis, proteins with specific "transport" functions in normal copper metabolism are known in other genera. Two such proteins are known which are responsible for insertion of copper cations into apoenzymes. An outer membrane protein, NosA, of *Pseudomonas stutzeri* inserts Cu into N_2O reductase [44], and the MelC1 protein of *Streptomyces* species has been shown to insert Cu into tyrosinase [45].

Enterococcus hirae contains two genes predicted from DNA sequence data to encode P-type ATPases [46,47]. These proteins (CopA, 727 amino acids; CopB, 745 amino acids) show 35% sequence similarity, and homology with other P-type ATPases. Their expression is regulated by copper concentration, and CopA is thought to be involved in uptake of copper, as disruption of the *copA* gene renders the cell copper-dependent. Disruption of the *copB* gene causes the cells to be hypersensitive to copper, suggesting a role in copper export. It is assumed that these gene products are required for copper homeostasis in *E. hirae.*

CopA shows 36% homology with CadA, the cadmium-exporting P-type ATPase from *Staphylococcus aureus*, which is responsible for cadmium resistance in this bacterium [48]. All P-type ATPases are cation-translocating membrane pumps, although the identity of the cation and the direction of translocation differ between proteins. The function of the CadA protein is well understood [7] and it contains an N-terminal Cys-x-x-Cys Cd^{2+}-binding motif. The sequence Met-x-x-x-x-Met-x-x-Met occurs three times in the corresponding region of the CopB ATPase, implying that this may be the copper-binding site in this presumptive copper-exporting ATPase.

Cupriavidus necator ("copper-loving killer") is a gram-negative predatory copper-resistant bacterium that requires copper to initiate cell growth [49]. It is reported to produce a peptide growth initiation factor (GIF) to obtain sufficient copper from its environment for initiation of bacterial growth, and production of the GIF is regulated by copper [50]. Further work on the biochemistry of this process is required to provide more concrete data. However, it is claimed [50] that the copper–GIF complex may be involved in killing prey bacteria; one can speculate that this may occur through the copper-dependent generation of hydroxyl radicals.

4. BACTERIAL SYSTEMS CONFERRING RESISTANCE TO EXCESS COPPER

Bacterial resistance to copper was recently described in several reviews [51–56]. Resistance to antibacterial agents is frequently the result of specific (and often extrachromosomal) genetic determinants [4,6,57]. Two such systems for copper resistance have been studied in detail: the *cop* determinant from *Pseudomonas syringae* pv. *tomato* and the *pco* determinant from *E. coli*. The understanding of the mechanisms of copper resistance is evolving rapidly and the current picture is quite confused, especially as apparently dissimilar mechanisms are conferred by genes showing great sequence similarity.

4.1. The *cop* Determinant from *Pseudomonas syringae*

Different copper-resistant strains of *P. syringae* that infect tomato in California carry a highly conserved plasmid, pTD23, that confers copper resistance [58] and from which the copper resistance genes were cloned [59]. Resistant strains of *P. syringae* can tolerate up to 2 mM Cu^{2+} in minimal media, in comparison to 0.1–0.2 mM Cu^{2+} for most sensitive strains of *P. syringae*. The mechanism of copper resistance involves copper accumulation, and copper-resistant colonies of *P. syringae* turn bright blue when grown with elevated levels of Cu^{2+} and can accumulate up to 0.3% dry weight as copper. This accumulation is *cop*-dependent and is specific for copper [60]. Aluminum also accumulates but can be removed by washing in 20 mM EDTA; whereas only about 50% of the copper is removed under these conditions.

The structural genes from the *cop* determinant were cloned [59] and their sequences were determined [61]. The sequence revealed four structural genes, *copABCD*, in the same orientation, giving rise to predicted gene products of 609, 328, 126, and 310 amino acids (Fig. 2). The *copB* gene contains five repeats of 24 base pairs, which would give

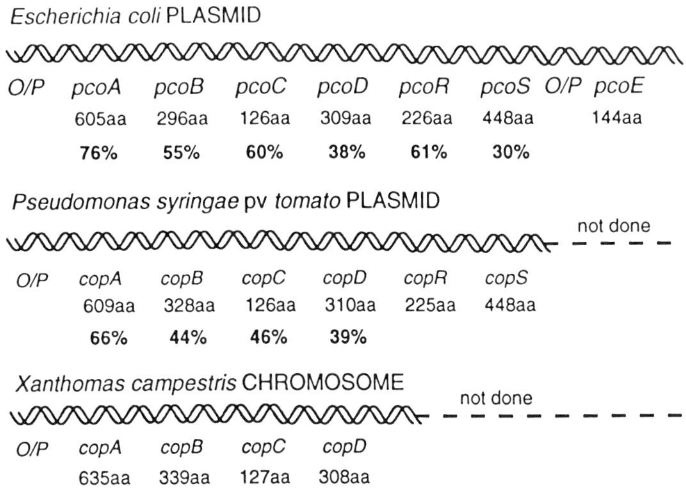

FIG. 2. Alignment of the sequenced regions of the *pco* determinant of *E. coli* and the *cop* determinants of *P. syringae* and *X. campestris* showing the predicted protein sizes and the percentage amino acid identities between similar pairs of proteins. O/P, operator–promoter region.

rise to five octapeptide repeat motifs Asp-His-Ser-Gln/Lys-Met-Gln-Gly-Met in the gene product. Four copies of a similar motif were observed in the CopA gene product but were less well conserved. Similar motifs have now been found in other copper resistance proteins (below) and in the *cutE* gene product from *E. coli* (Fig. 3), and these are hypothesized to be the sites of copper binding, although proof of this has not been obtained.

A series of specific deletion mutations showed that all four *cop* genes were required for expression of full resistance but that only the *copA* and *copB* genes are essential for partial resistance [61]. The DNA sequence of the genes also suggested that the proteins contained N-terminal signal sequences and were thus likely to be membrane-bound or extracellular. Processing of the signal sequences was demonstrated by direct amino acid sequence analysis and the subcellular locations of the proteins were determined using specific antibodies [62].

CopA is a periplasmic protein (M_r = 72,000) that binds about 11 copper atoms per polypeptide chain. It is expressed as about 3% of total protein following induction by $CuSO_4$. The sequence of CopA, in addi-

Pseudomonas syringae proteins:									
CopA	375-382	Asp	**His**	Gly	Ser	**Met**	Asp	Gly	**Met**
	408-415	Asp	**His**	Ser	Lys	**Met**	Ser	Thr	**Met**
	419-426	Asp	**His**	Gly	Ala	**Met**	Ser	Gly	**Met**
	427-434	Asp	**His**	Gly	Ala	**Met**	Gly	Gly	**Met**
CopB	36-43	Asp	**His**	Ser	Gln	**Met**	Gln	Gly	**Met**
	44-51	Asp	**His**	Ser	Lys	**Met**	Gln	Gly	**Met**
	52-59	Asp	**His**	Ser	Gln	**Met**	Gln	Gly	**Met**
	60-67	Asp	**His**	Ser	Lys	**Met**	Gln	Gly	**Met**
	68-75	Asp	**His**	Ser	Gln	**Met**	Gln	Gly	**Met**
CopA/B Consensus		Asp	His	Ser/ Gly	xxx	Met	xxx	Gly	Met
Escherichia coli proteins:									
PcoA	375-381	Gly	**His**	Asp	...	**Met**	Ala	Gly	**Met**
	382-389	Asp	**His**	Ser	Gln	**Met**	Gly	Gly	**Met**
	414-421	Asp	**His**	Ser	Ser	**Met**	Ala	Gly	**Met**
	422-429	Asp	**His**	Ser	Arg	**Met**	Ala	Gly	**Met**
PcoB	34-40	Gly	**His**	Asp	...	**Met**	Ser	Ala	**Met**
PcoE	26-33	**Met**	**His**	Asp	Gln	Val	Asn	Asn	Ala
	37-43	Ala	**His**	...	Gln	**Met**	Gln	Ser	Ser
	61-67	Ser	**His**	Asp	Gln	...	Ala	Ala	**Met**
	68-75	Ser	**His**	Asp	**Met**	**Met**	Gln	Asn	Gly
	79-86	Ala	**His**	Gln	Asp	**Met**	Ala	Glu	**Met**
	87-92		**His**	Lys	Lys	**Met**	**Met**	Lys	Ser
CutE	425-431	Gln	**His**	Phe	Gln	**Met**	Ala	Arg	**Met**

FIG. 3. Alignment of the presumptive copper-binding motifs in copper resistance proteins and in the chromosomal gene product CutE. Gaps (. . .) are to maximize alignment. Potential ligand-binding methionine and histidine residues are shown in boldface.

tion to four repeats related to the sequence Asp-His-x-x-Met-x-x-Met (Fig. 3), has been noted to contain a sequence near its C terminus showing similarity to the type 1 copper site of ascorbate oxidase from zucchini plants [63] (Fig. 4). Further similarities to the three remaining (one type 2 and two type 3) copper binding sites observed in ascorbate oxidase [64,65] have been identified in the CopA sequence [51] (Fig. 4). The binding of copper by the CopA protein is obviously complex and its exact role in copper resistance is unclear.

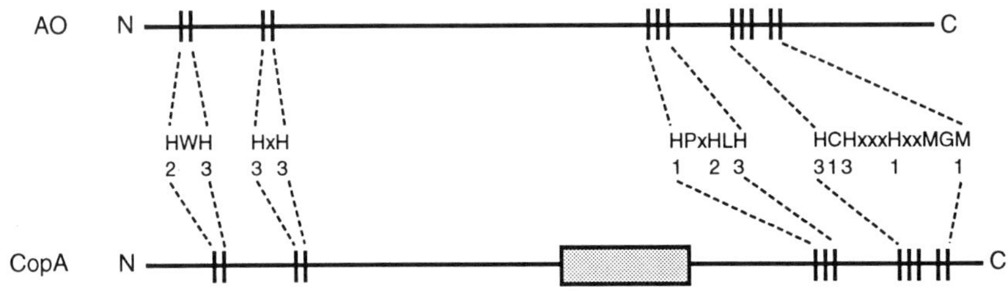

FIG. 4. Comparison of zucchini ascorbate oxidase (AO) and *P. syringae* CopA showing the presumed copper binding sites. The shaded box shows the position of the four repeated motifs (Asp His xx Met x Met) described in the text. Amino acids in the single letter code (H, histidine; W, tryptophan; P, proline; L, leucine; C, cysteine; M, methionine; G, glycine) are shown between the alignments, with x representing conservative differences. The amino acid ligands for copper atoms at the type 1, type 2, and type 3 copper binding sites in the AO crystal structure are shown as 1, 2, and 3, respectively. Details of these ligands are given in Refs. 64 and 65. The amino (N) and carboxyl (C) termini of the proteins are marked. (Taken with permission from Ref. 51.)

CopB is an outer membrane protein [62] and, as mentioned, it contains five repeated Asp-His-Ser-Gln/Lys-Met-Gln-Gly-Met repeats, but the copper binding stoichiometry of the protein has not been determined. Outer membrane preparations containing CopB are blue, indicating bound copper, but this copper is lost during purification of the CopB protein. The CopB protein shows no homology to the outer membrane copper-binding protein, NosA, of *P. stutzeri* [66].

CopC is a small periplasmic protein [62] and binds one Cu^{2+} per polypeptide. It is abundant in copper-induced resistant cells, reaching 1% of total protein. CopD is an inner membrane protein, but there is no evidence for its function. Because of its cellular location, it may be required for copper transport across the inner membrane. Expression of *copC* and *copD* in the absence of *copA* and *copB* results in hypersensitivity to copper [53]. This is reminiscent of hypersensitivity to mercuric salts conferred when the membrane transport functions are expressed in the absence of the detoxification functions [67,68] and suggests that copper is being transported into the cytoplasm in the absence of the presumptive sequestration proteins [53].

A model for copper resistance in *cop*-containing *P. syringae* proposed by Cooksey [53] is shown in Fig. 5. Excess copper outside the cell

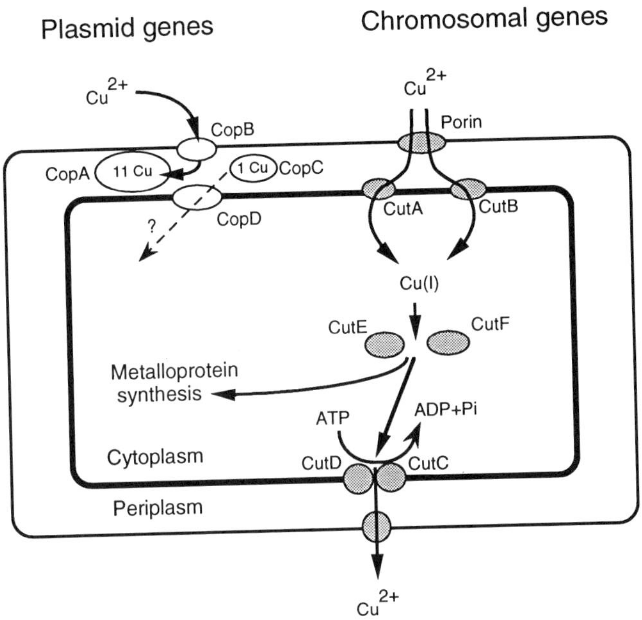

FIG. 5. Model for the *cop*-encoded resistance to copper salts in *Pseudomonas syringae*. The chromosomally encoded proteins are assumed to be related to those identified in *E. coli* (Fig. 1). The presumed roles of the CopA–D proteins are described in the text.

is bound by the outer membrane protein, CopB, which may facilitate transport of Cu^{2+} to the periplasmic compartment, where it is bound by CopA. The proteins CopC and CopD may constitute an uptake system that can provide sufficient copper for cell growth to the cytoplasmic compartment even in the presence of the strong sequestration activity of CopA. This model explains the observations that (1) CopA and CopB are essential to resistance, (2) CopC and CopD are required for maximal resistance (if one assumes that their absence would cause the high concentrations of CopA produced at high copper concentrations to limit copper uptake and growth), and (3) CopC and CopD alone cause hypersensitivity due to excessive uptake of copper. However, the accumulation of CopA with time does not mirror the accumulation of copper [62], suggesting that the mechanism is more complex than proposed. CopA may initially bind copper in the periplasm before delivering it to sequestration sites, perhaps on the membrane.

4.2. The *pco* Determinant from *E. coli*

An *E. coli* isolate from a piggery in which copper sulfate had been used as a food supplement contained a conjugative copper resistance plasmid, pRJ1004 [69]. When this plasmid was transferred to laboratory strains of *E. coli*, their tolerance of $CuSO_4$ in Luria–Bertani (LB) broth medium at pH 7.5 was increased from about 4 mM to about 20 mM $CuSO_4$. The real increase in resistance to free copper ions is probably more than fivefold, since the components of complex organic media bind metal ions [30,70]. Copper-resistant colonies appear dark brown on copper-containing LB agar plates and this was thought to be a precipitate of a "modified form" of copper. On incubation with $^{64}Cu^{2+}$ *E. coli* strains containing pRJ1004 accumulate less copper than the nonresistant parent, indicating that the mechanism of resistance includes either reduced uptake and/or increased export of copper [31,54]. These results are in complete contrast to those in *P. syringae* strains containing the apparently equivalent *cop* determinant, as resistant *Pseudomonas* cells accumulate copper and turn blue.

The copper resistance determinant, *pco* (for *p*lasmid-borne *co*pper resistance), responsible for this reduced accumulation was cloned from pRJ1004 into a low-copy-number plasmid and mutagenized by random insertion of the transposon Tn*1000* [31,54]. Copper resistance was lost in mutations in which the transposon was inserted within a 5.8-kb region of DNA. Four classes of mutation were identified by complementation analysis and designated *pcoARBC*. Subsequent DNA sequence analysis of this region has shown that the complementation group *pcoA* contains four genes with sequence similarities to *copABCD*. An additional region outside the 5.8-kb sequence shows homology to the repeated motifs in *copA* and *copB* (Fig. 3) [52]. These data led to a renaming of the putative genes in the *pco* determinant, and the seven genes *pcoABCDRSE* are now recognized [55] (Fig. 2).

The gene products of *pcoABCD* are similar to those of *copABCD* (Fig. 2). The sequence identity between CopA and PcoA is 76%; between CopB and PcoB, 54%; between CopC and PcoC, 60%; and between CopD and PcoD, 38% (S. R. Barrett, B. T. O. Lee, D. A. Rouch, and N. L. Brown, unpublished data). The PcoA protein retains all the sequence characteristics of the CopA protein in number and identity of potential copper binding sites, both of the octapeptide sequence and type 1, 2, and 3 copper binding sequences. Both proteins show similarity in arrangement of these binding sequences to the blue copper protein ascorbate oxidase. The PcoB protein contains only one of the octapeptide sequences instead of four in the N-terminal end of CopB. The C-terminal 80% of the proteins are well conserved and may have a function unrelated to copper binding. The PcoC protein and the CopB protein contain the motif Met-x-x-Met-x-x-His, which may have a role in copper recognition by these proteins. Both the PcoD and CopD proteins have potential transmembrane helices and conserved histidine and methionine groups are predicted to lie on the outer face of the cytoplasmic membrane. Until we have some indication of the role of each of these proteins in conferring resistance, we are unable to assess the significance of the similarities and differences between these proteins. The needed data are presently being obtained.

In view of the apparently different mechanisms of copper resistance conferred by *cop* and by *pco*, it would be rash to assume that the close evolutionary relationship between these proteins is reflected in identical or similar current functions.

The function of the *pcoR* and *pcoS* genes, which follow *pcoABCD*, is in regulation of expression of the resistance genes, as discussed in Sec. 4.3.

A presumptive gene, *pcoE* [52], was identified as a open reading frame in the DNA sequence with homology to the Asp-His-x-x-Met-x-x-Met repeat sequences found in CopA, CopB, PcoA, and PcoB (Fig. 3). A copper-inducible promoter has been identified upstream of *pcoE* (D. A. Rouch and N. L. Brown, unpublished data), indicating that the gene may be part of the copper resistance determinant. The sequence of the protein suggests that it is a periplasmic protein with an N-terminal leader sequence.

The model for copper resistance is based on physiological studies of the accumulation and export of $^{64}Cu^{2+}$ by *E. coli* K12 strain ED8739 (pRJ1004) and its derivatives lacking *pco* or containing chromosomal mutations [31,54]. Plasmid pPA173, which contains *pco* in a low-copy-number plasmid, does not confer resistance in the chromosomal enhanced-uptake *cutA* mutant and in the temperature-sensitive *cutD* efflux mutant at the nonpermissive temperature, but does confer resistance in the remaining *cut* mutants and in the wild-type parental strain. Resistance is also lost in the presence of arsenate or the *uncA* energy-uncoupling mutation. The tentative interpretation of these results is that the plasmid-encoded resistance mechanism includes an energy-dependent copper export process in which the *cutA* and *cutD* chromosomal gene products participate.

The presence of a brown precipitate around resistant colonies was thought to be due to the sequestration of copper in a complex that rendered it unavailable for subsequent uptake by cells [31,54]. More recent experiments [51] have shown that in liquid medium this precipitate is CuS, with associated amino acids, predominantly Glu/Gln, Gly, and Cys with some Thr, Ala, and Pro. A stable copper complex unique to the resistant strain could not be identified. Although the amino acids present suggest that glutathione may form part of the precipitate and be directly involved in resistance, chemical depletion of intracellular glutathione levels does not abolish resistance; it merely slows the growth of both resistant and parental strains [51].

Total copper accumulation in minimal medium by resistant and parental *E. coli* strains was measured by atomic absorption spectroscopy rather than $^{64}Cu^{2+}$ labeling, thus altering any effect of differential copper pool sizes [51]. The pRJ1004-containing resistant strain was shown to accumulate about 30% of the amount of copper compared with the parental strain in the presence of 0.3 mM CuSO4 and 100 μM each amino acid [51]. Under similar conditions *P. syringae* pv. *tomato* containing the copper resistance plasmid pCOP2 showed *increased* accumulation of copper [60]. This difference in accumulation, and therefore of mechanism of resistance, between the two systems is especially puzzling in view of the similarity of the genes in the two determinants. The different numbers of Asp-His-x-x-Met-x-x-Met octapeptide repeats found in the CopB and PcoB proteins (five and one, respectively; Fig. 3) may relate to this difference in mechanism.

4.3. Regulation of Copper Resistance Genes

Rouch *et al.* [31,54] investigated the regulation of the *pco* genes. The expression of the *pco* genes was shown to be copper-inducible by increased levels of hybridizing mRNA in the presence of 0.8 mM Cu^{2+}. Fusions of a transcriptional promoter from *pco* allowed copper-dependent expression of a *lacZ* reporter gene in the presence of *pco*. Tn*1000* insertion mutants of *pcoR* showed enhanced expression of the reporter gene, indicating that PcoR may function as a negative regulator of transcription. In the absence of *pco*, *lacZ* gene expression was regulated in a copper-dependent manner by one or more unidentified chromosomal gene(s). Inactivation of one of these chromosomal genes by Tn*10* mutagenesis allowed copper-independent expression of the promoter fusion; but reintroduction of *pcoR* into this mutant strain restored copper-dependent regulation. These data suggested that *pcoR* was a copper-dependent regulatory gene and a functional homolog of *pcoR* was present on the chromosome; this was designated *cutR*. The *cutE* gene is inducible by copper *in vivo* (S. D. Rogers and B. T. O. Lee, unpublished data), and may be regulated by *cutR*, although this has not been demonstrated directly. This gene is also regulated by Zn^{2+} (S. D. Rogers and B. T. O. Lee, unpublished data), thus providing a direct example of the interaction of these metals in bacteria, although the mechanism is unknown.

Two copper-inducible promoters have been identified in *pco*. One is responsible for expression of *pcoABCD*, and the other is immediately in front of the *pcoE* gene. The *pcoRS* promoter may be constitutive like the *Pseudomonas copRS* promoter [71]. Immediately following the *pcoE* gene is a sequence which shows the stem-loop structure and run of T residues indicative of a ρ-independent terminator of transcription [72]. The *pcoE* gene is likely to be transcribed as a monocistronic mRNA, whereas the *pcoABCD* operon is polycistronic. The two promoters show good -10 and -35 elements typical of eubacterial σ^{70} promoters [73].

Sequencing of *pcoR* showed that its gene product was similar to positive activators which are members of two-component regulator-sensor systems [74] and that the adjacent gene, designated *pcoS*, showed similarity to the sensor members of this family [52] (S. Barrett, D. A. Rouch, B. T. O. Lee, and N. L. Brown, unpublished). Interestingly, the *pcoS* gene spanned the physical position of two of the original genetic complementation groups, indicating that there may be more than one product from this gene. The exact interrelationship of the plasmid-borne regulatory genes and the chromosomal regulatory gene(s) is as yet unclear. However, the interaction of these two sets of genes is undoubtedly important in the homeostasis of intracellular copper, and in allowing cells to survive in both low and high external copper concentrations. A model for the regulation of the *pco* copper resistance genes from *E. coli* is shown in Fig. 6.

The expression of the *copABCD* genes was found to be inducible by copper and from a single promoter in front of the *copA* gene [75]. This promoter has not been fully characterized, but the proportion of *cop* mRNA to total RNA increases with $CuSO_4$ concentration between about 1 μM and 1 mM Cu^{2+} in minimal medium, and increases with time of induction, peaking at about 4 hr when induced with 100 μM $CuSO_4$. Recently, two trans-acting genes responsible for regulation of this promoter have been cloned and sequenced from downstream of the *copABCD* operon [71]; these genes, *copR* and *copS*, are related to the regulatory genes *pcoR* and *pcoS* of the *E. coli* copper resistance determi-

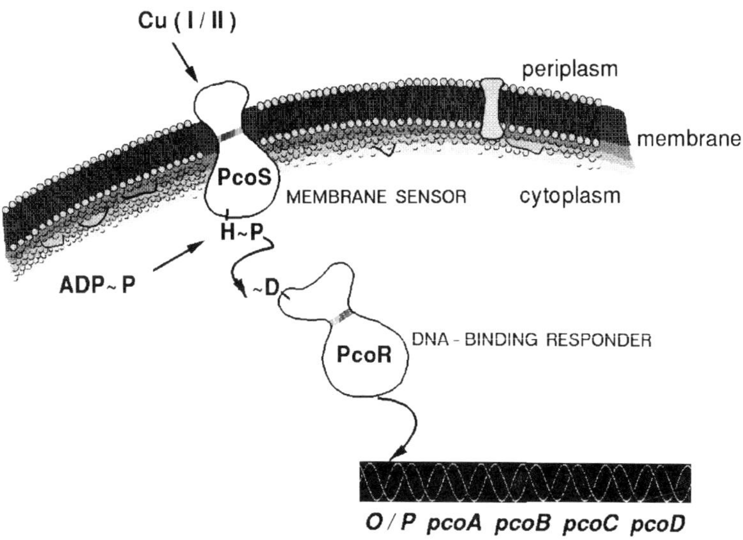

FIG. 6. Model for the regulation of the *pco* genes of *E. coli* by the PcoR and PcoS gene. The cartoon is derived from the known properties of other two-component systems in bacteria [74,87], in which the sensor, PcoS, becomes phosphorylated at a conserved histidine residue (H) in response to an environmental signal (here, increased external copper concentration); the sensor then phosphorylates the responder (or regulator), PcoR, at a conserved aspartate (D) residue, and phosphorylated PcoR binds DNA. The involvement of homologous chromosomal regulatory functions is not shown.

nant [52]. In the absence of *copRS*, the *copABCD* promoter is regulated by genes on the chromosome of *P. syringae* [71].

In the unrelated gram-positive bacterial species *Streptomyces lividans*, the chromosomal genes *cutR* (and *cutS*) regulate expression of the MelC1 protein, which is involved in copper metabolism [76]. The sequence of *cutR* and partial sequence of *cutS* indicate that these are also members of a two-component regulatory system. If the promoters for the *pcoABCD* and *pcoE* genes are cloned in front of a reporter gene (e.g., *lacZ* or *cat*), they are still regulated in *E. coli* in the absence of the remainder of the resistance determinant [31] (D. A. Rouch and N. L. Brown, unpublished data), thus providing genetic evidence for chromosomal homologs of the plasmid-borne *pcoRS* regulatory functions. Similar evidence has been obtained for chromosomal homologs of *copRS* [71]. DNA hybridisation studies indicate the presence of *pco*-related sequences on the *E. coli* chromosome [77] and of *cop*-related sequences in pseudomonads [78]. One may speculate that chromosomal regulation of copper uptake and transport functions, and therefore of copper homeostasis, is regulated by two-component systems in at least three bacterial genera, and possibly in a larger number.

4.4. Other Bacterial Copper Resistance Systems

Many bacteria adsorb heavy metals to extracellular polysaccharide, as described, for example, in Ref. 79. Strains which produce large amounts of extracellular polysaccharide may show much increased tolerance to heavy metals. If extracellular polysaccharide production is a response to exposure to heavy metals, then this may properly be deemed a resistance mechanism. The copper tolerance of many methanotrophs may be due to the requirement for copper in methane oxidation [80], as they are very sensitive to mercury and cadmium. The copper resistance of *C. necator* may be required because of its use of a copper–peptide complex as a bactericide against other bacteria [50]. Copper-resistant strains of *Alcaligenes denitrificans*, *Alcaligenes* species, and *Pseudomonas paucimobilis* have been isolated and shown to have an inducible resistance to copper [81].

Enterococcus hirae has been reported to show inducible resistance to copper [47]. More recent work [46] suggests that this apparent resistance is due to the genes for normal copper management in this species. The roles of the genes are described in Sec. 3.2. Interestingly, these gram-positive bacterial CopA, CopB proteins and the CadA cadmium export protein from *S. aureus* show similarity to the putative copper export ATPase encoded by the candidate gene for Menkes' disease in humans [82]. Menkes' disease is an X-linked lethal disorder of copper metabolism, in which copper-requiring proteins are deficient because of lack of copper; yet fibroblasts from Menkes patients accumulate excess copper. The putative copper-transporting ATPase has structural features similar to the CadA (see Fig. 7) and may be responsible for export of copper from normal cells. There is a sixfold repeated Cys-x-x-Cys motif, similar to bacterial Cd^{2+}- and Hg^{2+}-binding motifs, and a Cys-Pro-Cys-x-x-x-x-x-Pro pattern in the predicted transmembrane region, which is thought to be required for Cd^{2+} efflux in CadA [83].

DNA hybridisation studies have recently become an effective way of identifying the presence of specific resistance genes in a microorganism. Copper-resistant strains of a number of bacteria have been isolated from tomato plants or seeds and shown to contain sequences homologous to the *cop* determinant of *P. syringae* [78,84]. These strains include *Xanthomonas campestris* pv. *vesicatoria*, *P. cichorii*, *P. putida*, *P. fluorescens*, and other *Pseudomonas* strains. Of these only *X. campestris* has been further characterized genetically and shown to contain a copper resistance plasmid [85,86]. The *X. campestris cop* genes from a walnut pathogen in California were cloned and sequenced (Y. Lee and M. N. Schroth, personal communication). The *Xanthomonas* sequence contains genes with similarity to *pcoABCD* and *copABCD*; but, surprisingly, as *Xanthomonas* is a pseudomonad, these are more similar to the *E. coli pco* genes than to the *P. syringae cop* genes. The *Pseudomonas* strains identified as containing sequences with DNA homology to *cop* were shown to produce the CopB protein [78]. One new *Pseudomonas* strain accumulated 12% copper by dry weight [60]. Interestingly, copper-sensitive pseudomonads were shown to contain chromosomal sequences that hybridized to *cop*, indicating that the *cop* determinant may have arisen from chromosomal genes.

Additional DNA hybridisation and polymerase chain reaction studies have been performed on copper-resistant bacteria isolated from pig farms in the United Kingdom and in Australia, using sequences from the *E. coli pco* determinant as probes [77]. These studies identified

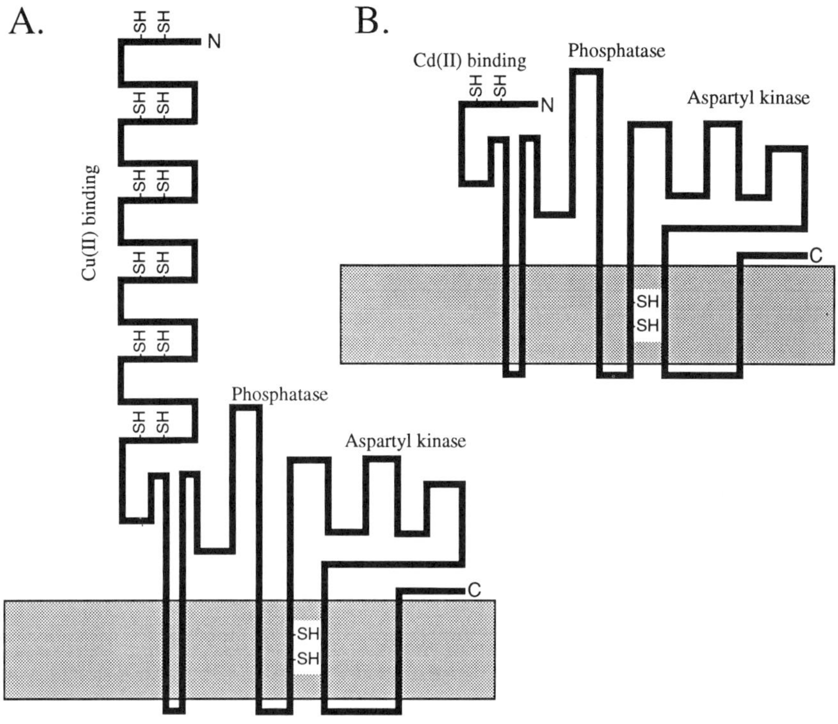

FIG. 7. The predicted topographical features and functional domains of (A) the human presumptive copper transport protein identified from the defect in Menkes' disease, and (B) the CadA cadmium resistance ATPase of *S. aureus* [83]. The shaded box represents a biological membrane (for CadA this is the cell membrane, with the cytoplasmic face uppermost). The cysteine thiol groups in each protein which are assumed to be involved in metal binding are shown as -SH. Other conserved residues (e.g., a conserved proline between the cysteine thiols in the membrane, a proline and conserved phosphorylated aspartate in the aspartyl kinase domain, etc.) are not marked, but are described in Refs. 82 and 83.

pco-homologous sequences in *E. coli*, *Citrobacter freundii*, and *Salmonella* species. Most of these resistances were inducible. The copper-resistant determinants from at least one *E. coli* and one *C. freundii* were transferrable to laboratory strains. Similarly, plasmid pRJ1004 can be transferred to, and confers copper resistance on, *Salmonella enteritidis*, *Salmonella typhimurium*, and *Klebsiella pneumoniae* strains. However, pRJ1004 could not be transferred or function in *Pseudomonas aeruginosa*, *Yersinia* species, *Aeromonas* species, *Shigella sarni*, *Enterobacter aerogenes*, *Proteus mirabilis*, or *Citrobacter diversi* under the experimental conditions used [31]. These negative results might be due to the narrow host range of the plasmid pRJ1004 or to differential expression of the *pco* genes in these species. The *cop* genes are not expressed when the genes are cloned and transformed into *E. coli*, indicating a failure at the level of transcription or in protein function.

5. CONCLUDING REMARKS

Unlike the well-characterized systems for bacterial resistance to mercury and cadmium, or the more recently characterized arsenate, arsenite, and antimonite resistance system, the mechanisms of copper resistance are poorly understood. Part of this is due to the studies being newer and being carried out in few laboratories. However, it may also be in part due to the intrinsic complexity of a system in which the resistance mechanism has to interact with the normal copper management systems of the nonresistant cell. Furthermore, our lack of a clear current understanding of the mechanism of copper resistance arises from the conundrum of having two systems (in *Pseudomonas* and *E. coli*) which are closely related in amino acid sequence, but which work by apparently opposite mechanisms, greater uptake and sequestration in *Pseudomonas* vs. reduced uptake and cellular exclusion in *E. coli*. We anticipate that this riddle will be solved in the near future. Undoubtedly more bacterial copper resistance systems remain to be discovered and characterized. A combination of genetic, physiological, and chemical techniques will have to be applied to the organisms, their genes and gene products, in order to elucidate the fundamental mechanisms whereby bacteria maintain homeostasis of this toxic but essential metal. One can predict that the normal copper management systems in the noncompartmentalized bacterial cell and the specific systems for resistance to high copper concentrations will sequester copper ions on proteins to prevent the formation of lethal hydroxyl radicals. The sequence similarities between the human presumptive copper transporter, which is defective in Menkes' disease, and bacterial metal resistance proteins will provide added impetus to this research.

ACKNOWLEDGMENTS

We are grateful to our colleagues and collaborators, particularly Siobhàn Barrett, Carol Bender, Jim Bryson, Jim Camakaris, Don Cooksey, Tom O'Halloran, Duncan Rouch, Sue Rogers, Milton Schroth, Mark Solioz, and Jill Williams, for communicating data prior to publication and for discussion and comment. Work from our laboratories reported here has been supported by grants from the Agricultural and Food Research Council, UK and the Medical Research Council, UK (to N.L.B.), from the Australian Research Council (to BTOL), and from the U.S. National Science Foundation (to S.S.). The sponsorship of The Royal Society, the British Council, the National Science Foundation, and the Australian Department of Industry, Trade and Commerce for travel between our laboratories is gratefully acknowledged.

DEFINITIONS AND ABBREVIATIONS

In accordance with standard bacterial genetic nomenclature, names in lower case italics (e.g., *pcoA*) represent genes and names in normal type with an initial capital letter represent the corresponding protein (e.g., PcoA).

lacZ is a gene encoding β-galactosidase. β-Galactosidase activity is often used to measure indirectly the activity of a promoter cloned in front of a *lacZ* "reporter gene".

cat is a gene encoding chloramphenicol acetyltransferase and is similarly used as a reporter gene.

A plasmid is an autonomously replicating DNA molecule within a bacterial cell in addition to the bacterial chromosome. Plasmids usually carry nonessential functions (such as resistance to antimicrobials).

ADP adenosine 5′-diphosphate
Ala L-alanine

Arg	L-arginine
Asp	L-asparagine
ATP	adenosine 5′-triphosphate
cop	*cop*per resistance
cut	*c*opper *u*ptake and *t*ransport
Cys	L-cysteine
EDTA	ethylenediamine-N,N,N′,N′-tetraacetate
GIF	growth initiation factor
Gln	L-glutamine
Glu	L-glutamate
Gly	L-glycine
GSH	reduced glutathione (γ-glutamylcysteinylglycine)
GSSG	oxidized glutathione
His	L-histidine
LB	Luria–Bertani medium
Lys	L-lysine
Met	L-methionine
NADPH	reduced nicotinamide adenine dinucleotide phosphate
NMR	nuclear magnetic resonance spectroscopy
pco	*p*lasmid-borne *co*pper resistance
Phe	L-phenylalanine
P_i	orthophosphate
Pro	L-proline
Ser	L-serine
SOD	superoxide dimutase
Thr	L-threonine
unc	energy-*unc*oupling mutation
x	any amino acid (in peptide sequences)

REFERENCES

1. T. C. Hutchinson, in *Copper in the Environment* (J. O. Nriagu, ed.), John Wiley and Sons, New York, 1979, p. 451 ff.
2. D. E. Baker, *Fed. Proc.*, *33*, 1188 (1974).
3. G. K. Davis, *Fed. Proc.*, *33*, 1194 (1974).
4. S. Silver and T.K. Misra, *Ann. Rev. Microbiol.*, *42*, 717 (1988).

5. A. O. Summers, *Trends Biotechnol.*, *3*, 122 (1985).
6. S. Silver (ed.), *Plasmid-Determined Metal Resistance Mechanisms, Plasmid Special Issue*, Vol. 27, Academic Press, New York. 1992.
7. S. Silver and M. Walderhaug, *Microbiol. Rev.*, *56*, 195 (1992).
8. T. Baasov and J. R. Knowles, *J. Bacteriol.*, *171*, 6155 (1989).
9. S. J. Pirt, *Principles of Microbe and Cell Cultivation*, Blackwell, Oxford, 1975.
10. H. Nar, A. Messerschmidt, R. Huber, M. van de Kamp, and G. W. Canters, *FEBS Lett.*, *306*, 119 (1992).
11. M. van de Kamp, G. W. Canters, S. S. Wijmenga, A. Lommen, C. W. Hilbers, H. Nar, A. Messerschmidt, and R. Huber, *Biochemistry*, *31*, 10194 (1992).
12. T. Matsubara, K. Frunzke, and W. G. Zumft, *J. Bacteriol.*, *149* , 816 (1982).
13. T. Kakutani, H. Watanabe, K. Arima, and T. Beppu, *J. Biochem. (Tokyo)*, *89*, 453 (1981).
14. R. A. Scott, W. G. Zumft, C. L. Coyle, and D. M. Dooley, *Proc. Natl. Acad. Sci. USA*, *86*, 4082 (1989).
15. W. G. Zumft, A. Dreusch, S. Lochelt, H. Cuypers, B. Friedrich, and B. Schneider, *Eur. J. Biochem.*, *208*, 31 (1992).
16. K. Lerch and L. Ettinger, *Eur. J. Biochem.*, *31*, 427 (1972).
17. K. Puget and A. M. Michelson, *Biochem. Biophys. Res. Commun.*, *58*, 830 (1974).
18. H. M. Steinman, *J. Bacteriol.*, *175*, 1198 (1993).
19. H. M. Steinman, *J. Biol. Chem.*, *257*, 10283 (1982).
20. J. J. R. Fraústo da Silva and R. J. P. Williams, *The Biological Chemistry of the Elements*, Clarendon Press, Oxford, 1991.
21. J. T. Trevors and C. M. Cotter, *J. Indus. Microbiol.*, *6*, 77 (1990).
22. R. B. Martin, in *Concepts on Metal Ion Toxicity*, Vol. 20 of *Metal Ions in Biological Systems*, (H. Sigel and A. Sigel, eds.), Marcel Dekker, New York, 1986, p. 21 ff.
23. S. Goldstein and G. Czapski, *J. Free Rad. Biol. Med.*, *2*, 3 (1986).
24. N. H. Stacey and C. D. Klaassen, *J. Toxicol. Environ. Health*, *7*, 139 (1981).
25. P. H. Beswick, G. H. Hall, A. J. Hook, K. Little, D. C. H. McBrien, and K. A. K. Lot, *Chem. Biol. Interact.*, *15*, 347 (1976).

26. J. H. Freedman, M. R. Ciriolo, and J. Peisach, *J Biol. Chem.*, *264*, 5598 (1989).

27. A. Alonso-Moraga, A. Bocanegra, J. M. Torres, J. Lopez-Barea, and C. Pueyo, *Mol. Cell. Biochem.*, *73*, 61 (1987).

28. C. J. Reed and K. T. Douglas, *Biochem. J.*, *275*, 601 (1991).

29. M. N. Hughes, in *Comprehensive Coordination Chemistry*, Vol. 6 (G. Wilkinson, R. D. Gillard, and J. A. McCleverty, eds.), Pergamon Press, Oxford, 1987, p. 541 ff.

30. M. N. Hughes and R. K. Poole, *Metals and Micro-organisms*, Chapman and Hall, London, 1989.

31. D. A. Rouch, Plasmid-mediated copper resistance in *E. coli*, Ph.D. thesis, University of Melbourne, 1986.

32. D. Rouch, J. Camakaris, and B. T. O. Lee, in *Metal Ion Homeostasis: Molecular Biology and Chemistry* (D. R. Winge and D. H. Hamer, eds.), Alan R. Liss, New York, 1989, p. 469 ff.

33. W. Ghandour, J. A. Hubbard, J. Deistung, M. N. Hughes, and R. K. Poole, *Appl. Microbiol. Biotechnol.*, *28*, 559 (1988).

34. B. P. Rosen, *Ann. Rev. Microbiol.*, *40*, 263 (1986).

35. M. G. Baldry and A. C. Dean, *Microbios*, *29*, 7 (1981).

36. W. B. Krueger and B. J. Kolodziej, *Microbios*, *17*, 141 (1976).

37. C. Cotter and J. T. Trevors, *Appl. Microbiol.*, *13*, 313 (1988).

38. M. N. Hughes and R. K. Poole, in *Metal–microbe interactions* (R. K. Poole and G. M. Gadd, eds.), IRL Press, Oxford, 1989, p. 1 ff.

39. S. D. Rogers, M. R. Bhave, J. F. Mercer, J. Camakaris, and B. T. O. Lee, *J. Bacteriol.*, *173*, 6742 (1991).

40. F. K. Gleason, C.-J. Lim, M. Gerami-Nejad, and J. A. Fuchs, *Biochemistry*, *29*, 3701 (1990).

41. A. Holmgren, *J. Biol. Chem.*, *264*, 13963 (1989).

42. D. McLaggan, T. M. Logan, D. G. Lynn, and W. Epstein, *J. Bacteriol.*, *172*, 3631 (1990).

43. J. T. Greenberg and B. Demple, *J. Bacteriol.*, *168*, 1026 (1986).

44. K. Mokhele, Y. J. Tang, M. A. Clark, and J. L. Ingraham, *J. Bacteriol.*, *169*, 5721 (1987).

45. L. Y. Chen, W. M. Leu, K. T. Wang, and Y. H. Lee, *J. Biol. Chem.*, *267*, 20100 (1992).

46. A. Odermatt, H. Suter, R. Krapf, and M. Solioz, *J. Biol. Chem., 268*, 12775 (1993).
47. A. Odermatt, H. Suter, R. Krapf, and M. Solioz, *Ann. N. Y. Acad. Sci., 936*, 484 (1992).
48. G. Nucifora, L. Chu, T. K. Misra, and S. Silver, *Proc. Natl. Acad. Sci. USA, 86*, 3544 (1989).
49. N. S. Makkar and L. E. Casida, Jr., *Int. J. System. Bacteriol., 37*, 323 (1987).
50. L. E. Casida, *Appl. Environ. Microbiol., 53*, 1515, (1987).
51. J. W. Bryson, T. V. O'Halloran, D. A. Rouch, N. L. Brown, J. Camakaris, and B. T. O. Lee, in *Bioinorganic Chemistry of Copper* (K. D. Karlin and Z. Tylekar, eds.), Chapman and Hall, London, 1993, p. 101ff.
52. N. L. Brown, D. A. Rouch, and B. T. O. Lee, *Plasmid, 27*, 41 (1992).
53. D. A. Cooksey, *Mol. Microbiol., 7*, 1 (1993).
54. D. Rouch, B. T. O. Lee, and J. Camakaris, in *Metal Ion Homeostasis: Molecular Biology and Chemistry* (D. R. Winge and D. H. Hamer, eds.), Alan R. Liss, New York, 1989, p. 439 ff.
55. S. Silver, B. T. O. Lee, N. L. Brown, and D. A. Cooksey, in *Chemistry of Copper and Zinc Triads* (A. J. Welch and S. K. Chapman, eds.), Royal Society of Chemistry, London, 1993, p. 38ff.
56. J. T. Trevors, *Microbiol. Sci., 4*, 29 (1987).
57. T. J. Foster, *Microbiol. Rev., 47*, 361 (1983).
58. C. L. Bender and D. A. Cooksey, *J. Bacteriol., 165*, 534 (1986).
59. C. L. Bender and D. A. Cooksey, *J. Bacteriol., 169*, 470 (1987).
60. D. A. Cooksey and H. R. Azad, *Appl. Environ. Microbiol., 58*, 274 (1992).
61. M. A. Mellano and D. A. Cooksey, *J. Bacteriol., 170*, 2879 (1988).
62. J. S. Cha and D. A. Cooksey, *Proc. Natl. Acad. Sci. USA, 88*, 8915 (1991).
63. C. Ouzounis and C. Sander, *FEBS Lett., 279*, 73 (1991).
64. A. Messerschmidt and R. Huber, *Eur. J. Biochem., 187*, 341 (1990).
65. A. Messerschmidt, A. Rossi, R. Ladenstein, R. Huber, M. Bolognesi, G. Gatti, A. Marchesini, R. Petruzelli, and A. Finazzi-Agrò, *J. Mol. Biol., 206*, 513 (1989).

66. H. S. Lee, A. Abdelal, M. A. Clark, and J. L. Ingraham, *J. Bacteriol.*, *173*, 5406 (1991).
67. P. A. Lund and N. L. Brown, *Gene*, *52*, 207 (1987).
68. H. Nakahara, S. Silver, T. Miki, and R.H. Rownd, *J. Bacteriol.*, *140*, 161 (1979).
69. T. J. Tetaz and R. K. J. Luke, *J. Bacteriol.*, *154*, 1263 (1983).
70. S. Ramamoorthy and D. J. Kushner, *Microb. Ecol.*, *2*, 162 (1975).
71. S. D. Mills, C. A. Jasalavich, and D. A. Cooksey, *J. Bacteriol.*, *175*, 1656 (1993).
72. T. Platt, *Ann. Rev. Biochem.*, *55*, 339 (1986).
73. D. K. Hawley and W. R. McClure, *Nucl. Acids Res.*, *11*, 2237 (1983).
74. R. Gross, B. Aricò, and R. Rappuoli, *Mol. Microbiol.*, *3*, 1661 (1989).
75. M. A. Mellano and D. A. Cooksey, *J. Bacteriol.*, *170*, 4399 (1988).
76. H. C. Tseng and C. W. Chen, *Mol. Microbiol.*, *5*, 1187 (1991).
77. J. R. Williams, A. Morgan, D. A. Rouch, N. L. Brown, and B. T. O. Lee, *Appl. Environ. Microbiol.*, *59*, 2531 (1993).
78. D. A. Cooksey, H. R. Azad, J. S. Cha, and C. K. Lim, *Appl. Environ. Microbiol.*, *56*, 431 (1990).
79. M. D. Mullen, D. C. Wolf, F. G. Ferris, T. J. Beveridge, C. A. Flemming, and G. W. Bailey, *Appl. Environ. Microbiol.*, *55*, 3143 (1989).
80. J. P. Bowman, L. I. Sly, and A. C. Hayward, *Lett. Appl. Microbiol.*, *10*, 85 (1990).
81. C. Dressler, U. Kues, D. H. Nies, and B. Friedrich, *Appl. Environ. Microbiol.*, *57*, 3079 (1991).
82. C. Vulpe, B. Levinson, S. Whitney, S. Packman, and J. Gitschier, *Nature Genetics*, *3*, 7 (1993).
83. S. Silver, G. Nucifora, and L. T. Phung, *Mol. Microbiol.*, *10*, 7 (1993).
84. D. A. Cooksey, *Appl. Environ. Microbiol.*, *56*, 13 (1990).
85. C. L. Bender, D. K. Malvick, K. E. Conway, S. George, and P. Pratt, *Appl. Environ. Microbiol.*, *56*, 170 (1990).
86. S. Garde and C. L. Bender, *Appl. Environ. Microbiol.*, *57*, 2435 (1991).
87. C. W. Ronson, B. T. Nixon, and F. M. Ausubel, *Cell*, *49*, 579 (1987).

Author Index

Numbers in parentheses are reference numbers and indicate that an author's work is referred to although his name may not be cited in the text. Underlined numbers give the page on which the complete reference is listed.

A

B

D

E

G

H

I

J

L

M

N

O

P

Q

R

S

T

W

Subject Index

A

D

F

M

N

Q

R

S

U

V

W

X

Y

Z